Wissenschaftsethik und Technikfolgenbeurteilung
Band 26

Schriftenreihe der Europäischen Akademie zur Erforschung
von Folgen wissenschaftlich-technischer Entwicklungen
Bad Neuenahr-Ahrweiler GmbH
herausgegeben von Carl Friedrich Gethmann

B. v. Maydell · K. Borchardt · K.-D. Henke · R. Leitner · R. Muffels ·
M. Quante · P.-L. Rauhala · G. Verschraegen · M. Żukowski

Enabling Social Policy

Springer

Series Editor
Professor Dr. Dr. h.c. Carl Friedrich Gethmann
Europäische Akademie GmbH
Wilhelmstraße 56, 53474 Bad Neuenahr-Ahrweiler, Germany

For the Authors
Bernd v. Maydell
Siebengebirgsstraße 58a, 53757 St. Augustin, Germany

Editing
Friederike Wütscher
Europäische Akademie GmbH
Wilhelmstraße 56, 53474 Bad Neuenahr-Ahrweiler, Germany

ISBN 978-3-642-06747-1 e-ISBN 978-3-540-29772-7

Bibliographic information published by Die Deutsche Bibliothek.
Die Deutsche Bibliothek lists this publication in the Deutsche Nationalbibliographie;
detailed bibliographic data is available in the Internet at http://dnb.ddb.de

Springer-Verlag is a part of Springer Science+Business Media
springeronline.com

Softcover reprint of the hardcover 1st edition 2006

Coverdesign: deblik, Berlin

Europäische Akademie

zur Erforschung von Folgen wissenschaftlich-technischer Entwicklungen
Bad Neuenahr-Ahrweiler GmbH

The Europäische Akademie

The Europäische Akademie zur Erforschung von Folgen wissenschaftlich-technischer Entwicklungen GmbH is concerned with the scientific study of consequences of scientific and technological advance for the individual and social life and for the natural environment. The Europäische Akademie intends to contribute to a rational way of society of dealing with the consequences of scientific and technological developments. This aim is mainly realised in the development of recommendations for options to act, from the point of view of long-term societal acceptance. The work of the Europäische Akademie mostly takes place in temporary interdisciplinary project groups, whose members are recognised scientists from European universities. Overarching issues, e. g. from the fields of Technology Assessment or Ethic of Science, are dealt with by the staff of the Europäische Akademie.

The Series

The series "Wissenschaftsethik und Technikfolgenbeurteilung" (Ethics of Science and Technology Assessment) serves to publish the results of the work of the Europäische Akademie. It is published by the academy's director. Besides the final results of the project groups the series includes volumes on general questions of ethics of science and technology assessment as well as other monographic studies.

Acknowledgement

The project *European Social Policy* was supported by the Ministerium für Wissenschaft, Weiterbildung, Forschung und Kultur of Rhineland-Palatinate.

Preface

The Europäische Akademie is concerned with the scientific study of the consequences of scientific and technological advance for the individual, society and the natural environment. It intends to contribute to find a rational way for society to deal with the consequences of scientific progress, by proposing recommendations for options of actions with long-term social acceptance. The work of the Europäische Akademie mostly takes place in temporary interdisciplinary project groups, whose members are recognised scientists from European universities and other independent research institutes.

The study at hand is the result of the Europäische Akademie's project group 'European Social Policy', in which experts coming from different European countries and representing various disciplines – economics, law, sociology, political science, philosophy – worked for more than two years on the multifarious aspects of the topic, providing a thorough analysis and a series of comprehensive recommendations for policy making.

In the light of the recent enlargement of the European Union, the ever contended issue of a European social policy gets of increasingly pressing importance. The differences between member states are widening, the international competition grows within the Union as well as between the EU and its global competitors, the further economic integration restrains the scope of national decision-making. These developments aggravate the tension between the pressures national welfare systems face and the distribution of competences between the national and the supra-national level of governance in matters of social policy. The twofold question arising from this situation is on the one hand how *European*, and on the other how social European social policy should be in the decades to come. 'Enabling Social Europe' is an attempt to answer the question in both regards. Firstly, by assessing the role a European policy should play taking into account both the potentialities and limitation of the 'European social model' as a normative framework as well as the relations between economic and social policies and the future challenges of the European welfare systems. Secondly, by advocating the paradigm of the 'enabling welfare state' as a new perspective for social policy aimed at raising the personal autonomy, individual responsibility and social inclusion of people and at enabling them to manage and balance their life courses in a better way.

I would like to thank the members of the project group Dr Katja Borchardt, Professor Dr Klaus-Dirk Henke, Professor Dr Ruud Muffels, Professor Dr Michael Quante, Professor Pirkko-Liisa Rauhala, PhD, Dr Gert Verschraegen, Professor Dr Maciej Żukowski, and in particular the chair Professor Dr Bernd von Maydell and the project coordinator Dott Rupert Leitner, for their commitment and the excellent work in this study. Special thanks is due to Friederike Wütscher for the editorial work in preparing the text for print.

Bad Neuenahr-Ahrweiler, November 2005 Carl Friedrich Gethmann

Foreword

The term European social policy has become a polarising item of contention. On the one hand, a social policy framed by the European Union is feared to pose a threat to national social and labour market policies; on the other, the absence of the Union's clearly defined competences in this field is held responsible for the citizen's lacking identification with the Community. Both of these – mutually contradictory – lines of argumentation have been invoked to explain the rejection of the Constitutional Treaty by the citizens of France and the Netherlands. Presumably, controversial statements on European social policy also have to do with the different understandings of the term. The need to clarify an issue of such elementary significance to the further development of the European Union was no doubt one of the reasons for the Europäische Akademie in Bad Neuenahr to entrust an interdisciplinary project group with the investigation of this problem.

The work commenced in the second half of 2002 with the establishment of a working party which sought to identify questions thought relevant. This was followed by the formation of the actual project group, which did not include all members of the working party. Through their contributions to the discourse, the party members who withdrew enriched the subsequent work of the project group – for which we record our gratitude. Personnel changes also occurred as the project work progressed. Professor Dr Weyma Lübbe (University of Leipzig) and Professor Dr Jos Berghman (Catholic University of Leuven) left the group and had to be replaced by new members. With their valuable ideas and contributions, Ms Lübbe and Mr Berghman helped map out the path of the project group. Professor Berghman moreover continued to accompany the discussions, notably through his participation in the mid-term meeting of June 2004.

The multifaceted nature of the subject matter was accommodated by the interdisciplinary and international composition of the project group. Its members were recruited from very different regions of Europe and represented various disciplines. From the abundance of possible themes, the group singled out and focussed on a set of specific problems. Thus one task was to deal with a modern conception of social policy whose objectives are perceived as dynamic. This aspect is expressed by the title of the book. Another approach emerged from the diversity of national social systems. In the light of this diversity, one must ask what is meant by the often cited European Social Model and how systemic diversity can be utilised to advance the European Union.

The intensive discussion within the group was facilitated and enhanced through the reliable forum presented by the Europäische Akademie. Of especial importance here was the cooperation of Dott Rupert Leitner, who served as coordinator and, in addition, delivered a substantive input as co-author.

The project group mooted the initial concept and the main points of emphasis in two events that included external colleagues. Besides the project group members, the following experts attended the kick-off meeting in March 2003:

- Professor Dr Klaus-Dirk Henke, Technische Universität Berlin (economics),
- Professor Dr Stephan Leibfried, University of Bremen (social sciences),
- Professor Dr Franz Marhold, University of Graz (legal science), and
- Professor Dr Kieke Okma, Queen's University Kingston (health policy).

Shortly after the kick-off meeting Professor Dr Klaus-Dirk Henke joined the project group. Along with the complete project group and its former member Professor Dr Jos Berghman, the following participants attended the mid-term meeting in June 2004:

- Professor Dr Beatrix Karl, University of Graz (legal science),
- Professor Dr Stephan Leibfried, University of Bremen (social policy),
- Professor Lutz Leisering, PhD, University of Bielefeld (sociology),
- Professor Dr Paul Schoukens, Katholieke Universiteit Leuven (law),
- Professor Dr Ulrich Steinvorth, University of Hamburg (philosophy), and
- Dr Werner Tegtmeier, retired state secretary (economics).

The project group is indebted to all of them. With their advice and critical comments they spurred on our work and rendered it productive.

The completion of the joint project left all members regretful that the intense exchange of thought within the group, in which we had managed to find a common form of expression, was now ending. It is to be hoped that the product of our work will perpetuate the discourse in an expanded circle interested in social policy and European issues.

Sankt Augustin, November 2005 Bernd v. Maydell

List of Authors

Bernd Baron v. Maydell, Professor Dr., born in 1934 in Reval/Estonia, studied law and economics in Marburg and Berlin (Freie Universität). After the two state examinations in law and his doctorate in public international law, he wrote his post-doctoral thesis on civil law ('Geldschuld und Geldwert' – Monetary Debt and Monetary Value), on the basis of which he acquired professorial status in Bonn in the fields of civil law, labour law and social (security) law. Following professorships in Berlin (1974–1981) and Bonn (1981–1992), he accepted the appointment as Director of the Max Planck Institute for Foreign and International Social Law in Munich (1992 until conferral of emeritus status in 2002). The research fields in which Bernd v. Maydell has been engaged range from civil law and labour law to social law, although in recent years his chief interest has been in international and European social law. In addition to his academic publications, Professor v. Maydell participated in numerous national and international bodies. Thus, he belonged to the Permanent Deputation of the German Law Congress (Deutscher Juristentag) from 1990 to 2002, and for over 20 years, he was a member of the Committee of Experts on the Application of Conventions and Recommendations of the International Labour Organisation in Geneva. His expertise in German, foreign and international labour and social law was enlisted by many, often interdisciplinary, expert commissions. For instance, he was a member of the Bundestag Commission of Enquiry on demographic change and took part in a number of task forces dealing, inter alia, with old-age security for women and the harmonisation of Germany's various pension systems. Under the leadership of Professor v. Maydell, the Max Planck Institute in Munich elaborated concepts for the transformation of social security systems in Central and Eastern Europe on the basis of comparative law methodology and was able to verify these concepts in performing actual advisory functions.

Katja Borchardt, Dr., born 1975, studied economics at the Technical University in Berlin, Germany and at the Institut Etudes Européennes in Brussels, Belgium. From July 2000 until middle of May 2005, she worked as teaching and research assistant at the Department Public Finance and Health Economics chaired by Prof. Dr. Klaus-Dirk Henke. Since May 17th, 2005, she started working for the German Pharmaceutical Association as Head Affairs, Health Policy. She worked as teaching assistant in Public Finance, Economic

Policy and Health Economics. Borchardt's main research area is the analysis of social security systems, with a special focus on health care systems, migration of health professionals, social policy in the EU and EU health policy. Her doctoral thesis in 2005 is about 'The migration of physicians from and to Germany – an empirical and theoretical approach' and will be published in autumn 2005. Working as teaching and research assistant Borchardt contributed to a number of different research projects at the Department. During summer 2004, she stayed for a research period in Copenhagen at the World Health Organisation, Regional Office for Europe working on migration of health professionals in the European region.

Klaus-Dirk Henke, Professor Dr., studied economics and social sciences at the University of Cologne, Germany, and at the London School of Economics from 1963 to 1968. After studies at the University of Michigan, Ann Arbor, USA, he received his doctoral degree from the University of Cologne in 1970. From 1970 to 1976, he held assistant professorships at the Universities of Cologne and Marburg and was visiting scholar at the Brookings Institution, Washington, D. C. He received his habilitation from the University of Marburg in 1976. From 1976 to 1995 he held a professorship for economics at the University of Hannover, Germany. He was visiting professor at the University of Vienna and the University of Bristol. Since 1995, he is Professor of Economics (Public Finance and Health Economics) at the Technical University of Berlin. From 1996 to 2000, he has also been director at the European Centre for Comparative Government and Public Policy, Berlin. Professor Henke received scholarships from the German Academic Exchange Service (DAAD), from the German Ministry of Foreign Affairs, and from the German National Science Council. From 1987 to 1998, Professor Henke was a member of the Advisory Board for the Concerted Action in Health Care in Germany of which he was Chairman from 1993 to 1998. Since 1984, he has served as a member of the Advisory Board to the German Ministry of Finance. From 1989 to 1991, he was president of the European Health Care Management Association. Professor Henke is editor of the book series Europäische Schriften zu Staat und Wirtschaft (since 1999) and of the Journal of Public Health / Zeitschrift für Gesundheitswissenschaften (since 1993).

Rupert Leitner, Dott., born 1966 in Brixen/Italy, studied business economics at the Universitá Commerciale "L. Bocconi" in Milan and at the York University in Toronto. In 1995, he graduated in business economics with a thesis on the recycling problem of complex mass products investigated with reference to the automotive sector. After graduation, he was research assistant and lecturer at the Department of Business Sciences at the University of Applied Sciences in Merseburg. At the same time, he began his studies of philosophy, sociology and psychology at the Fern-Universität in Hagen. Dott. Leitner worked as management consultant with the Accenture GmbH in Hamburg before joining, in 2002, the Europäische Akademie GmbH in Bad Neuenahr-Ahrweiler. As a member of the scientific staff there, he is coordinating the project group 'European Social Policy' and preparing for graduation in philosophy at the University of Duisburg-Essen. The topic of his thesis is the interrelation between ignorance of and responsibility for consequences of technological innovation. Besides his research activities, he was teaching business administration at the University of Applied Sciences in Remagen.

Ruud Muffels, Professor Dr., is Professor of Socio-Economics (labour market and social security science) at Tilburg University (NL). He is also director of the Tilburg Institute of Social and Socio-Economic Research (TISSER) and one of the research directors at the Institute for Labour Studies (OSA) at Tilburg University. In 2005, he also became a fellow at the NETSPAR research Institute at the Economic Faculty. He has been involved in a large number of comparative research projects for the European Science Foundation and the European Commission on income distribution, poverty, comparative analyses of European welfare states, social protection and ageing. His primary interests concern labour market economics, income dynamics, comparative analysis of the welfare state, panel methodology and socio-economic policies. Professor Muffels has published in a wide range of socio-economic and interdisciplinary journals (Journal of Population Economics, Social Indicators Research, Journal of Public Policy, Journal of Socio-Economics, Journal of European Social Policy, Vocational Training, Acta Sociologica, and others) and in a large number of international academic volumes. Among his recent (English) books are 'Social Exclusion in European Welfare States' in 2002 (with P. Tsakloglou and D. Mayes, Edward Elgar, 2002), 'Solidarity and Care in the European Union' (with R. ter Meulen and W. Arts, Kluwer Academic Publishers, 2001) and 'The Real Worlds of Welfare Capitalism' in 1999 (with Robert E. Goodin and Bruce Headey, Cambridge University Press, 1999).

Michael Quante, Professor Dr., studied philosophy and Germanistik (German language and literature studies) in Berlin and Münster, where he passed his 1. state examination in 1989 and was conferred a PhD in 1992. From 1993 to 1995, he was a fellow in the DFG (German Research Council) research project on 'Ethical Aspects of Organ Transplantation' and was awarded a habilitation scholarship from the DFG in 1995. From 1996 to 2001, Professor Quante was scientific assistant at the department of philosophy of the Universität Münster and qualified as university lecturer in 2001. From 1993 to 2001 he was a permanent guest lecturer at the department of design of the University of Applied Sciences in Münster. Since 2004, Professor Quante has been Professor for Practical Philosophy at the Universität Duisburg-Essen, and since autumn 2005 at the University of Cologne. His areas of research combine historical and systematic questions on the one hand and cover both practical and theoretical philosophy on the other hand. He has done research on German Idealism (especially G.W.F. Hegel), Philosophy of Mind (especially theory of action and personal identity), ethics, meta-ethics, social philosophy and biomedical ethics. Besides, he is the author of books on Hegel, the ethics of organ transplantation, personal identity and ethics. Furthermore, since 2004 Professor Quante has been co-editor of the journal "Ethical Theory and Moral Practice" and since 1996 member of the Akademie der Ethik in der Medizin (AEM).

Pirkko-Liisa Kristiina Rauhala, Professor PhD, born 1951, studied social policy and public health sciences at the University of Jyväskylä, and at the University of Tampere, Finland. Her habilitation thesis was on 'The Development of the Social Care Services in Finland.' As a member of a Scandinavian study group she studied the gender as a factor of social care services. During the years 1997–2001, Professor Rauhala was a senior researcher and vice-leader of the welfare research group at the National Research and Development Centre for Welfare and Health, Helsinki, Finland. Her study interests have focused on the Beveridgean social policy, on local social policies and social care, and she has published analyses concerning the development of the immaterial social protection. The social development of the Baltic Sea region is emphasised in her research work, too. Since 1998, she is senior reader of social policy at the University of Helsinki, and since 2001 university lecturer of social work at the University of Helsinki. Since 2001, she holds a Guest-professorship of Social Policy and Social Work at the University of Tartu, Estonia. Professor Rauhala is member of the Welfare Research Committee of Nordic Council of Ministers (2001–2006), and acted as a reader and guest-editor of the Haworth Press Inc., New York. She is publishing fiction essays, too, and in 1999, she was invited as a member of the Finnish Literature Society.

Gert Verschraegen, Dr., born 1973, studied sociology, philosophy and political economy at the Universities of Leuven (Belgium) and Hull (Great Britain). He received a Phd in the social sciences (2000) from the University of Leuven, where he is currently appointed as a postdoctoral fellow of the Research Foundation Flanders (FWO-Vlaanderen). From 2002 to 2003, he was assistant professor at Ceps/Instead (Luxemburg). Dr. Verschraegen's main research areas are international and comparative social policy, sociology – especially sociology of law and human rights – and political theory. Dr. Verschraegen wrote several publications in these areas and is the co-editor of 'Rawls. Een inleiding in zijn werk' (2002); 'De verleiding van de ethiek. Over de plaats van morele argumenten in de huidige maatschappij' (2003), 'Internationale Rechtvaardigheid' (2005) and 'Between Cosmopolitan Ideals and State Sovereignty: Studies in Global Justice' (2006). He is currently working on a co-authored book on the post-war development of the Belgian welfare state.

Maciej Żukowski, Professor Dr., born 1959, studied economics at the University of Economics in Poznań, Poland, where he is currently Professor for Economic Sciences, holding a chair in Labour and Social Policy. Since 1996, he also holds a professorship in banking at the University of Applied Sciences of Banking (WSB) in Poznań. Professor Żukowski's main research area is the analysis of social security systems, with a special focus on old-age security systems, labour market, labour migration, social policy in the EU and in the transforming countries. His doctoral thesis in 1991 was on 'The Relationships between Pension- and Labour-Income and their Determinants – in Poland, the Federal Republic of Germany and Great Britain.' His translation into Polish of Nicholas Barr's textbook, 'The Economics of the Welfare State' was published in 1993. In addition to several other research appointments abroad (among others at the Vienna University of Economics and at the London School of Economics), from 1994 to 1996 Professor Żukowski was engaged at the Centre for Social Policy of the University of Bremen. There, as a scholarship holder from the Alexander-von-Humboldt-Foundation, he worked on his postdoctoral thesis on the subject of 'Multi-Tier Old-Age Pension Systems in the European Union and Poland. Between State and Market.' In 1997 he qualified as a university lecturer in economic sciences and was awarded a year later a first grade individual research prize from the Polish Minister of Education for his book.

Table of Contents

Executive Summary

With the recent enlargement of the European Union, the ever-present issue of whether there is a need for a European social policy will be debated even more. Disparities between member states are widening, international competition within the Union as well as between the EU and its global competitors is growing, and the road to further economic integration is restricting the scope for national decision-making. These developments aggravate the pressures national welfare systems have to cope with and question the distribution of competences between national and supranational levels of governance with respect to social policy. The twofold question arising from this situation is, on the one hand, how *European* and, on the other, how *social* European social policy should be in the decades to come. The study 'Enabling Social Europe' is an attempt to answer the question in both regards.

This book is the result of a research project on the challenges and future of European social policy undertaken by the Europäische Akademie in Bad Neuenahr-Ahrweiler from fall 2002 to spring 2005. The interdisciplinary project group consisted of experts coming from various European countries and active in the fields of economics, law, sociology, political science and philosophy. The study objectives were to assess the groundwork and scope for national and European social policy in the coming decades with a view to the pressures and challenges member states face in the wake of the aforementioned contextual developments. Its aim is to provide a thorough analysis of the relationship between national and European social policy, culminating in a number of more or less far-reaching recommendations for policymaking and policy actions.

The study consists of four parts. Part one gives a general overview of social policy in Europe in the 21st century. It begins with the historical development of the various social structures in Europe, going on to outline the problems and challenges confronting them today. Subsequently, it describes the ethical foundations of the 'European social model', eventually leading up to the paradigm of the 'enabling welfare state' as an innovative approach to social policy in Europe. Part two investigates the significance of the European Union: first, as a framework for national policies, which are influenced by changes in migration, labour market and employment patterns, and by the process of economic integration and supranational legislation; and second, as an agent for social policy. Part three analyses four important areas of social policy – health care, old-age security, family policy and poverty prevention – by conducting comparative case studies of pairs of countries belonging to different welfare regimes: Germany and the United Kingdom, Poland and Germany, Finland and Estonia, Belgium and Denmark. The case studies are aimed at: the comparison of regimes and practices, thus seeking to gain more insight into the potentialities for designing 'enabling' policies; the relevance of policies at the Union level compared to the national level; and the interactions between nation states, especially with regard to policy learning and coordination. Part four brings together the conclusions of the previous analyses and assesses the potential added-value of the paradigm of the 'enabling welfare state' for the modernisation or renewal of social policy in Europe. Accordingly, the study ends with recommendations for strategies of action and instruments for the renewal of social policies at Union and nation-state level, with a specific view to the investigated policy areas.

Social Policy in the 21st Century

The first chapter of part one gives a synopsis of the development of national welfare institutions in Europe from their origins in the nineteenth century up to their current state in today's Europeanised and globalised environment. In doing so, some key ideas and concepts, which are then referred to throughout the rest of the book, are likewise identified in a socio-historical context. Finally, the present routes of social policy development in the European Union are discussed along with the most pressing challenges confronting such a policy.

European welfare states emerged in the last third of the nineteenth century as a consequence of the so-called 'social question'. The Industrial Revolution had led to the formation of a working class whose incipient organisation was starting to threaten the political predominance of the elites, thereby leading to a growing involvement of the state. In that setting, the welfare state pursued a number of diverse goals. Thus it attempted to coordinate social and economic policy by adjusting the primary distribution of income by means of a state-sponsored form of secondary distribution. At the same time, the socialisation of risks and the enhancement of income security for workers aimed to increase the legitimacy of the state and safeguard the social order and political stability. The welfare state was also a project of nation building and inclusion: nation-wide solidarity was institutionalised by defining the rules and conditions for entitlements to new social rights and by linking them to citizenship.

After World War II, social policy turned from a workers' policy into a general (re)distributive policy. The expansion of the social welfare system in post-war Europe was driven by four main factors: (i) the progressive extension of coverage under welfare state benefit schemes and of services to the entire population; (ii) the broadening of the income protection function to cover risks endured over the entire life course; (iii) the growing size of the welfare state and the enhanced role of the state acting as an employer; and (iv) the shortening of employment spells over the entire life course.

Looking at those trends in welfare state development from the perspective of today's pressures and challenges, it is clear that by now – actually, from the beginning of the nineteen-seventies – these trends seem to have reversed as limits to growth and expansion have become more relevant, if not decisive. Today, welfare systems are challenged from many sides. Growing international competitive pressure resulting from the acceleration of economic globalisation, with the ever-closer integration of financial and product markets, affects welfare states worldwide because their systems of taxation and regulation heavily impact the international competitiveness of their economies. Developments towards post-industrial service economies with declining rates of productivity growth do not only transform the general potential for economic growth, but also reconfigure the distribution of income between different business sectors. The ongoing extension and accelerated pace of innovation in information and communication technologies alter the requirements for occupational qualifications, leading to an increasing and continuous demand for higher education and life-long learning. Persistently high rates of long-term unemployment – also amongst the young – tighten the financial restraints on social policy in general and demand new, modernised approaches in welfare state policies, particularly in the labour market. Growing levels of migration cause tensions between different ethnic and religious groups in our multicultural societies. The problems of regional differentiation in the EU – with disparities growing both within as well as between nation states, especially after the latest enlargement – question the distribution of competences between national and supranational levels of governance. The aging of society increasingly exerts its influence on all aspects of social policy and especially impacts expenditure on health care and old-age security. Changes in traditional family structures, the rising importance of the dual breadwinner model and growing female participation in the labour market require innovations in employment, family and poverty prevention policies.

If these are the main challenges facing social policy in Europe, then the Europeanisation of social policy – as a dynamic interaction process between national welfare regimes, taking place simultaneously at two levels of governance, the national and the supranational – may be seen as part of the challenges and, at the same time, part of the solutions. It depends on whether the diversity of welfare regimes will prove a hindrance to coordination or rather a source of enhancing competition and policy learning. Which of these two lines will be followed, depends on the different actors involved – the Union, with its economic and monetary policies; the Councils and the European Court of Justice, with their decisions; and the nation states, with their social policy measures and their interaction via the Open Method of Coordination (OMC). It depends on whether they succeed in managing the diversity, in coordinating the – albeit not formally – overlapping competences, in integrating social and economic policy, and in ultimately building some form of social citizenship in Europe.

The questions raised in the first chapter set the scene for the following analyses. Yet, before the paradigm of the 'enabling welfare state' can be advocated as a new approach to a European social policy in response to the depicted challenges, an explication and systematic treatment of the general ethical principles and values underlying social policy in Europe is needed – also to ensure that the innovative perspective does not contrast with these basic prerogatives.

In the second chapter of part one, an attempt is made to critically reconstruct the norms, values and principles which frame the normative self-understanding of social policy in Europe. The thesis put forward concerning social policy is that there is a European consensus on two normative principles: personal autonomy and social inclusion of the individual. ('Consensus' here is meant in a broad sense and used as synonymous with 'common acknowledgement of the basic principles and values'.) In asking how the two principles should be understood, reference is made to the distinction between negative and positive freedom, affirming that the consensus on social policy in Europe relies on the richer and more demanding notion of positive freedom. The capability approach developed by Sen (1999) is referred to as a theory of freedom, which takes into account that state interference primarily seeks to promote and realise those capacities necessary for personal autonomy. Sen's conception of freedom (1992) allows for an understanding of the state's role only as an enabling and activating one. However, since human beings can exercise their autonomy only within a community, autonomy is impossible without social inclusion, which requires active participation and recognition just as much as it does material resources. The dynamic aspects of a life-course perspective are integrated into the normative framework by widening the notion of personal autonomy over the temporal structure of a human person's existence.

Building upon these assumptions, it is possible to further investigate the relationship between the two principles of autonomy and inclusion, on the one hand, and the principles on which a good or just society is built, on the other, thereby asking what that means for the relationship between the individual and the community. With reference to the four models serving as a normative basis for social policy in Europe, namely libertarianism, (weak and strong) moderate liberalism, liberal communitarianism and anti-liberal communitarianism, the European consensus remains underdetermined and only allows for a negative delimitation, excluding the extreme poles of the spectrum. The paradigm of the enabling welfare state does not, however, require a narrower scope of determination at the level of principles. Although in the understanding of the state's role, (strong) moderate liberalism may best serve as a normative starting point, it is not necessary for the purposes of this book to decide between moderate liberalism and liberal communitarianism. Since the analysis here is restricted to the level of principles and values, it leaves room for further specifications on the levels of aims and instruments.

Underlying the two principles of personal autonomy and social inclusion, the concepts of subsidiarity and sustainability are regarded only as being of instrumental value. Subsidiarity, when referring to interaction between different levels of governance, is analysed in its twofold significance as an interdiction to interfere and as a duty to ensure. Sustainability becomes an issue in matters of intrapersonal, interpersonal and intergenerational effects of social policy. However, since social policy always has to (re)distribute goods and services between individuals and age groups, a third principle has to be incorporated into the normative framework: the principle of distributive justice. In the light of the principle of justice, other options for the justification of the distribution of goods through social policy, such as informed self-interest and solidarity, appear too weak to pay regard to the sense of obligation actually supported within European societies. Although a basic universal moral claim, the principle of distributive justice in conjunction with social policy in Europe must be specified for contexts of action in order to account for national and cultural differences, and to avoid overburdening individuals and social institutions. Similarly, the tension between universal moral claims and the spatial and temporal limits of a European social policy has to be eased by weighing up the restriction to Europe against the consideration of third countries and future generations.

Given this normative framework, the third chapter of part one focuses on the basic goals and main tasks of an enabling social policy. Compared with the traditional tools of social policy, like income replacement and support but also the 'making work pay' policies of the 'activating welfare state', the paradigm of the 'enabling welfare state' implies a shift to a more preventive logic and a new distribution of responsibilities and roles between different societal actors. As a preventive approach to social policy, it places more emphasis on fostering active participation and enhancing the quality of human capital. Welfare policies have to be designed to 'enable' more people to work by focusing on human investment and skill development. Innovative practices and ideas can give citizens more 'free choice' by opening up opportunities to all and by enabling people to accept responsibility for managing their lives. The life course may be taken as a starting point for defining new routes to social policy through life-long learning and a better reconciliation or tuning of work and private life.

Sen's capabilities approach (1992, 1999) can serve as a conceptual tool for the development of a basic understanding of the principles and goals of the enabling welfare state. Therefore, a brief overview of his theory and its significance to a new social policy approach is given. Sen understands capabilities as opportunities to achieve functionings, as potential options of choice, as freedoms to get hold of what produces welfare and contributes to well-being. This belief can be translated into a reformulation of welfare policy goals in terms of 'freedom to act' instead of 'freedom from want', aimed at enabling people – not as passive recipients but as acting individuals – to maintain or raise their human and social capital and to prevent their capabilities from becoming obsolete or redundant. Since Sen's view is questioned by Dworkin's (2000) claim for equality of resources rather than capabilities, an attempt is made to integrate both approaches by further specifying the understanding of the state's and the individual's role for ensuring capabilities and resources as well as for achieving well-being.

Albeit the welfare state always has taken into account and even institutionalised the life course of citizens, the paradigm of an enabling welfare state demands a more dynamic life-course perspective. The point here is that the life course has become more diverse, involving many different and partly overlapping stages with less clear-cut transitions between them. Accordingly, welfare arrangements should aim at ensuring a more balanced spread of resources and time during the individual's life, providing a continuous flow of support to the development of capabilities and a better combination of various phases of the life course, as well as transitions between them.

Since the paradigm of the 'enabling welfare state' actually boils down to the notion of a 'welfare society', the distribution of roles and responsibilities between the various social actors in welfare regimes must likewise be addressed. To this end, five regime types are distinguished: corporatist, liberal, social-democratic, southern, and eastern or transitional. The roles of the government, the market, the community, the family and the individual vary significantly across welfare regimes. Work, welfare and family dependencies are structured and balanced in different ways, and social policy goals like income, employment and social integration are disparately weighted and prioritised. These differences in the structures of welfare state dependencies and their impact on the autonomy of individuals and households also affect the degree to which regimes are open to social arrangements that are less governmental and contractual and thus more individual, informal and market-related, as envisioned by the paradigm of the 'enabling welfare state'.

European Union and Social Policy

Based on the challenges, normative framework and the enabling approach to European social policy explained in part one, part two proceeds with empirical facts and research into the welfare state. Two chapters deal with the significance of the European Union as a framework for national social policy and as a policymaking actor in its own right.

The first chapter of part two is divided into four sections dealing with migration, labour markets, economic integration and the institutional framework of the Union. The first section investigates the effects of EU enlargement on the mobility flows of workers and other persons moving within the Union, and their impact on the national social systems as well as on the Union's coordination policies and institutions. Free movement is not only granted as one of the 'four freedoms' forming a particular trait of the Union; historically, it was also one of the first and main causes for the need of supranational coordination. After a short sketch of the present pattern of EU-wide migration, its influential factors and various forms, the section concentrates on the expected effects of EU enlargement. The main questions concern: overall migration potential, duration of migration and other forms of mobility, push and pull factors, characteristics of the majority of migrants, transitional rules governing migration from new to old member states, immigration flows from outside the European Union, and the future need for new efforts of coordination. Migration may impact the age structure, labour potential and sustainability of social systems in both the emigration and immigration countries. Meanwhile, however, migration potential seems to have become highly restricted, and in the near future – also after the transitional period – the pressure to migrate might even decrease owing to the effects of faster economic growth, direct investments and benefits of the EU Structural Fund for the new member states. Hence, in view of the limited migration flows likely to ensue from EU enlargement, national and international migration policies will presumably be affected only on a marginal scale, at least in the short run. As to the long-run consequences, these depend on the extent of ongoing European integration, which in turn ultimately depends on the economic success of the accessing countries in the decades to come.

The second section addresses employment policy, an area in which the EU is already active, and illustrates the effects of supranational policies on national and international labour markets. A brief outline of the European Employment Strategy is followed by some figures reflecting the extent to which various national labour markets are capable of attaining the goals agreed by the EU member states at the Lisbon summit, in 2000 and 2003 respectively. Here again, the welfare regime type approach is applied as a means of stressing the different labour market outcomes. In this way, the role policy regimes may play for explaining the different country-specific outcomes is assessed. Member states differ not only as regards the performance of their national labour markets but – in view of the European Employment Strategy guidelines – also as regards the pressures they face in striving to attain performance goals. The main challenge consists in balancing and combining the goals of work and income security with the goals of labour market flexibility and mobility, in both geographical and occupational terms, thereby not only meeting social requirements but also enhancing economic productivity and competitiveness. Improving the adaptability and employability of the labour force and increasing labour market participation and work security necessitates an extension of opportunities. These are furnished through innovative life-course oriented arrangements for a better combination of work, care and education, as well as through investments in human capital via life-long learning and practices of active aging. The EU can play an important role in promoting these policy changes by supporting and recommending national innovations in labour market policy and by improving the efficacy of the coordination process at EU level through a better alignment and fine-tuning of the decisions of the Economic and Financial Affairs Council (ECOFIN) with the work of the Employment, Social Policy, Health and Consumer Affairs Council (ESPHCA). The Open Method of Coordination (OMC) is designed to enhance compliance with ambitious targets and innovative policy approaches, although its effectiveness is often greatly restricted, especially in periods of economic downturn and on account of the principle of subsidiarity.

The third section describes the consequences of ongoing and widening economic integration for the national social systems. Economic integration – especially through the European Monetary Union (EMU) and the Stability and Growth Pact (SGP) – increasingly influences and limits the scope of action of the member states in their efforts to respond to economic adversity and growing pressure on social expenditures. This forces them to launch incisive retrenchment and/or reform programmes in their welfare systems. Social policies thus face pressure from two sides: mounting expenditures owing to long-term trends like aging, individualisation and unemployment on the one front, and the EMU- and SGP-induced inability to use deficit spending policies to deal with adverse economic shocks, on the other. Since a relaxation of Community rules or a prolongation of their implementation do not seem feasible solutions in the long term, the only option available to nation states is the reformation of their labour markets and social security systems. Looking in more general terms at the relationship between economic and social policy in the EU, it is easy to see the steady rise in tensions between the growing unity of economic policy at supranational level and the widening heterogeneity of national welfare systems at state level. A convergence of different social policy regimes is still a long way off; instead, there is increased policy competition between nation states within a Europe of different speeds. Whether this situation will give rise to more competition or more policy learning between the member states, and whether these interaction processes will lead to a race to the bottom or rather to the top, remain open questions. Yet the OMC could contribute to the modernisation of social protection systems along the lines of the 'enabling welfare state'. Its chances will be better the more it remains restricted to the formulation of general aims, or to a framework in the form of non-binding guidelines on the basis of which the member states have sufficient discretion to work out specific national targets and arrangements, leaving the implementation of the guidelines to national and even regional and local tiers of government.

The fourth section analyses the significance of the Union's coordination mechanisms and legal framework for national policies, showing how the former impact the latter. Although the Treaty on European Union cites as an important objective the tasks to improve social progress, assure a high level of employment and strengthen social cohesion, it does not confer upon the EU any competences in the social policy sphere. Nor is the Constitution going to change this; instead, it expressly demands 'concern' for the principle of subsidiarity. Nevertheless, as a legal union, the Union affects national policies through Community law and the activities of the Community institutions. The main impact is exerted through the 'fundamental freedoms', the coordination rules governing migrant workers and the numerous judgments of the European Court of Justice related therewith. In addition, some provisions of Community law are also of direct and indirect relevance to social policy. Finally, the effects of Community competition law will meet with growing interest as clear-cut distinctions between public systems of social security and private-sector arrangements become increasingly blurred.

The second chapter of part two deals with the European Union as an agent in the area of social policy – which in many ways it is, regardless of the principle of subsidiarity. In this respect, although social policy remains a matter of member state competence in principle, manifold activities of the Community institutions in this area have constantly enhanced the social dimension of the Union. The role and relative importance of the Councils, the Commission, the European Court of Justice and the Parliament nevertheless differ significantly. This also holds true for the instruments available for action at Union level, comprising legislation, structural promotion and political activities. The diverse objectives of efforts undertaken at Union level range from coordination to convergence and the securing of minimum standards.

National and Supra-national Social Policy: Comparative Case Studies

The analyses in part one and two give no in-depth consideration to specific policy areas or countries. National particularities are taken into account only with a view to their attribution to different welfare regimes, but no attention is drawn to single member state features. In part three, the focus of the study shifts to more concrete levels of policymaking. The four main chapters of part three are thus devoted to four areas of social policy, namely health care, old-age security, family policies and poverty prevention. Each of these areas is scrutinised by way of a concrete comparison of the policies and institutional arrangements of two selected countries. The selection of the country pairs is based on the welfare regimes attributed to each of them, which together cover most of the earlier mentioned regime types, including the transition or Eastern regime type. Following a comprehensive introduction, the first chapter contrasts healthcare policies in Germany, as a corporatist welfare regime, with those existing in the United Kingdom, representing the liberal model. The next chapter comparatively evaluates policies of old-age security in Germany and Poland, the latter representing a transition country.

The third chapter discusses the two different approaches to family policy in Finland and Estonia, a social-democratic and another transition state respectively. Finally, the fourth chapter deals with the political and institutional arrangements adopted to prevent poverty in Belgium and Denmark, a corporatist and a social-democratic welfare regime. The exemplary analysis of the chosen policy areas is representative of the main issues and the variety of social policies encountered in Europe, and makes it possible to draw conclusions needed for general evaluations and recommendations. The same holds true for the countries investigated and compared. Thus, the intention of the study is not to give a full picture of all EU member state policies, but – by undertaking policy-specific comparisons of two countries representing more or less distinct welfare regimes – to demonstrate differences and similarities from both a static comparative and a dynamic developmental, reform-oriented perspective.

Proceeding from the core elements embodied in the concept of the enabling welfare state, the capabilities approach, the life-course perspective and sustainability (being the main condition for the maintenance and efficiency of any social system), the analysis of the four social policy areas brings to the fore some major problems and questions. Family policies centre on childhood and parenthood, while provisions for old age delineate, structure and support specific phases of the life course, enhancing the capabilities of beneficiaries during these phases. Health care and poverty prevention systems afford protection against two basic risks of life, illness and poverty, which are not directly connected to specific life phases but can more or less temporarily and randomly interrupt and distort life plans and security expectations. By emphasising preventive rather than compensative measures in both these fields, priority is again given to individual capability in the sense of, say, health awareness, provident investment in the maintenance and improvement of productive and social skills or employability – especially if poverty prevention is combined with activation policies. The funding of healthcare and pension systems addresses issues relating to their sustainability and the re-distribution of means – between age cohorts in terms of society as a whole and between life phases of the individual. Corresponding problems involve expectations of levels of assistance and security, which in turn affect the distribution of responsibility between the citizen, the state and other social institutions.

Based on the common lines of analysis, the four chapters share the same structure. Each begins with a review of the main issues of the respective social policy area. This is followed by a description of the policy approaches taken in each of the two countries selected for the given area, illustrating their particular institutions, methods, developments, challenges and policy goals. Subsequently, the country pairs are compared, stressing similarities and peculiarities, strengths and weaknesses. The comparison is geared to the two basic elements of the enabling welfare state, namely the capability approach and the life-course perspective, and additionally takes account of the concept of sustainability. In this way, the two countries are compared and evaluated not only with respect to each other, but also as regards the degree to which they have either already implemented measures in support of the paradigm, or are open to reforms and changes needed to realise the underlying elements. Another section highlights the European dimensions and elements of each policy area. On the one hand, the focus is on the interaction between the supranational and the national level – that is, the role of the Union with a view to its various institutions and policies in other domains that significantly influence the specific social policy area and the national policies of the two countries under consideration. On the other hand, attention is given to the relations between member states – that is, the effects of national policies on other member states through direct interaction, comparison and competition, as well as through institutions and mechanisms at Union level, above all the OMC. Finally, each chapter finishes by drawing conclusions and formulating recommendations. The final remarks seek to evaluate the systems with a view to the following: their capacity to respond to the challenges they face; their needs and their potential for innovation and reform towards the enabling welfare society; the extent to which they respect ethical principles of personal autonomy, social inclusion and distributive justice; their possibilities for reciprocal policy learning and adaptation of best practices; the options available for action at the supranational level; and the expected effects of the most promising interventions.

The overall picture resulting from the detailed studies of the four policy areas and seven countries is characterised above all by heterogeneity. Although the main purpose of the comparisons is not to confront good with bad practices, some conclusions can be drawn regarding favourable and less favourable examples. Thus it is possible to identify models that can best be recommended for policy learning and innovation – of course, without neglecting either the difficulties of transferring measures and methods from one context to another or the specificity of some urgent challenges facing individual states.

In the case of health care, for instance, the comparison between Germany and the United Kingdom reveals that the two systems differ in many, often fundamental respects. But to say that the former outperforms the latter as regards quality standards, responsiveness and equal access does not automatically mean that the British system should be adapted to the purported German model. From the perspective of the enabling paradigm, it is more important to elucidate in how far the two systems face different challenges and how urgent these challenges are. That is decisive for the priorities they give to cost containment, efficiency and long-term sustainability (Germany), or enhancement of equal access, extension of patients' freedom of choice and self-responsibility (United Kingdom), in order to implement an enabling approach to health care. Accordingly, the two systems are affected differently through the European Union: While the jurisdiction of the Court of Justice addresses the British waiting lists problem, Community competition law and the common market are of growing relevance to German public health insurance as healthcare provisions are increasingly excluded from public schemes and covered instead by supplementary private plans. Both countries, however, will be impacted by the strengthening of the markets of goods and services, healthcare providers and insurers, as well as by the application of the OMC in the field of health care.

In the area of old-age security, Poland and Germany both aim at adequate and sustainable public pensions and both face similar challenges arising from demographic pressures and labour market trends. One of the main issues is how public pension systems can be adjusted to ensure sustainability and supplemented by private provisions for the sake of building up satisfactory and adequate coverage. The question is: to what degree should public systems eventually suffice only as a source of minimum retirement income, while supplementary private provisions serve to maintain the previous standard of living? A greater degree of private provisioning does not only induce individuals to assume more responsibility for themselves but also enables them to choose from a wider range of options. Whether the Polish three-pillar system could serve as a model here depends not only on its success but also on specific preconditions for the implementation of more or less radical reforms. The extraordinary transition situation in Poland, under which the financial sustainability of the former pension schemes could not be guaranteed long-term, posed particular challenges but also opened exceptional options for political action. Of an entirely different nature is the present situation in Germany, whose specific requirements are better met by measured policy shifts and gradual system changes. Yet the move from pure pay-as-you-go systems to their combination with funded schemes, regardless whether they are voluntary or obligatory and more or less subsidised, seems unavoidable for all European countries. In fact, since they all face the same challenges, the OMC has been launched for the field of pensions with the aim of coordinating efforts and exchanging information about best practices to ensure adequate and sustainable pensions in the EU. Unlike the coordination rules for migrant workers – the only form of 'Europeanisation' of old-age security existing in the past – the OMC represents a shift in supranational policy. Even so, old-age security systems will probably remain more nationally oriented than other areas of social policy, at least in the near future. The more old-age provisioning is covered by private insurance schemes, however, the greater the influence of Community competition law and the common market.

In the area of family policy, the conclusions drawn from the pertinent comparison seem more clear-cut. In Finland, family policy is definitely more advanced in the direction of the enabling paradigm and, hence, may serve as a model for policy learning. But the comparison between Finland and Estonia also demonstrates how fundamentally divergent aims can be connected to family policies. Thus the main goals of Scandinavian family policy seek to reconcile family and work, also with a particular view to fathers, and to protect and enhance childhood over a comparatively long time span. This raises the question whether Estonian policies, given the dramatic demographic situation in the smaller country, should also strive for the same goals. The differing Estonian priorities, which are clearly pro-natality, notably in support of births and the first year of child life, likewise appear acceptable and reasonable – at least as long as present circumstances persist. Notwithstanding these differences, the political aims of family policy in Estonia and Finland can be considered similar from a fundamental perspective: easing the parental burden of childcare expenditure, support of family formation in the early phases of newborn life, and measures towards the reconciliation of work and family. On a very general level, both strategies can be described as 'enabling' in the sense that they support sensitive phases in the life courses of children as well as (young) parents striving to fulfil the roles of employees in the labour market and carers within the family. So far, the EU has not directly intervened in the area of national family policy. However, its supranational policies have an indirect effect, for instance through employment policies aimed at reconciling work and family and at strengthening the role of fathers, as well as through legal efforts in the field of gender equality.

At first glance, the most promising policy learning situations seem to exist in Denmark and Belgium. While both countries are quite successful in keeping financial poverty at low levels, Denmark performs better on the labour market owing to its distinctly active labour policy centred on human capital investment, a more flexible labour market and social inclusion, especially with a view to woman, youths and older persons. Some of the concrete measures implemented by Denmark with great success include the reduction of early retirement options and unemployment benefit duration as well as the subsidisation of low-skilled workers. Although such obviously successful policy measures are clearly identified, there are nonetheless significant institutional and political barriers and path dependencies that obstruct the implementation of policy learning practices and the transferability of arrangements that have proven successful elsewhere. In the case of Belgium, the labour market remains rather inflexible owing to early retirement options, the considerable tax burden on labour, long entitlement periods and a high level of market regulation, all of which tend to impede social inclusion. So although the Danish experience suggests ways for pursuing an enabling policy of poverty prevention – by placing more emphasis on activation programmes, boosting service employment and linking social spending to investment in human and social capital – its transferability to a different context remains difficult, with prospective success neither straightforward nor easy to attain. In the light of these comparative results, the idea of improving policy convergence at the European level through the use of 'soft' law instruments and OMC practices is highly ambitious. Nevertheless, promoting social inclusion and fighting poverty has become one of the key objectives of the European Union, and the application of the OMC in this policy area has progressed further and with greater success than in others. It has to be kept in mind, however, that policy learning does not simply mean transferring elements of one distinct system to another, but involves rendering support to the development of novel ideas within a given national context.

As to the supranational level, the analysis of the four social policy areas produces a rather unbalanced and complex picture. The influence of European policy on national policies is growing both directly and indirectly. The importance of the supranational level therefore has to be taken into account comprehensively at both levels of policymaking. Yet as long as the principle of subsidiarity regulates competences in the social sphere, the soft instrument of the OMC seems the only feasible alternative. Even so, it still remains unclear what purposes the OMC may, can or should serve best. The findings on the success of the OMC process in the four policy areas under scrutiny point in the same direction.

Elements of a European Social Policy

In part four, the intermediary results obtained from the analysis of the exemplified policy areas and member states are brought together with the central issues examined in parts one and two. The summary of arguments in favour of basing social policy on the paradigm of the 'enabling welfare state' and the overview of current conditions, prospects and circumstances for a paradigm shift create the basis for the recommendation of action strategies aimed at an enabling social policy in the European Union. Recommendations and suggestions are thus given on two counts. The one focuses on concrete measures of an enabling social policy in the specific policy areas analysed above, and the other highlights prospects and instruments necessary for streamlining and integrating a European social policy at Union level.

Adopting a European social policy raises two central questions. The first asks how 'social' European policy should be and thus concerns the relationship between macro-economic, employment and social policy. It reverts to the idea of social protection conceived as a productive factor that has been translated into the capability and life-course policy approaches. This question also addresses the interventions and mechanisms available for putting the paradigm of the enabling welfare state into practice in the four policy areas of health care, old-age security, family policy and poverty prevention. Accordingly, social policy should abandon the fear that employment and competitiveness are impeded by too high social costs and overly burdensome regulations. Instead, it should focus on the goal of attracting mobile capital by offering a high-skilled labour force made available through active policies of human capital investment. Such policies are increasingly considered as instruments for activating both economic growth and innovation, and for rendering opportunities to all.

The second question asks how 'European' national social policy should be. The principle of subsidiarity sets forth that Europe is not to play a role in the social domain unless commonly agreed social goals are unattainable without its interference. In its positive meaning, however, subsidiarity also implies that any European interference should entail beneficial effects for the member states as such and for Europe as a whole. It is in this sense that the Union's commitment to the social domain at national level appears to constitute part of the European Social Model. Even so, the Union cannot assume any tasks of an institutional or productive welfare state but can only play a purely regulative role, exercising an ensuring and enabling influence on social policies at member state level. An enabling European social policy should aim at the coordination of national policies and strengthen their orientation towards the paradigm of the enabling welfare state through regulative interventions. To what extent this ensuring role of the Union will require more regulation depends on the success of 'soft coordination' or OMC processes in the various domains. Union intervention does not necessarily imply that the ultimate convergence of the different systems must be striven for. Coordination can also consist in the management of diversity, meaning that systems define similar tools to attain shared goals, while the specific design of these tools differs from one country to the next. It can also mean that the tools themselves or their implementation may differ, while the outcomes in terms of the targeted goals prove more or less similar. Furthermore, the idea of managing diversity is also compatible with the setting of minimum social standards, which might differ at national level if they are considered relative and thus adjusted to national conditions. And that again would comply with the subsidiarity rule governing European social policy.

It may seem a paradox to plead for the European Union's stronger interference with national social policies in times when member states tend to view these policies as a predominantly national concern – whether because of the commonly agreed principle of subsidiarity or the pressure of meeting international economic challenges. And it seems even more paradoxical to demand greater commitment on the part of social actors, in particular the state, when most governments are seeking to reduce or at least contain current levels of social spending. However, both paradoxes can be solved from a broader perspective. As to the first, it must be stressed that international competition between the world's large economic regions has become much more severe than in the past. The competition the EU faces comes above all from the United States, China and other Asian economies. In order to compete successfully on a global scale, Europe could profit from reducing competition within its own borders, thereby acting as a single economic block rather than a cluster of economies competing in regard to wages, social costs and taxation levels. To this end, the Union's stronger involvement in economic and social matters might be highly beneficial. As to the second paradox, it should be kept in mind that the paradigm of the enabling welfare society demands a more active role on the part of social actors, including the state, but does not necessarily entail more government spending. Welfare-state or social policy needs to shift its emphasis from a compensatory to a preventive logic, placing less stress on income support to people out of work and, instead, fostering active participation in the labour force by enhancing the quality of human capital. This policy aims at giving citizens more freedom of choice and enabling them to assume greater responsibility for managing their lives. Simultaneously, it takes account of the exigencies of time, money and education that go hand in hand with flexible and heterogeneous life plans. The role of the state thereby shifts from providing protection and compensation through the redistribution of financial resources towards exercising more control, management and coordination of the services rendered either by its public institutions or other private or non-state agencies. This shift presupposes increased responsibility on the part of the market and societal organisations in the provision of social services, for example through private market initiatives, agencies of civil society, charitable institutions, non-governmental organisations, and individual and informal networks, notably relatives and friends.

Building on these conclusions, the study ends by recommending strategies of action for defining concrete policy measures, at both national and European level – measures that fit into the framework and goals of the enabling policy perspective. Some of these measures have already been enacted and mirror recent changes underway in some welfare states as well as at Union level; others are still in debate and form part of the trend towards modernising and innovating social systems in Europe. These findings are identified for the four areas of social policy analysed in part three of this study, namely health care, old-age security, family policy and poverty prevention. In addition, recommendations are also given for the employment policies and labour market situations analysed in part two from a European perspective. Subsequently, separate and specific recommendations are summarised for actions and instruments at the supranational level. In the following, only the main points of the recommendations are outlined.

Health care – Based on the principles of equal access, fair financing and evidence-based medicine, nationally adjusted minimum standards should be defined as a safety net to preserve adequate levels of health care also in the poorer regions of the Union. In order to reach a clear division between allocation and distribution within the funding structure, a shift towards the financing of re-distributive elements through general taxes is required. By introducing obligatory private insurance schemes for the whole population, with public support for low-income groups, more room should be given to market solutions. Additionally, a transition to partially funded arrangements in public health insurance schemes might enhance the sustainability and efficiency of the entire system, thus responding to demographic challenges in keeping with the principle of intergenerational justice. Through the further implementation of the EC Treaty with respect to the free movement of labour, capital, and goods and services, restrictions on cross-border health care could be reduced to widen consumers' range of choices between providers, insurers (public and private) and forms of treatment. Finally, a stronger emphasis on preventive health care constitutes an indispensable prerequisite for investments in personal capabilities and their maintenance over the life course.

Old-age security – Improvements to old-age security systems aim at their long-term budgetary sustainability and the maintenance of adequate levels of income replacement upon retirement. Enhancing the role of funded and private schemes seeks to accomplish a better balance between various old-age protection models, thereby improving the sustainability of public schemes as well as supporting new and more flexible life-course patterns. More free choice, diversity and flexibility should be realised through: the individualisation of public systems (individual saving accounts, notional defined contributions); the move from defined-benefit to defined-contribution plans, or the development of fully funded private systems; and flexible retirement-age arrangements and the introduction of partial pensions. The latter may also help to activate older workers and prolong working life. In any case, however, the factual retirement age must be postponed, also by restricting early retirement options. Old-age protection must be balanced with other policy domains like primary health care, long-term care, social services and housing, which also affect the standard of living in old age. In order to remove persisting barriers to mobility and enhance policy learning, the OMC should focus more on indicators and aim at streamlining various areas of importance to old-age security. The mechanisms for coordinating social security systems within the European Union should be adjusted further to improve the cross-border portability of occupational pension rights.

Family policy – The reconciliation of work and family life is not only central to fostering personal autonomy and enhancing individual capabilities, but is even more important for the well-being of children. This goal of social policy can be attained by extending options for parental leave, strengthening fathers' role in child care and offering more flexible day-care facilities. Additionally, broadening the emphasis of family policy to include families with school-age children seeks to address the growing problem of early school drop-out through: prolonged flexible part-time and job-sharing arrangements; more flexible school hours; and better after-school care facilities, which could also be provided by private organisations or agencies of 'civil society' (e.g. non-governmental organisations). The rights of children must be strengthened, for instance in the form of the right to a decent childhood or the right to be raised by both parents. Moreover, such rights need to be underpinned by combating child poverty, supporting single parents and offering equal education opportunities to all children. A fine-tuning of family and employment policies as well as the enhancement of children's rights are feasible only if they rest on agreements that are enforceable Europe-wide, either by framing rights and minimum standards or by extending the OMC process to the area of family policy. While the aging of society leads to a growing proportion of older people becoming dependent on care, the actual supply of care does not rise proportionally. Hence, family policies should provide assistance to middle-aged women and men willing to take care of elderly parents. The responsibility for this task should not be left entirely to the market or the family; needed instead is a form of co-responsibility shared by public institutions and the state.

Minimum protection and poverty prevention – The shift of minimum protection and poverty prevention policies from a compensatory to an enabling and activating approach requires a change in the definition of poverty prevention from an income-based measure to a multidimensional concept. Accordingly, static indicators such as, say, low income at a given point in time have to be supplemented by longitudinal indicators, which take account of housing conditions, health, education, and social and cultural participation over time, in order to provide a comprehensive and dynamic description of social exclusion patterns. As long-term unemployed persons, single mothers and children constitute the most vulnerable groups, poverty policies must primarily focus on: prevention of long-term unemployment; reintegration into the labour market; reconciliation of work and family care, particularly on behalf of single parents; employment promotion, especially female labour participation; and safeguarding sufficient levels of minimum protection. The accomplishment of these objectives, notably minimum income protection, requires a definition of European – alongside national – minimum income and/or poverty standards, which could then be adapted to the specific circumstances or conditions prevailing in individual member states. Social inclusion and poverty prevention policies need to be linked to active labour market policies, particularly where entitlement to social assistance is made to depend on the willingness to work. This integrative approach should be taken into consideration also in applying the OMC to these closely interrelated areas.

Employment and labour market policies – Enabling labour market policies should focus on two overriding objectives: first, the activation of the unemployed and second, the flexibilisation of labour markets. According to the 'flexicurity' approach, policies should generally aim at facilitating changes of employment by reducing the attendant risks. This aim can be pursued by: supporting investments in human capital formation through life-long learning, education, training and skill acquisition; providing better opportunities for occupational advancement; and preventing people from 'falling down the career ladder' or being excluded from the labour market owing to obsolete skills. Public intervention could be directed towards a less regulated labour market, with lower levels of employment protection and less union power. Combined with policies to invest in the capabilities of people, such measures could ultimately increase rather than diminish employment opportunities. An activating labour market policy designed to raise employability and flexibility includes: the decentralisation and regionalisation of labour market interventions; a system of tax credits geared to employment rather than wage subsidies; and more transitional arrangements and opportunities regarding retirement leave, education or study leave, and leave for caring duties and family obligations. Both poverty prevention and employment policy possess a genuine European dimension because they are inextricably bound with the economic policies of the Union and the member states. For this reason, poverty reduction and employment targets, adequate minimum protection standards for the unemployed, and minimum wages for those in work should be determined and coordinated at European level, taking into account the member states' specific national and regional distinctions.

At EU level, the most important instruments for coordinating and guiding national social policies towards an enabling welfare society include: a greater degree of acknowledgement by the competent Community institutions, the closer integration and streamlining of policy areas, the structuring of competence and interaction across the various levels of governance, and the application of the OMC in the social domain.

Enhancement of institutional awareness – Coordinating and guiding European social systems towards the paradigm of the enabling and activating welfare state requires an improved awareness of the meaning of European social policy and its impact on the entire policy of the Union. This initially pertains to the awareness of the social dimension of policies at supranational level. Thus it must explicitly be acknowledged that the political decisions and measures adopted by the European Union, irrespective of the particular area concerned – whether the basic freedoms, the internal market, monetary and macro-economic policies, or structural and social policies – all have a fundamental social bearing and impact. It is therefore, first and foremost, necessary that the significance of the social domain be perceived and valued at the highest level within the EU, above all by the Commission and Council of Ministers. The awareness of the importance of social issues could profit from the setting of clear and ambitious quantitative targets for the envisaged aims. This is especially the case where these targets are announced and put up for serious evaluation. Institutional awareness seems to be heightened mainly through a strong top-down communication process – from the highest organisational and hierarchical levels of the Union's institutions down to the political representatives of all tiers of government at national level, and eventually the broad public. On the other hand, the paradigm of the enabling welfare state must be emphasised and reinforced at all levels of political decision-making. In this respect, it is not the social dimension itself that has to be recognised, but rather the explicit role the enabling welfare state perspective could play when it comes to elaborating and implementing the required reforms.

Integration of social policy across various policy domains – Relations between the various domains of European socio-economic policy require a stronger degree of interaction and more dialogue between the institutions involved in the decision-making processes across the various policy areas. A full-fledged implementation of the targeted fundamental freedoms, the internal market, the EMU, the SGP, as well as structural and agricultural policies – along with the removal of barriers to their attainment – will have a significant impact on the outcomes of European and national policies in the social domain. This holds especially true with a view to the rising importance of market solutions in the social domain, for instance healthcare and pension systems. Thus, only a joint treatment of different policy domains makes it possible to determine overall target levels that might be achieved – given conflicting goals and priorities. Although not all of these influences can be aligned, coordinated and integrated, opportunities do exist to improve the fine-tuning of economic, employment and social policies by restructuring, integrating or streamlining the governance process. The most powerful institutions are the Councils, above all the EU Council and ECOFIN. Since the decisions of these Councils carry the most political weight while reflecting the will of the member states, it is obvious that social policy interaction with the other political domains will take place at this hierarchical level. The alignment and integration of these domains occurs in numerous different forms. A feasible option may consist in the occasional or permanent extension of the ECOFIN or other Councils, either through joint sessions or participation of the social affairs and employment ministers in ECOFIN, or the finance and economic affairs ministers in ESPHCA. Additionally, a rescheduling of the entire decision-making process by taking better account of the sequence and timing of decisions in other domains, as well as an improved interlocking of procedural schedules and agendas, could improve possibilities for the mutual recognition and alignment of each other's decisions.

Definition of competences and interaction between levels of governance – As long as any form of supranational social policy seems to contrast with the principle of subsidiarity and the sovereignty of nation states in this area, any plea for a European social policy appears an unrealistic scenario without much chance of success for achieving significant coordination. Limitations built into the theoretical principle on the one hand, and practical requirements on the other, seem to provoke endless debate and dispute over competence. These conflicts pertain not only to relations between the supranational and the national level, but also to those between the various tiers of member state government, for instance regional and municipal bodies. Hence, there is a need for a clear assignment and distribution of authority between the various levels of competence. That is all the more necessary as the picture becomes increasingly blurred through the multifarious forms of interaction between supranational and regional levels, for instance through the Union's regional and interregional programmes. Except for particular circumstances, a further transfer of competences to the European level will not solve the competence disputes. Instead, the interplay between the different tiers of governance has to be improved by shifting the focus from national and sub-national levels, which are already tightly regulated, to these levels' interaction with the Union.

The Open Method of Coordination – The OMC should serve three purposes: the setting of general goals, the mutual exchange of learning experiences between member states and the benchmarking of achievements. Although the social systems of most European member states face similar challenges and pressures, the concrete measures taken to solve them can be very dissimilar owing to differing national conditions, institutional settings and historical roots. The limitation of the OMC process to the formulation of and agreement on goals is appropriate to the intent of leaving the member states plenty of room for choosing and adapting nation-specific tools for the achievement of these goals. Although, admittedly, there are no formal sanctions in the case of insufficient results, the effect of public 'naming, shaming and blaming' campaigns is likely to increase with the growing awareness of social policy benchmarking among the general public, the electorate as well as politicians. Possibly, this could lead to a cycle of self-reinforcement. Enhanced awareness boosts the effects of rankings and public campaigns, which in turn will be used more intensely, thus heightening public attention to social topics from an international perspective and ultimately strengthening the binding nature of goals defined under the OMC. In this way, the setting of ever more ambitious goals becomes feasible. Moreover, it seems reasonable to expect that the more market elements are introduced into the welfare systems the more transparent comparisons will become, so increasing the pressure on poorer performers. In matters of policy learning and the mutual exchange of experiences, there have been only few examples of cross-national cooperation and policy transferability so far. In order to initiate and implement an effective process of mutual learning through the OMC, it is not conducive merely to put pressure on a country's achievement through benchmarking exercises; the more important task is to take due account of the differences and particularities of national policies within that process.

Indeed, the idea of mutual policy learning should not be misunderstood as a learning strategy for policy transferability – that is, for simply transferring successful practices from one member state to another.

Current experiences with the OMC as well as prospects for its application in the near future illustrate that some effort is still needed to make it an effective instrument for the purposes postulated here. In the areas of social inclusion and employment, shared objectives accomplished and common sets of indicators have been found. In the field of pensions, the agreed goals are very fundamental and the learning process has not progressed beyond the identification of best practices and innovative approaches. In the healthcare sector, preference is currently given to a slim version of the OMC, serving only as a platform for the exchange of information and the determination of comparison indicators. An application of the OMC to family policies is not yet in sight.

Zusammenfassung

Mit der letzten Erweiterung der Europäischen Union verschärft sich die fortwährende Diskussion um die Frage, ob es einer Europäischen Sozialpolitik bedarf. Die Unterschiede zwischen den Mitgliedstaaten werden größer, der internationale Wettbewerb wächst nicht nur innerhalb der Union, sondern auch zwischen der EU und ihren globalen Konkurrenten, und die fortschreitende wirtschaftliche Integration engt den nationalen Entscheidungsspielraum ein. Diese Entwicklungen verstärken den Druck auf nationale Wohlfahrtsysteme und stellen die Verteilung von Kompetenzen zwischen der nationalen und der supranationalen Regelungsebene im Bereich der Sozialpolitik in Frage. Aus dieser Situation ergibt sich die doppelte Fragestellung: wie *europäisch* einerseits und wie *sozial* andererseits soll die europäische Sozialpolitik in den nächsten Jahrzehnten sein? Die Studie ‚Enabling Social Europe' ist ein Versuch, diese beiden Fragen zu beantworten.

Dieses Buch ist das Ergebnis eines Forschungsprojektes über die zukünftigen Herausforderungen und Optionen europäischer Sozialpolitik, das an der Europäischen Akademie in Bad Neuenahr-Ahrweiler von Herbst 2002 bis Frühjahr 2005 durchgeführt wurde. In der interdisziplinären Projektgruppe arbeiteten Experten der Wirtschaftswissenschaften, des Rechts, der Soziologie, der Politikwissenschaften und der Philosophie aus verschiedenen europäischen Ländern zusammen. Die Zielsetzung des Projektes war die Bestimmung der Grundlagen und der Optionen für nationale und europäische Sozialpolitik in den kommenden Jahrzehnten mit Blick auf die Herausforderungen und Belastungen, mit denen die Mitgliedstaaten aufgrund oben genannter Entwicklungen konfrontiert sind. Im Ergebnis werden mit der Studie eine eingehende Untersuchung der Beziehung zwischen nationaler und europäischer Sozialpolitik und darauf aufbauend Handlungsempfehlungen für politische Entscheidungen und Strategien unterschiedlicher Reichweite vorgelegt.

Die Studie setzt sich aus vier Teilen zusammen. Der erste Teil gibt einen allgemeinen Überblick über Sozialpolitik in Europa am Anfang des 21. Jahrhunderts. Zu Beginn werden die historischen Entwicklungslinien sowie die gegenwärtigen Probleme und Herausforderungen der verschiedenen Sozialsysteme in Europa betrachtet. Im Anschluss daran werden die ethischen Grundlagen des „Europäischen Sozialmodells" beschrieben und schließlich das Paradigma eines ‚enabling welfare state', eines ‚befähigenden Wohlfahrtstaates', als ein innovativer Ansatz für Sozialpolitik in Europa vorgestellt. Im zweiten Teil wird die Bedeutung der Europäischen Union untersucht: erstens als Rahmen für nationale Politik, welcher von Veränderungen der Migrationsströme, von Entwicklungen im Bereich der Beschäftigung und des Arbeitsmarktes, von der wirtschaftlichen Integration und der europäischen Gesetzgebung bestimmt ist; zweitens als eigenständiger Akteur im Bereich der Sozialpolitik. Im dritten Teil werden vier wichtige sozialpolitische Handlungsfelder – Gesundheitsversorgung, Alterssicherung, Familienpolitik, Armutsprävention – anhand von vergleichenden Fallstudien von jeweils zwei Ländern untersucht, die verschiedene Typen von Wohlfahrtsregimen repräsentieren: Deutschland und Großbritannien, Polen und Deutschland, Finnland und Estland, Belgien und Dänemark. Die vergleichende Analyse der einzelnen Fallbeispiele stellt Regime, Praktiken und Verfahren gegenüber, um daraus einen besseren Überblick über die Gestaltungsmöglichkeiten einer ‚befähigenden' Sozialpolitik, über die Bedeutung der Unionsebene im Verhältnis zu den Politiken auf nationaler Ebene und über die Interaktionen zwischen den Mitgliedstaaten insbesondere hinsichtlich politischer Lern- und Koordinierungsprozesse zu gewinnen. Im vierten Teil werden die Schlussfolgerungen aus den vorherigen Untersuchungen zusammengeführt, um das Wirkungspotential des Paradigmas des ‚enabling welfare state' für die Modernisierung und Erneuerung der Sozialpolitik in Europa abzuschätzen. Abschließend werden Handlungsempfehlungen gegeben bezüglich der Strategien und Instrumente für die Erneuerung der Sozialpolitik auf Unions- und nationaler Ebene unter besonderer Berücksichtigung der untersuchten Politikfelder.

Sozialpolitik im 21. Jahrhundert

Das erste Kapitel von Teil Eins gibt einen Überblick über die Entwicklung der nationalen Wohlfahrtsinstitutionen in Europa von ihren Anfängen im 19. Jahrhundert bis zu ihrer heutigen Lage in einem europäisierten und globalisierten Umfeld. Dabei werden auch einige grundlegende Ideen und Konzepte, auf die im weiteren Verlauf der Studie zurückgegriffen wird, in ihren gesellschaftlichen und historischen Entstehungskontext eingebettet. Schließlich werden die gegenwärtigen Entwicklungstendenzen und die derzeit größten Herausforderungen für eine Europäische Sozialpolitik thematisiert.

Die europäischen Wohlfahrtsstaaten entstanden im letzten Drittel des neunzehnten Jahrhunderts in Folge der so genannten ‚Sozialen Frage'. Die Industrielle Revolution führt zur Bildung einer Arbeiterklasse, die sich zu organisieren und damit die politische Vorherrschaft der Eliten zu bedrohen beginnt, was eine wachsenden Einflussnahme seitens des Staates provoziert. Mit der Bildung des Wohlfahrtstaats werden verschiedene Ziele verfolgt. Es wird versucht, soziale und Wirtschaftspolitik dadurch zu koordinieren, dass die primäre Einkommensverteilung durch staatliche Sekundärverteilung korrigiert wird. Durch die Sozialisation von Risiken und die Verbesserung der Einkommenssicherung für Arbeiter wird aber auch die Legitimation staatlicher Intervention zur Aufrechterhaltung der sozialen Ordnung und der politischen Stabilität erhöht. Letztendlich trägt der Wohlfahrtsstaat auch zur Bildung und Stärkung der Nationalstaaten und ihrer Abgrenzung nach außen bei, denn nationale Solidarität wird dadurch institutionalisiert, dass die Regeln und Bedingungen für die Inanspruchnahme neuer sozialer Rechte definiert und an die Staatsangehörigkeit gebunden werden. Nach dem Zweiten Weltkrieg wandelt sich die Sozialpolitik von einer Arbeiterpolitik zu einer allgemeinen Umverteilungspolitik. Die Expansion der Wohlfahrtssysteme im Nachkriegseuropa lässt sich hauptsächlich auf vier Faktoren zurückführen: erstens die schrittweise Ausdehnung der Absicherung durch wohlfahrtsstaatliche Leistungen und Zuwendungen auf die gesamte Bevölkerung, zweitens die Erweiterung der Einkommenssicherung gegen Risiken über den gesamten Lebenslauf, drittens das Wachstum der Institutionen der sozialen Sicherung und ihre zunehmende Bedeutung – zusammen mit anderen öffentlichen Einrichtungen – als Arbeitgeber, und viertens die Verkürzung der Lebensarbeitszeit.

Betrachtet man diese Entwicklungslinien aus dem Blickwinkel des Drucks und der Herausforderungen, denen sich Wohlfahrtsstaaten heute gegenübergestellt sehen, dann ist klar ersichtlich, dass diese Trends mittlerweile – und tatsächlich schon seit den 70er Jahren – eine Umkehrung erfahren haben und die begrenzenden Faktoren vorrangig, wenn nicht sogar bestimmend, geworden sind. Gegenwärtig gerät der Sozialstaat von mehreren Seiten unter Druck. Die Beschleunigung der wirtschaftlichen Globalisierung führt zu einer immer engeren Verflechtung der Finanz- und Produktmärkte weltweit und damit zu einer Erhöhung des internationalen Wettbewerbsdrucks. Davon sind auch Wohlfahrtssysteme direkt betroffen, weil die Steuersysteme und Regulierungen insbesondere des Arbeitsmarktes die internationale Wettbewerbsfähigkeiten der entsprechenden Wirtschaftssysteme stark beeinflussen. Der fortschreitende Strukturwandel in Richtung post-industrieller Dienstleistungsökonomien mit sinkenden Wachstumsraten der Produktivität beeinträchtigt das gesamtwirtschaftliche Wachstumspotential und verändert die Einkommensverteilung zwischen unterschiedlichen Wirtschafts- und Beschäftigungssektoren. Die weitere Verbreitung und die Beschleunigung der Innovationszyklen der Informations- und Kommunikationstechnologien verändern die Anforderungen an berufliche Qualifikationen. Der Bedarf nach besserer Bildung und neuen Qualifikationen steigt kontinuierlich und verlangt lebenslanges Lernen. Die andauernd hohen Quoten der Langzeitarbeitslosigkeit auch unter jungen Arbeitssuchenden schränken den finanziellen Handlungsspielraum für die gesamte Sozialpolitik ein und verlangen deshalb nach innovativen und modernen Ansätzen der Wohlfahrts- und insbesondere der Arbeitsmarktpolitik. Die ansteigenden Migrationsströme erhöhen die Spannungen zwischen unterschiedlichen ethnischen und religiösen Gruppen in multikulturellen Gesellschaften. Angesichts der regionalen Differenzierung in der EU, wo die Disparitäten – insbesondere seit der letzten Erweiterung – sowohl innerhalb als auch zwischen den Mitgliedsstaaten zunehmen, wird die Verteilung von Kompetenzen zwischen den nationalen und supranationalen Regierungs- und Steuerungsebenen in Frage gestellt. Die Alterung der Gesellschaft beeinflusst alle Bereiche der Sozialpolitik, in finanzieller Hinsicht vor allem aufgrund der steigenden Ausgaben für das Gesundheitswesen und die Alterssicherung. Die Veränderung traditioneller Familienstrukturen, die steigende Bedeutung des Doppelverdienerhaushalts und die wachsende Teilnahme von Frauen am Arbeitsmarkt verlangen nach neuen Lösungen in den Bereichen der Beschäftigungs- und Familienpolitik sowie der Armutsprävention.

Wenn dies die wichtigsten Herausforderungen für Sozialpolitik in Europa sind, dann kann die Europäisierung der Sozialpolitik – als ein dynamischer Interaktionsprozess zwischen nationalen Wohlfahrtssystemen, der gleichzeitig auf zwei politischen Steuerungsebenen, der nationalen und der supranationalen, vollzogen wird – sowohl als Teil der Herausforderungen, als auch als Teil der Lösungen gesehen werden. Was der Fall sein wird, hängt davon ab, ob sich die Unterschiede zwischen den Wohlfahrtsregimen eher als ein Hindernis für ihre Koordinierung oder vielmehr als ein Ausgangspunkt für konstruktiven Wettbewerb und wechselseitiges politisches Lernen erweisen. Welcher dieser beiden Wege eingeschlagen wird, hängt weiterhin davon ab, inwieweit die verschiedenen beteiligten Akteure – die Union mit ihrer Wirtschafts- und Währungspolitik, die Räte und der Europäische Gerichtshof (EuGH) mit ihren Entscheidungen, die Mitgliedstaaten mit ihren sozialpolitischen Maßnahmen und ihrer Zusammenarbeit über die Offene Methode der Koordinierung (OMK) – erfolgreich darin sein werden, die Vielfalt zu regeln und zu lenken, die zwar nicht formaliter aber realiter überlappenden Kompetenzen zu koordinieren, die Wirtschaftspolitik mit der Sozialpolitik zu integrieren und letztendlich eine Form von sozialer Bürgerschaft in Europa zu etablieren.

Die im ersten Kapitel aufgeworfenen Fragestellungen bereiten den Boden für die weitern Untersuchungen der Studie. Im Folgenden wird das Paradigma des ‚enabling welfare state', eines ‚befähigenden Wohlfahrtsstaates', als ein neuer Ansatz für Sozialpolitik in Europa sowie als Antwort auf die genannten Herausforderungen dargestellt und begründet. Vorher bedarf es allerdings noch einer Explizierung und systematischen Rekonstruktion der allgemeinen ethischen Prinzipien und Werte, auf die sich Sozialpolitik in Europa gründet. Damit soll gewährleistet werden, dass der innovative Ansatz nicht in Widerspruch zu den grundlegenden normativen Voraussetzungen steht.

Im zweiten Kapitel von Teil Eins wird versucht, jene Normen, Werte und Prinzipien kritisch zu rekonstruieren, die das normative Selbstverständnis von Sozialpolitik in Europa bestimmen. Es wird die These aufgestellt, dass in Europa in Bezug auf soziale Politiken ein Konsens über zwei ethische Prinzipien besteht: die personale Autonomie und die soziale Inklusion des Individuums. („Konsens" ist dabei in einem weiten Sinn zu verstehen als synonym mit „allgemeiner Anerkennung der grundlegenden Prinzipien und Werte".) Zur Klärung des näheren Verständnisses der beiden Prinzipien wird auf die Unterscheidung zwischen positiver und negativer Freiheit zurückgegriffen und aufgezeigt, dass der sozialpolitische Konsens in Europa auf die inhaltsreichere und anspruchsvollere Auffassung der positiven Freiheit gründet. Sens (1999) Ansatz der ‚capabilities' bietet sich diesbezüglich an als eine Theorie von Freiheit, die davon ausgeht, dass Eingriffe des Staates primär auf die Ausbildung und Unterstützung jener Fähigkeiten ausgerichtet sind, die für die Ausübung der individuellen Autonomie notwendig sind. Laut Sens Konzept von Freiheit (1992) muss die Rolle des Staates als eine befähigende und aktivierende verstanden werden. Da Individuen ihre Autonomie nur innerhalb einer Gemeinschaft ausüben können, ist Autonomie nicht möglich ohne soziale Inklusion, für die es aktiver Partizipation, Anerkennung und materieller Ressourcen bedarf. Geht man darüber hinaus von einem erweiterten Verständnisses von personaler Autonomie aus, das sämtliche Abschnitte eines Lebenslaufs umfasst, dann lassen sich auch die dynamischen Aspekte einer Lebenszeitperspektive in die normativen Grundlagen integrieren.

Von diesen Grundannahmen ausgehend, wird das Verhältnis zwischen den beiden Prinzipien der Autonomie und der Inklusion weiter bestimmt sowie die Frage erörtert, welche Prinzipien einer guten und gerechten Gesellschaft zugrunde liegen, und was dies für die Beziehung zwischen Individuum und Gemeinschaft bedeutet. In Bezug auf die vier Modelle des Libertarianismus, des (schwachen und starken) moderaten Liberalismus, des liberalen Kommunitarismus und des anti-liberalen Kommunitarismus als normative Grundlage für Sozialpolitik in Europa, bleibt der europäische Konsens unterbestimmt und erlaubt nur eine negative Abgrenzung, welche die beiden extremen Formen (Libertarianismus und anti-liberaler Kommunitarismus) ausschließt. Doch für das Paradigma des ‚enabling welfare state' ist eine engere Bestimmung auf der Ebene der Prinzipien auch nicht notwendig. Obwohl hinsichtlich des Verständnisses der Rolle des Staates der (starke) moderate Liberalismus am besten als normativer Ausgangspunkt geeignet scheint, ist für die Zwecke der vorliegenden Studie eine Entscheidung zwischen moderatem Liberalismus und liberalem Kommunitarismus nicht notwendig. Da sich die Untersuchung hier auf die Ebene der Prinzipien und Werte beschränkt, bleibt Raum für nähere Spezifizierungen auf den Ebenen der Ziele und der Instrumente.

Im Gegensatz zu den zwei Prinzipien der personalen Autonomie und der sozialen Inklusion, wird den Konzepten der Subsidiarität und der Nachhaltigkeit nur instrumenteller Wert beigemessen. In der Interaktion zwischen verschiedenen Steuerungsebenen wird Subsidiarität in seiner zweiseitigen Bedeutung betrachtet: sowohl als Verbot, in die Kompetenzen der unteren Ebene einzugreifen, als auch als Pflicht, diese in der Umsetzung ihrer Kompetenzen zu unterstützen. Nachhaltigkeit wird im Zusammenhang mit intrapersonellen, interpersonellen, und intergenerationellen Auswirkungen sozialpolitischer Maßnahmen thematisiert. Da Sozialpolitik immer eine Umverteilung von Gütern und Dienstleistungen zwischen Individuen und Altersgruppen mit sich bringt, muss das Prinzip der Verteilungsgerechtigkeit als drittes Prinzip in die normative Rahmenstruktur eingeführt werden. Für die Rechtfertigung der Verteilung von Gütern durch sozialpolitische Entscheidungen scheinen andere Optionen, wie aufgeklärtes Eigeninteresse oder Solidarität, im Vergleich zum Gerechtigkeitsprinzip nicht hinreichend zu sein, weil sie den tatsächlich in den europäischen Gesellschaften akzeptierten Sinn der Verpflichtung zu und der Verbindlichkeit von Umverteilung nicht erfassen. Wenngleich das Prinzip der Verteilungsgerechtigkeit einen grundlegenden universellen moralischen Anspruch impliziert, muss es im Bereich der Sozialpolitik in Europa in Relation zu spezifischen Handlungsfeldern kontextualisiert werden, um nationale und kulturelle Unterschiede berücksichtigen und um eine mögliche Überforderung von Individuen und Institutionen vermeiden zu können. In gleicher Weise muss die Spannung zwischen dem universellen moralischen Anspruch und den räumlichen und zeitlichen Grenzen einer Europäischen Sozialpolitik dadurch aufgelöst werden, dass die Begrenzung auf das gegenwärtige Europa gegen die Belange von Drittstaaten und zukünftigen Generationen abgewogen werden.

Vor dem Hintergrund dieses normativen Rahmens werden im dritten Kapitel von Teil Eins die Hauptziele und -aufgaben einer ‚befähigenden' Sozialpolitik dargelegt. Im Vergleich sowohl zu den traditionellen sozialpolitischen Instrumenten der Einkommensabsicherung und -unterstützung als auch zu den Beschäftigungsanreizen und -maßnahmen des aktivierenden Sozialstaates, verlangt das Paradigma das ‚enabling welfare state' eine verstärkt vorbeugende Ausrichtung sowie eine Änderung der Verteilung von Zuständigkeiten und Rollen unter den gesellschaftlichen Akteuren. Im Sinne eines präventiven sozialpolitischen Ansatzes liegt der Schwerpunkt bei der Förderung aktiver Partizipation und bei der qualitativen Verbesserung des Humankapitals. Das Sozialwesen muss so gestaltet werden, dass es durch Investitionen in die Entwicklung von Fähigkeiten und Qualifikationen mehr Menschen zur Arbeit befähigt. Innovative Praktiken und Ideen können mehr Wahlmöglichkeiten schaffen, indem für alle Menschen Chancen eröffnet werden, die es ihnen ermöglichen, ihr Leben mit einem höheren Grad an Eigenverantwortlichkeit zu planen und zu führen. Dabei muss vom gesamten Lebenslauf ausgegangen werden, damit in der Sozialpolitik neue Wege eingeschlagen werden, die lebenslanges Lernen und eine bessere Vereinbarkeit und Abstimmung von Arbeits- und Privatleben ermöglichen und unterstützen.

Da Sens Ansatz der ‚capabilities' (1992, 1999) als konzeptuelles Instrument für die Entwicklung des Verständnisses der wesentlichen Prinzipien und Ziele des ‚enabling welfare state' dienen kann, wird ein kurzer Überblick über dessen Grundlagen und die Bedeutung für einen neuen Ansatz der Sozialpolitik gegeben. Sen versteht ‚capabilities' (Fähigkeiten, Vermögen) als Gelegenheiten und Möglichkeiten, ‚functionings' (Grundfunktionen) zu verwirklichen, als tatsächlich verfügbare Wahloptionen, als die Freiheit und Fähigkeit das zu erlangen, was zu Wohlbefinden und Zufriedenheit führt und zu Wohlfahrt beiträgt. Aus dieser Sichtweise lassen sich sozialpolitische Zielsetzungen neu bestimmen im Sinne von ‚Freiheit zu Handeln' an Stelle von ‚Freiheit von Bedürftigkeit'. Ziel einer solchen Sozialpolitik ist es, Menschen – nicht als passive Empfänger, sondern als handelnde Individuen – dazu zu befähigen, ihr humanes und soziales Kapital zu erhalten und zu steigern sowie zu vermeiden, dass ihre Fähigkeiten obsolet oder redundant werden. Sens Ansatz wird zwar von Dworkins (2000) Forderung nach Gleichheit von Ressourcen anstatt Fähigkeiten in Frage gestellt, doch lassen sich beide Positionen dadurch in Einklang bringen, dass das Verständnis der Rolle des Staates und der Individuen sowohl für die Gewährleistung von Fähigkeiten als auch für das Erreichen von Wohlbefinden näher spezifiziert wird.

Der Lebenslauf wird von Sozialpolitik immer schon beeinflusst und zu einem Teil auch bestimmt und institutionalisiert, doch im Sinne des Paradigmas des ‚enabling welfare state' ist eine dynamischere Lebenszeitperspektive notwendig, die der Tatsache Rechnung trägt, dass Lebensläufe vielfältiger werden und eine größere Anzahl unterschiedlicher, zum Teil überschneidender Phasen beinhalten, zwischen denen oft keine klaren Übergänge stattfinden. Dementsprechend müssen soziale Sicherungssysteme darauf abzielen, eine ausgewogene Verteilung von Ressourcen und Zeit für unterschiedliche Aktivitäten über den gesamten individuellen Lebenslauf zu gewährleisten, um die Entwicklung von Fähigkeiten kontinuierlich zu unterstützen und sowohl die Vereinbarkeit von parallelen als auch die Übergänge zwischen sukzessiven Lebensphasen zu erleichtern.

Da das Paradigma des ‚befähigenden' Wohlfahrtsstaates letztlich auf das Konzept einer ‚enabling welfare society', einer ‚befähigenden' Wohlfahrtsgesellschaft hinausläuft, muss auch die Verteilung von Rollen und Verantwortlichkeiten zwischen den gesellschaftlichen Akteuren in verschiedenen Wohlfahrtssystemen in Frage gestellt werden. Zu diesem Zweck werden fünf Typen von Sozialstaaten unterschieden: der korporatistische, der liberale, der sozial-demokratische, der südliche und der östliche (oder transitionale). Die Rollen, die Regierung, Markt, Gesellschaft, Familie und Individuum in den verschiedenen Wohlfahrtsregimen einnehmen, unterscheiden sich zum Teil sehr stark. Die Abhängigkeitsverhältnisse von Arbeitseinkommen, sozialer Sicherung, Familie und Zivilgesellschaft sind unterschiedlich strukturiert und gewichtet; die sozialpolitischen Zielsetzungen wie Einkommenssicherung, Beschäftigung und soziale Integration werden mit unterschiedlichen Schwerpunkten und Prioritäten verfolgt. So wie die Abhängigkeitsbeziehungen und ihr Einfluss auf die Autonomie von Individuen und Haushalten in den einzelnen Systemen variieren, unterscheidet sich auch die Offenheit der Wohlfahrtsregime für die Umsetzung von weniger staatsgebundenen, weniger verrechtlichten und mehr individuellen, informellen und marktbezogenen sozialen Regelungen und Einrichtungen, die im Sinne eines ‚enabling welfare state' stärker in den Vordergrund rücken sollten.

Europäische Union und Sozialpolitik

Nachdem in Teil Eins die Herausforderungen, die normativen Grundlagen und der Ansatz des ‚enabling welfare state' für Europäische Sozialpolitik dargestellt wurden, beginnt mit Teil Zwei die empirische Untersuchung der gegenwärtigen Situation der Sozialpolitik in Europa. In zwei Kapiteln wird die Bedeutung der Europäischen Union zunächst in ihrer Relevanz für die Rahmenbedingungen der nationalen Sozialpolitiken und im Anschluss daran in ihrer Funktion als eigenständiger politischer Akteur hinterfragt.

Das erste Kapitel von Teil Zwei ist in vier Sektionen zu den Themen Migration, Arbeitsmarkt, wirtschaftliche Integration und institutioneller Rahmen der Union unterteilt. In der ersten Sektion werden die Auswirkungen der Erweiterung der EU auf die Migrationsströme von Arbeitern und anderen Personen innerhalb der Union und deren Einfluss sowohl auf die nationalen Sozialsysteme als auch auf die Koordinierungspolitik auf Unionsebene untersucht. Die Freizügigkeit von Personen ist nicht nur als eine der vier Grundfreiheiten garantiert und damit ein spezifisches Merkmal der EU, sondern sie ist aus historischer Sicht auch eine der ersten und wichtigsten Ursachen für die Notwendigkeit einer supranationalen Koordinierung gewesen. Nachdem zuerst die gegenwärtige Situation, die Einflussfaktoren und die unterschiedlichen Formen der Migration in der EU kurz umrissen werden, folgt eine Untersuchung der zu erwartenden Effekte der Erweiterung. Die wichtigsten Fragen betreffen die Einschätzung des Gesamtpotentials an Migranten, die Dauer von Ein- und Auswanderung und anderer Mobilitätsformen, die förderlichen und hinderlichen Einflussfaktoren, die Eigenschaften der größten Gruppen der Migranten, die Übergangsregelungen der Wanderung zwischen alten und neuen Mitgliedstaaten, die Immigrationsströme von außerhalb der Europäischen Union und den zukünftigen Bedarf an neuen Formen und Einrichtungen der Koordinierung. Wanderung kann die Altersstruktur, das Arbeitsangebotspotential und die Nachhaltigkeit der Sozialsysteme sowohl in den Immigrations- als auch in den Emigrationsländern beeinflussen. Doch bisher deuten alle Einschätzungen darauf hin, dass das Migrationspotential eher beschränkt ist und dass die Anreize dafür mittelfristig – auch nach Ablauf der Übergangsfristen – aufgrund des höheren wirtschaftlichen Wachstums, der direkten Investitionen und der Zuwendungen durch den EU-Strukturfonds in den neuen Ländern sogar schwächer werden könnten. Angesichts des begrenzten Umfangs der durch die Erweiterung zu erwartenden Wanderungsbewegungen werden die Auswirkungen auf die nationale und supranationale Migrationspolitik zumindest kurz- bis mittelfristig gering sein. Welche Konsequenzen langfristig auftreten werden, hängt hauptsächlich davon ab, wie weit und wie schnell die Europäische Integration voranschreitet, was seinerseits letztendlich vom wirtschaftlichen Erfolg der Beitrittsländer in den kommenden Jahrzehnten abhängt.

In der zweiten Sektion werden, zumal die EU im Bereich der Beschäftigungspolitik bereits aktiv ist, die Auswirkungen der supranationalen Politik auf den nationalen und internationalen Arbeitsmarkt beschrieben. Nach einer kurzen Darstellung der Europäischen Beschäftigungsstrategie wird anhand einiger aussagekräftiger Zahlen aufgezeigt, zu welchem Grad die verschiedenen Mitgliedsstaaten tatsächlich in der Lage sind, auf dem Arbeitsmarkt jene Ziele zu erreichen, die mit der Lissabon-Strategie auf den Gipfeltreffen 2000 und 2003 vereinbart wurden. Auch hier wird auf die Einteilung in fünf Wohlfahrtsregimetypen zurückgegriffen und die Arbeitsmarktergebnisse werden danach klassifiziert, um die Bedeutung abzuschätzen, die den Regimetypen und den entsprechenden Politiken für die Erklärung der unterschiedlichen Resultate zukommt. Die Mitgliedsstaaten unterscheiden sich nicht nur hinsichtlich ihrer Leistungsmerkmale auf den nationalen Arbeitsmärkten, sondern auch hinsichtlich des Drucks, dem sie ausgesetzt sind, um die in den Leitlinien der Europäische Beschäftigungsstrategie gesetzten Ziele zu erreichen. Die größte Herausforderung besteht darin, die Ziele der Arbeits- und Einkommenssicherheit mit den Zielen der – geographischen und beruflichen – Arbeitsmarktflexibilität und -mobilität aufeinander abzustimmen und zusammen zu bringen, um damit nicht nur sozialstaatlichen Anforderungen gerecht zu werden, sondern auch die wirtschaftliche Produktivität und Wettbewerbsfähigkeit zu steigern. Um sowohl die Einsatzfähigkeit und Vermittelbarkeit der Arbeitskräfte zu verbessern als auch die Teilnahme und Sicherheit am Arbeitsmarkt zu steigern, ist eine Erweiterung der Wahlmöglichkeiten und Chancen notwendig. Dazu bedarf es innovativer, lebenslauf-orientierter Regelungen zur besseren Vereinbarkeit von Arbeit, Familie und Bildung, der Steigerung des Humankapitals durch lebenslanges Lernen und der Förderung eines aktiven Alterns. Die EU kann eine wichtige dabei Rolle spielen, diese politischen Veränderungen voranzutreiben, indem sie nationale Neuerungen in der Arbeitsmarktpolitik befürwortet und unterstützt und indem sie die Effektivität des Koordinierungsprozesses auf Unionsebene erhöht. Letzteres kann durch einen besseren Abgleich und eine feinere Abstimmung der Entscheidungen des Rats der Wirtschafts- und Finanzminister (ECOFIN) mit der Arbeit des Rats für Beschäftigung, Sozialpolitik, Gesundheit und Verbraucherschutz (ESPHCA) erreicht werden. Auch die OMK ist darauf ausgerichtet, die Erreichung und Einhaltung anspruchsvoller Zielsetzungen sowie die Umsetzung innovativer Politikansätze zu fördern, wenngleich ihre Wirksamkeit insbesondere in Zeiten schwacher oder negativer konjunktureller Entwicklung und aufgrund des Subsidiaritätsprinzips eng begrenzt ist.

In der dritten Sektion werden die Auswirkungen der fortwährenden und -schreitenden wirtschaftlichen Integration der EU auf die nationalen Sozialsysteme beschrieben. Die wirtschaftliche Integration – insbesondere durch die Europäische Währungsunion (EWU) und den Stabilitäts- und Wachstumspakt (SWP) – beeinflusst und begrenzt in zunehmendem Maße den Handlungsspielraum der Mitgliedstaaten. Dies gilt vor allem für deren Bestrebungen, auf eine schwache wirtschaftliche Entwicklung und auf den steigenden Druck auf die Sozialausgaben zu reagieren, und zwingt sie zu einschneidenden Kürzungs- und/oder Reformprogrammen im sozialen Bereich. Sozialsysteme stehen von zwei Seiten unter Druck: Einerseits wachsen die Ausgaben aufgrund langfristiger Entwicklungstrends wie Alterung, Individualisierung und Arbeitslosigkeit, andererseits erscheint eine Erhöhung der öffentlichen Ausgaben durch Verschuldung als Antwort auf die konjunkturelle Lage aufgrund der EWU und des SWP nicht mehr machbar. Da eine Aufweichung der EU-Regelungen oder eine zeitliche Verzögerung ihrer Umsetzung auf lange Sicht keine tragfähigen Lösungen sind, bleibt den einzelnen Staaten als Ausweg nur die Reformierung ihrer Arbeitsmärkte und sozialen Sicherungssysteme. Betrachtet man das Verhältnis zwischen Wirtschafts- und Sozialpolitik in der EU im Allgemeinen, dann wird ersichtlich, dass die Spannung zwischen einer zunehmend vereinheitlichten Wirtschaftspolitik auf supranationaler Ebene einerseits und einer immer breiter werdenden Vielfalt nationaler Wohlfahrtssysteme andererseits ständig ansteigt. Eine Konvergenz der verschiedenen Politikregime ist zurzeit weder in Sicht- noch in Reichweite, stattdessen wächst der politische Wettbewerb zwischen den Nationalstaaten in einem Europa unterschiedlicher Geschwindigkeiten. Ob diese Situation zu noch mehr Wettbewerb oder zu mehr Politiklernen unter den Mitgliedstaaten führen und ob sich aus diesen Interaktionsprozessen ein Abwärts- oder ein Aufwärts-Wettlauf ergeben wird, bleibt offen. Die OMK kann aber trotzdem dazu beitragen, dass sich die Modernisierung der sozialen Sicherungssysteme den Leitlinien des ‚enabling welfare state' entsprechend vollzieht. Die Chancen dafür stehen umso besser, je mehr die Methode auf die Festsetzung allgemeiner Zielvorgaben und die Bestimmung der Rahmenbedingungen in Form von unverbindlichen Leitlinien beschränkt wird, weil auf dieser Grundlage den Einzelstaaten genügend Entscheidungsspielraum erhalten bleibt sowohl für die Formulierung spezifischer nationaler Ziele und Regelungen als auch für die Umsetzung der Leitlinien auf nationaler, regionaler und kommunaler Ebene.

In der vierten Sektion werden die Bedeutung und der Einfluss der Koordinierungsmechanismen und des rechtlichen Rahmens der EU für nationale Politiken ermittelt. Obwohl die Förderung des sozialen Fortschritts, die Erhöhung des Beschäftigungsniveaus und die Stärkung des sozialen Zusammenhalts wichtige Ziel der Union sind, überträgt der Vertrag der Europäischen Union derselben keine Kompetenzen im Bereich der Sozialpolitik. Auch mit der Verfassung wird sich daran nichts ändern, weil sie verstärkt die Beachtung des Subsidiaritätsprinzips verlangt. Dennoch beeinflusst die EU als eine rechtliche Union die nationale Politik durch das Gemeinschaftsrecht und die Institutionen der Gemeinschaft. Der größte Einfluss wird durch die Grundfreiheiten, das Koordinierungsrecht für Wanderarbeitnehmer und die damit verbundenen Entscheidungen des EuGH ausgeübt. Darüber hinaus haben auch andere gemeinschaftsrechtliche Regelungen teils mittelbare, teils unmittelbare sozialpolitische Relevanz. Schließlich ist zu beobachten, dass die Bedeutung des Wettbewerbsrechts zunimmt, weil die klare Abgrenzung zwischen öffentlichen Systemen der sozialen Sicherung und privatwirtschaftlichen Versicherungsangeboten mehr und mehr verschwimmt.

Im zweiten Kapitel von Teil Zwei wird die Europäische Union als eigenständiger Akteur auf dem Feld der Sozialpolitik, was die Union unabhängig vom Subsidiaritätsprinzip in vieler Hinsicht ist, betrachtet. Obwohl Sozialpolitik im Prinzip eine Angelegenheit der Mitgliedstaaten bleibt, wird durch eine Reihe von Aktivitäten der gemeinschaftlichen Institutionen im sozialen Bereich die soziale Dimension der EU beständig verstärkt. Dabei sind die Funktionen und das relative Gewicht der Räte, der Kommission, des EuGHs und des Parlaments sehr unterschiedlich ausgeprägt. Dies gilt auch für die auf Unionsebene verfügbaren Instrumente, die von Rechtssetzung über Strukturförderung bis zu weiteren politischen Aktivitäten reichen. Die verschiedenen Ziele der Bemühungen auf Unionsebene umfassen Koordinierung, Konvergenz und die Sicherung von Mindeststandards.

Nationale und supranationale Soziapolitik: vergleichende Fallstudien

In den Ausführungen in Teil Eins und Zwei werden weder spezifische Politikfelder noch einzelne Länder in Betracht gezogen. Nationale Besonderheiten rücken nur im Zusammenhang mit der Untersuchung unterschiedlicher Wohlfahrtsregime ins Blickfeld, aber darüber hinaus bleiben spezifische Eigenschaften der einzelnen Mitgliedstaaten unbeachtet. In Teil Drei stehen nun konkrete Beispiele und Fälle politischen Handelns im Mittelpunkt. In den vier Hauptkapiteln von Teil Drei werden vier sozialpolitische Handlungsfelder anhand des Vergleichs der konkreten politischen Maßnahmen und institutionellen Einrichtungen von je zwei Ländern behandelt. Die vier Felder umfassen Gesundheitsversorgung, Alterssicherung, Familienpolitik und Armutsprävention. Die exemplarische Auswahl der Länder ist durch ihre Zugehörigkeit zu verschiedenen Typen von Wohlfahrtsregimen bestimmt: Zusammen decken die Beispiele den größten Teil des Spektrums inklusive der Transitionsländer ab. Nach einer umfassenden Einleitung zu Teil Drei werden im ersten Kapitel die Gesundheitspolitiken von Deutschland, einem korporatistischen Sozialstaat, und Großbritannien, einem liberalen Wohlfahrtsregime, gegenübergestellt. In der Folge werden die Alterssicherungssysteme von Deutschland und Polen, einem Transitionsland, verglichen. Im dritten Kapitel werden die Ansätze zur Familienpolitik in Finnland und Estland, respektive einem sozial-demokratischen und einem weiteren Transitionsstaat, erörtert. Schließlich werden im vierten Kapitel die politischen und institutionellen Regelungen und Einrichtungen behandelt, die Belgien und Dänemark, ein korporatistisches und ein sozial-demokratisches Wohlfahrtsregime, im Bereich der Armutsprävention anwenden. Mit der exemplarischen Analyse der ausgesuchten Felder wird versucht, repräsentativ die wichtigsten Themen und die Vielfalt der Sozialpolitiken in Europa abzudecken und daraus die Schussfolgerungen zu ziehen, die für allgemeine Einschätzungen und Handlungsempfehlungen notwendig sind. Dasselbe gilt für die vergleichsweise untersuchten Länder. Der Studie liegt nicht die Absicht zugrunde, einen Gesamtüberblick der Politiken aller Mitgliedstaaten der EU zu geben. Indem je zwei Ländern, die mehr oder weniger exakt verschiedene Sozialstaatstypen repräsentieren, in vier zentralen Bereichen der Sozialpolitik gegenübergestellt werden, sollen vielmehr die Differenzen und Gemeinsamkeiten sowohl aus einer vergleichenden statischen als auch aus einer dynamischen reformorientierten Entwicklungsperspektive aufgezeigt werden.

Ausgehend von den zentralen Elementen des Paradigmas des ‚enabling welfare state', dem ‚capabilities'-Ansatz und der Lebenslaufperspektive, sowie der Nachhaltigkeit (als einer Grundbedingung für die Beständigkeit und Wirksamkeit jedweden Sozialsystems), bringt die Untersuchung der vier sozialpolitischen Handlungsfelder die wichtigsten Probleme und Fragestellungen ans Licht.

Familienpolitik (ausgerichtet auf Kindheit und Elternschaft) und Alterssicherung differenzieren, strukturieren und unterstützen spezifische Phasen des Lebenslaufs und fördern die ‚capabilities' der Leistungsempfänger während dieser Lebensabschnitte. Armutsprävention und Gesundheitsversorgung funktionieren als Vorkehrungen gegen zwei grundlegende Lebensrisiken, Armut und Krankheit, die nicht an spezifische Lebensphasen gebunden sind, sondern mehr oder weniger zufällig und vorübergehend Lebenspläne unterbrechen oder Sicherheitserwartungen beeinträchtigen können. Indem in beiden Feldern die Prioritäten auf präventive anstatt kompensatorische Maßnahmen gelegt werden, können wiederum Vermögen und Fähigkeiten der Menschen verbessert werden: durch Gesundheitserhalt durch Vorbeugung, durch Investitionen in den Erhalt und die Verbesserung von produktiven und sozialen Qualifikationen und der Arbeitsfähigkeit, insbesondere wenn Armutsprävention mit beschäftigungspolitischen Aktivierungsmaßnahmen kombiniert wird. Probleme der Finanzierung des Gesundheitswesens und der Rentensysteme verweisen auf die Frage der Nachhaltigkeit und der Umverteilung von Mitteln – gesellschaftlich zwischen Altersgruppen, individuell zwischen Lebensabschnitten. Eng damit verbunden sind Fragen zur Höhe von Sicherheits- und Fürsorgeerwartungen und – daran anschließend – der Verteilung von Verantwortung zwischen den Bürgern, dem Staat und den gesellschaftlichen Institutionen.

Aufgrund der einheitlichen Gesamtplans der Untersuchung sind alle vier Kapitel gleich gegliedert. Die Ausgangslage bildet jeweils ein Überblick über die wichtigsten Aspekte des sozialpolitischen Handlungsbereichs. Darauf folgt eine Beschreibung der politischen Ansätze in den Ländern, die Institutionen, Methoden, Entwicklungen, Herausforderungen und Ziele beider Staaten in dem Bereich umfasst. Anschließend werden in einem Vergleich Ähnlichkeiten und Eigenarten, Stärken und Schwächen hervorgehoben. Die Kriterien für den Vergleich sind zum einen die beiden Grundelemente des ‚enabling welfare state', der Ansatz der ‚capabilities' und die Lebenslaufperspektive, zum anderen das Konzept der Nachhaltigkeit. Auf diese Weise werden die beiden Länder nicht nur im Bezug zueinander verglichen und beurteilt, sondern auch hinsichtlich des Grades, zu dem sie entweder bereits Maßnahmen umsetzen, die die Elemente des Paradigma realisieren, oder offen sind für die dafür notwendigen Veränderungen und Reformen. In einer weiteren Sektion werden die europäischen Aspekte und Elemente des jeweiligen Politikfeldes herausgearbeitet. Einerseits wird dabei der Fokus auf die Interaktion zwischen der supranationalen und der nationalen Ebene gerichtet, mit besonderer Berücksichtigung der Union, die mit ihren Institutionen und politischen Maßnahmen in anderen Bereichen ggf. einen wichtigen Einfluss auf den untersuchten sozialpolitischen Bereich und die von den beiden Ländern darin umgesetzten nationalen Politiken ausübt. Andererseits wird besonderes Augenmerk auf die Interaktion zwischen den Staaten gelegt, d.h. auf die Auswirkungen nationaler Politik auf andere Mitgliedstaaten entweder durch direkte Wechselbeziehungen, Vergleich und Wettbewerb oder indirekt über die Institutionen und Mechanismen auf Unionsebene, vor allem die OMK. Abschließend werden die beiden nationalen Systeme in mehrfacher Hinsicht beurteilt. Dabei werden die folgenden Kriterien berücksichtigt: die Fähigkeit, die zukünftigen Herausforderungen zu meistern; der Bedarf und das Potential für Neuerungen und Reformen in Richtung einer ‚befähigenden' Wohlfahrtsgesellschaft; das Ausmaß, in dem die ethischen Prinzipien der personalen Autonomie, der sozialen Inklusion und der Verteilungsgerechtigkeit respektiert werden; die Möglichkeiten für wechselseitiges Politiklernen und für die Adaptierung von ‚best practices'; die verfügbaren Optionen für Maßnahmen auf Unionsebene und die zu erwartenden Auswirkungen der meistversprechenden Eingriffe.

Das Gesamtbild, das sich aus der detaillierten Analyse der vier Bereiche und der sieben Staaten ergibt, zeichnet sich vor allem durch Heterogenität aus. Obwohl die Vergleiche nicht vorrangig darauf ausgerichtet sind, gute und schlechte Praktiken gegenüberzustellen, ermöglichen sie eine Beurteilung von besseren und schlechteren Beispielen und damit die Ermittlung von empfehlenswerten Modellen für Innovation und wechselseitiges Lernen. Dabei darf allerdings nicht vergessen werden, wie schwierig es ist, Maßnahmen und Methoden von einem nationalen Kontext in einen anderen zu übertragen, und wie spezifisch einige der dringlichsten Herausforderungen der einzelnen Staaten sind.

Beispielsweise ist es im Fall der Gesundheitsversorgung, in dem Deutschland und Großbritannien verglichen werden und sich beide Systeme in vielen auch grundlegenden Aspekten voneinander unterscheiden, nicht sinnvoll, aus der Tatsache, dass Deutschland Großbritannien hinsichtlich Qualitätsstandards, Wahlfreiheit, Reaktionsfähigkeit und Zugangsgleichheit übertrifft, den trügerischen Schluss zu ziehen, dass das britische System dem deutschen Modell entsprechend verändert und angepasst werden sollte. Aus Sicht des Paradigma des ‚enabling welfare state' ist es vielmehr wichtig aufzuzeigen, inwieweit sich die beiden Systeme verschiedenen Herausforderungen mit unterschiedlicher Dringlichkeit stellen. Denn davon hängt es ab, wie die Prioritäten zwischen Kosteneindämmung, Effizienz und langfristiger Nachhaltigkeit einerseits (Deutschland), Verbesserung der Zugangsgleichheit, Erweiterung der Wahlmöglichkeiten und Eigenverantwortung der Patienten andererseits (Großbritannien) festgelegt werden müssen, um einen ‚befähigenden' Ansatz in der Gesundheitsversorgung zu verwirklichen. Darüber hinaus werden die beiden Systeme auf unterschiedliche Weise durch die Europäische Union beeinflusst. Während die Rechtssprechung des EuGH das britische Problem der Wartelisten behandelt, ist die gesetzliche Krankenversicherung in Deutschland, je mehr sie Gesundheitsleistungen vom gesetzlichen Versicherungskatalog ausschließt und stattdessen mit zusätzlichen privaten Versicherungen abdeckt, dem Wettbewerbsrecht und dem Gemeinsamen Markt ausgesetzt. Außerdem werden beide Länder von der weiteren Ausdehnung des Marktes auf Güter und Dienstleistungen, Anbieter und Versicherer im Bereich der Gesundheit und durch die Anwendung der OMK auf die Gesundheitsversorgung betroffen sein.

Auf dem Gebiet der Alterssicherung sind Polen und Deutschland in gleicher Weise um angemessene und nachhaltige Renten bemüht und durch die Entwicklungen in der Altersstruktur und auf dem Arbeitsmarkt ähnlichem Druck ausgesetzt. Eine der wichtigsten Fragen ist, wie öffentliche Pensionssysteme im Sinne der Nachhaltigkeit angepasst und mit privater Vorsorge ergänzt werden können, um ein zufrieden stellendes und angemessenes Rentenniveau zu garantieren. Der Streitpunkt liegt darin, in welchem Maße öffentliche Renten ggf. nur ein Mindesteinkommen im Alter absichern sollen, während der Erhalt des vor dem Ruhestand erreichten Lebensstandards durch zusätzliche private Vorsorge erbracht werden soll. Wobei die Versicherten für letztere nicht nur selbst verantwortlich wären, sondern auch eine größere Wahlfreiheit hätten. Ob in dieser Hinsicht das polnische Drei-Säulen-Modell als Vorbild dienen kann, hängt sowohl von seinem Erfolg, als auch von den spezifischen Ausgangsbedingungen für die Umsetzung mehr oder weniger radikaler Reformen ab. Die außergewöhnliche Transformationssituation in Polen, in der die finanzielle Nachhaltigkeit der vormals existierenden Alterssicherung langfristig nicht gewährleistet werden konnte, forderte, aber ermöglichte auch Optionen für politisches Handeln, die sich von denen der gegenwärtigen Situation in Deutschland klar unterscheiden. Deren spezifischen Anforderungen scheint ein langsames Vorgehen mit graduellen Veränderungen der Alterssicherung eher gerecht zu werden. Dennoch ist abzusehen, dass der Übergang von umlagefinanzierten

Rentensystemen zu Kombinationen mit kapitalgedeckter Vorsorge, unabhängig ob freiwillig oder obligatorisch und mehr oder weniger subventioniert, für alle europäischen Länder unvermeidbar ist. Da tatsächlich alle in ähnlicher Weise unter Handlungsdruck stehen, wurde die Anwendung der OMK auch auf den Bereich der Renten erweitert, um Bemühungen zu koordinieren und Informationen über ‚best practices' auszutauschen, mit dem Ziel, angemessene und nachhaltige Renten EU-weit zu gewährleisten. Im Vergleich mit der einzigen bereits vorher existierenden Form einer Europäisierung der Alterssicherung – dem Koordinierungsrecht für Wanderarbeitnehmer – stellt die OMK eine innovative Komponente der supranationalen Politik dar. Dennoch werden die Rentensysteme auch in absehbarer Zukunft stärker national ausgerichtet bleiben als andere Bereiche der Sozialpolitik. Allerdings wird sich mit dem steigenden Anteil privater Altersvorsorge der Einfluss der EU über das Wettbewerbsrecht und den gemeinsamen Markt erhöhen.

Auf dem Feld der Familienpolitik scheinen die Folgerungen aus dem Vergleich klarer zu sein. Die Familienpolitik in Finnland ist eindeutig weiter fortgeschritten im Sinne des ‚enabling welfare state' und kann deshalb als ein Modell für Politiklernen dienen. Doch der Vergleich zwischen Finnland und Estland zeigt auch, wie grundlegend verschieden mit Familienpolitik verbundene Zielsetzungen sein können. Betrachtet man die wichtigsten Ziele der Familienpolitik in dem skandinavischen Land, die Vereinbarkeit von Familie und Beruf insbesondere auch für Väter sowie der Schutz und die Unterstützung der Kindheit über einen vergleichsweise langen Zeitraum, dann stellt sich angesichts der dramatischen demographischen Situation in dem kleinen Land die Frage, ob die estische Politik die selben Ziele anstreben sollte. Die spezifische Prioritätensetzung in Estland mit ihrer klaren Ausrichtung auf die Steigerung der Natalität, derzufolge hauptsächlich Geburten gefördert und das erste Lebensjahr des Kindes unterstützt werden, kann ebenso annehmbar und angemessen sein – zumindest solange die besonderen Bedingungen fortbestehen. Letztendlich sind in grundlegender Hinsicht die familienpolitischen Ziele in den beiden Ländern trotz der Unterschiede ähnlich: beide erleichtern es den Eltern, die Kosten und den Aufwand für Kinder zu tragen, beide unterstützen die Bildung der Familie in den frühen Lebensphasen des neugeborenen Kindes und beide entwickeln Maßnahmen zur besseren Vereinbarkeit von Arbeit und Familie. Allgemein betrachtet können beide Strategien als ‚befähigend' beschrieben werden, da sie kritische Abschnitte des Lebenslaufs von Kindern und (jungen) Eltern unterstützen und damit die Bemühungen letzterer, den doppelten Anforderungen – als Berufstätige am Arbeitsmarkt und als Erzieher und Pfleger in der Familie – gerecht zu werden. Die EU nimmt bisher auf dem Gebiet der Familienpolitik keinen direkten Einfluss. Dennoch hat die supranationale Politik indirekte Auswirkungen auf nationale Familienpolitiken vor allem durch beschäftigungspolitische Maßnahmen zu Förderung der Vereinbarkeit von

Arbeit und Familie und zur Stärkung der Rolle der Väter sowie durch gesetzliche Anstrengungen im Bereich der Geschlechtergleichstellung.

Auf den ersten Blick scheint die Situation in Belgien und Dänemark die besten Chancen für Politiklernen anzubieten. Obwohl beide Länder ziemlich erfolgreich die finanzielle Armut auf einem niedrigen Niveau halten, erzielt Dänemark bessere Ergebnisse auf dem Arbeitsmarkt. Dies wird durch eine ausgeprägt aktive Beschäftigungspolitik erreicht, die auf Investitionen in das Humankapital, auf die Flexibilisierung des Arbeitsmarktes und auf die soziale Partizipation insbesondere von Frauen, jungen und alten Menschen ausgerichtet ist. Zu den erfolgreich umgesetzten Maßnahmen zählen im Konkreten die Einschränkung von Vorruhestandsregelungen, die Kürzung der Bezugsdauer der Arbeitslosenunterstützung und die Subventionierung von niedrig qualifizierten Arbeitern. Trotzdem bestehen auch in diesem Fall, in dem Erfolg versprechende Maßnahmen klar identifiziert werden können, signifikante institutionelle und politische Hindernisse und Pfadabhängigkeiten, die ein wechselseitiges Politiklernen und die Übernahme von anderswo bewährten Regelungen und Praktiken erschweren. In Belgien bleibt der Arbeitsmarkt aufgrund der Frühverrentungsoptionen, der hohen steuerlichen Belastung der Löhne, der langen Anspruchzeiten für Sozialleistungen und der umfassenden Regulierung weiterhin unflexibel und damit ein Hindernis für die soziale Inklusion der Menschen. Obwohl die Erfahrungen in Dänemark einen Weg aufzeigen für die Umsetzung einer ‚befähigenden' Politik im Bereich der Armutsprävention – durch den breiteren Einsatz von Aktivierungsprogrammen, durch die Erhöhung der Beschäftigung im Dienstleistungsbereich, durch die Verknüpfung von Sozialausgaben mit Investitionen in Human- und Sozialkapital – bleibt die Übertragung auf einen anderen Kontext problematisch und der erhoffte Erfolg schwer zu erreichen. Angesichts der Ergebnisse des Vergleichs scheint die Zielsetzung, die Politikkonvergenz auf der Unionsebene durch die Anwendung „weicher" rechtlicher Instrumente und der OMK zu verbessern, sehr hoch gesteckt. Dennoch zählen die Förderung der sozialen Inklusion und die Bekämpfung der Armut zu den wichtigsten Zielen der EU und die Anwendung der OMK ist in diesem Politikfeld weiter fortgeschritten und erfolgreicher als in anderen Bereichen. Dabei darf allerdings nicht übersehen werden, dass Politiklernen nicht einfach die Übernahme von fertigen Elementen aus einem System in ein anderes bedeutet, sondern vielmehr als Unterstützung für die Entwicklung neuer Ideen im eigenen nationalen Kontext zu verstehen ist.

Bezüglich der europäischen Ebene ergibt die Untersuchung der vier sozialpolitischen Felder ein ziemlich unausgewogenes und komplexes Bild. Der Einfluss der Unionspolitik auf nationale Politiken steigt sowohl auf direktem als auch auf indirektem Wege, weshalb die Bedeutung der Union auf beiden politischen Handlungsebenen in umfassender Weise berücksichtigt werden muss. Trotzdem scheint, solange das Subsidiaritätsprinzip die Kompetenzverteilung im sozialen Bereich bestimmt, nur das „weiche" Instrument der

OMK zur Verfügung zu stehen. Dabei bleibt aber die Frage, welche Zwecke mittels der OMK am besten verfolgt werden können und sollen, weitgehend offen – wie auch die Schlussfolgerungen zum Erfolg der OMK in den vier untersuchten Bereichen zeigen.

Elemente einer europäischen Sozialpolitik

In Teil Vier werden die Zwischenergebnisse aus den exemplarischen Analysen der Politikfelder und Beispielländer mit den zentralen Fragestellungen aus Teil Eins und Zwei zusammengeführt. Die Argumente für eine Orientierung der Sozialpolitik am Paradigma des ‚enabling welfare state' werden zusammengefasst und die gegenwärtigen Ausgangsbedingungen und zukünftigen Aussichten eines Paradigmenwechsels aufgezeigt. Auf dieser Grundlage lassen sich Handlungsempfehlungen und Strategien einer ‚befähigenden' Sozialpolitik in der Europäischen Union formulieren. Die Empfehlungen und Vorschläge werden in zweierlei Hinsicht gegeben: einerseits zu konkreten Maßnahmen einer ‚befähigenden' Sozialpolitik in den verschiedenen, vorher untersuchten Politikfeldern, andererseits zu den Optionen und Instrumenten, die für die Rationalisierung („streamlining") und Integrierung einer europäischen Sozialpolitik auf Unionsebene notwendig sind.

Das Thema einer europäischen Sozialpolitik bringt eine zweifache Fragestellung mit sich. Die erste Frage ist, wie sozial eine europäische Politik sein soll. Sie betrifft die Beziehungen zwischen Wirtschafts-, Beschäftigungs- und Sozialpolitik und verweist auf ein Verständnis der sozialen Sicherung als eines Produktivfaktors; ein Verständnis, das durch den Ansatz der ‚capabilities' und die Lebenslaufperspektive in politisches Handeln übertragen wird. Diese Frage betrifft auch das Problem, durch welche Interventionen und Mechanismen das Paradigma des 'enabling welfare state' in den vier Politikbereichen Gesundheitsversorgung, Alterssicherung, Familienpolitik und Armutsprävention in die Praxis umgesetzt werden kann. Sozialpolitik muss sich lösen von der Befürchtung, dass Beschäftigung und Wettbewerbsfähigkeit beeinträchtigt werden, wenn Sozialkosten zu hoch und Regulierungen zu aufwendig sind, um sich dem Ziel zuzuwenden, mobiles Kapital durch das Angebot einer hoch qualifizierten Arbeitskraft anzuziehen, welche durch eine aktive Politik der Investition in Humankapital bereitgestellt werden kann. Eine derartige Politik wird zunehmend als ein Mittel sowohl zur Steigerung von wirtschaftlichem Wachstum und Innovation als auch zur Erweiterung der Chancen für die gesamte Bevölkerung angesehen.

Die zweite Frage ist, wie europäisch nationale Sozialpolitik sein soll. Dem Subsidiaritätsprinzip zufolge kommt der Union im sozialen Bereich keine Funktion zu, außer wenn gemeinsam beschlossene soziale Ziele ohne Einwirkung seitens der Union nicht erreicht werden können. Dagegen verlangt Subsidiarität in der aktiven Lesart des Begriffs, dass die Union ihren Einfluss auf die einzelnen Mitgliedstaaten als solche und auf die Union als Ganzes positiv geltend machen soll. In diesem Sinne scheint im Sozialbereich eine Europäische Einwirkung auf die nationale Ebene Teil des ‚Europäischen Sozialmodells' zu sein. Aber die EU kann keine Aufgaben und Leistungen eines institutionellen Wohlfahrtsstaats, sondern nur eine regulierende Funktion übernehmen und damit einen gewährleistenden und fördernden Einfluss auf die Sozialpolitiken auf Mitgliedsstaatsebene ausüben. Das Ziel einer ‚befähigenden' Europäische Sozialpolitik sollte die Koordinierung der nationalen Politiken und die Stärkung ihrer Orientierung am Paradigma des ‚enabling welfare state' sein und dies sollte mit regulierenden Interventionen verfolgt werden. In welchem Maße eine solche gewährleistende Rolle der Union mehr rechtlicher Regelung bedarf, hängt vom Erfolg der „weichen" Koordination mittels der OMK in den verschiedenen Bereichen ab. Jedenfalls impliziert die Einflussnahme der Union nicht zwingend, dass am Ende eine Konvergenz der verschiedenen Systeme erreicht werden soll. Koordinierung kann auch darin bestehen, die Vielfalt zu regeln und zu lenken, was bedeutet, dass die verschiedenen Länder zwar ähnliche Instrument verwenden, um gemeinsame Ziele zu erreichen, die Ausgestaltung der konkreten Anwendung dieser Instrumente aber von Land zu Land unterschiedlich ausfällt. Es kann auch bedeuten, dass nicht nur die Umsetzungsweisen voneinander abweichen, sondern auch unterschiedliche, spezifische Instrumente zum Einsatz kommen, um mehr oder weniger ähnliche Resultate zu erzielen. Darüber hinaus ist die Lenkung und Regelung von Vielfalt auch mit der Setzung von Mindeststandards vereinbar, insofern diese relativ bestimmt und den nationalen Bedingungen angepasst werden. Zudem wären national differenzierte Mindeststandards auch mit dem Subsidiaritätsprinzip kompatibel.

Es mag paradox erscheinen, für eine stärkere Einmischung der EU in nationale Sozialpolitik zu argumentieren, zu einem Zeitpunkt, zu dem die Mitgliedstaaten angesichts des durch internationale wirtschaftliche Veränderungen steigenden Wettbewerbsdrucks und bekräftigt durch das Subsidiaritätsprinzip, Sozialpolitik mehr denn je als vorrangig nationale Domäne ansehen. Und noch mehr scheint es paradox, einen größeren Einsatz der sozialen Institutionen, insbesondere des Staates zu fordern, wenn die meisten Regierungen versuchen, ihre Sozialausgaben zu kürzen oder wenigstens auf den bestehenden Niveaus einzufrieren. Doch beide Paradoxa lassen sich aus einem breiteren Blickwinkel auflösen. Zum ersten Punkt muss hervorgehoben werden, dass der internationale Wettbewerb zwischen den großen Wirtschaftsregionen der Welt viel stärker als in der Vergangenheit geworden ist. Deshalb ist die EU zuallererst dem Wettbewerb mit den USA, China und den anderen asiatischen Länder ausgesetzt. Um in diesem Wettbewerb der globalen Wirtschaft erfolgreich zu sein, kann es für Europa von Vorteil sein, den Wettbewerb innerhalb seiner Grenzen abzuschwächen und nach außen als ein einheitlicher wirtschaftlicher Block zu agieren, anstatt als eine Vielzahl von Ökonomien, die in wechselseitiger Konkurrenz hinsichtlich Lohnhöhe, Sozialkosten und Steuerlast zueinander stehen. Zu diesem Zweck kann eine stärkere Einbindung der Union nicht nur in wirtschaftlichen, sondern auch in sozialen Belangen angebracht sein. Zum zweiten Punkt muss daran erinnert werden, dass die Vorstellung einer ‚enabling welfare society' eine aktivere Rolle der sozialen Akteure einschließlich des Staates verlangt, dies aber nicht zwingend zu höheren Staatsausgaben führen muss. Der Schwerpunkt der Sozialpolitik muss weg von Kompensation hin zu mehr Prävention verlagert werden, die Einkommenssicherung für die Nicht-Beschäftigten muss zweitrangig werden im Vergleich zur Stärkung der aktiven Teilnahme am Arbeitsmarkt durch die Erhöhung der Qualität des Humankapitals. Ziel einer solchen Politik ist es, den Bürgern mehr Wahlmöglichkeiten zu eröffnen und sie zu befähigen, mehr Eigenverantwortung für ihre Lebensplanung und -führung zu übernehmen. Dabei muss auch der über flexible und heterogene Lebensläufe wechselnde Bedarf an Zeit, Geld und Bildung berücksichtigt werden. Die Funktion des Staates verschiebt sich von Schutz und Sicherung durch Umverteilung von finanziellen Ressourcen zu mehr Kontrolle, Koordinierung und Steuerung von Leistungen, die entweder vom Staat oder von anderen, privaten und nicht-staatlichen Organisationen erbracht werden. Diese Verschiebung setzt eine erweiterte Zuständigkeit des Marktes und anderer gesellschaftlicher Organisationsformen für die Erbringung sozialer Leistungen voraus: private marktwirtschaftliche Unternehmen, bürgerliche und zivile Einrichtungen, Wohltätigkeits- und karitative Institutionen, nicht-staatliche Organisationen, individuelle informelle Netzwerke, insbesondere Verwandte und Freunde.

Die Studie endet – ausgehend von diesen Schlussfolgerungen – mit der Empfehlung von Handlungsstrategien für die Definition von konkreten politischen Maßnahmen auf nationaler und Europäischer Ebene, die den Grundlagen und den Zielsetzungen eines ‚befähigenden' sozialpolitischen Modells entsprechen. Einige dieser Maßnahmen werden bereits umgesetzt und reflektieren seit kurzem anlaufende Veränderungsprozesse in einigen Ländern und auf Unionsebene. Andere stehen zurzeit zur Debatte und sind Teil der Modernisierungs- und Innovationstrends der Sozialsysteme in Europa. Die Empfehlungen werden in Bezug auf die vier im dritten Teil der Studie untersuchten sozialpolitischen Handlungsfelder formuliert: Gesundheitsversorgung, Altersicherung, Familienpolitik und Armutsprävention. Zusätzlich werden im Rückblick auf den zweiten Teil, in dem – in Anbetracht ihrer übergreifenden Bedeutung – Beschäftigungs- und Arbeitsmarktpolitik aus einer Europäischen Perspektive betrachtet wurden, Vorschläge auch für diesen Bereich gemacht. Im Anschluss daran werden spezifische Empfehlungen für Vorgehensweisen und Instrumente auf Unionsebene gegeben. Im Folgenden werden nur die wichtigsten Punkte der Vorschläge skizziert.

Gesundheitsversorgung – Auf der Grundlage der Prinzipien der Zugangsgleichheit, der sozial gerechten Finanzierung und der evidenz-basierten Medizin sollten national angepasste Mindeststandards als ein Sicherheitsnetz definiert werden, durch das auch in den ärmeren Gebieten der Union ein angemessenes Niveau der Gesundheitsversorgung gewährleistet wird. Um eine klare Trennung von Allokation und Distribution in den Finanzierungsstrukturen zu realisieren, müssen re-distributive Elemente in größerem Umfang über das allgemeine Steuersystem finanziert werden. Durch die Einführung von obligatorischen privaten Versicherungssystemen für die gesamte Bevölkerung – mit öffentlicher Unterstützung für Menschen mit geringem Einkommen – sollte mehr Raum für marktwirtschaftliche Angebote geschaffen werden. Zusätzlich könnte die Umstrukturierung der gesetzlichen Krankenversicherung in Richtung Teilkapitaldeckung die Nachhaltigkeit und Effizienz des gesamten Systems verbessern, wodurch sowohl den demographischen Herausforderungen begegnet als auch dem Prinzip der intergenerationellen Gerechtigkeit entsprochen werden könnte. Im Zuge der weiteren Umsetzung des EU-Vertrags insbesondere der Personenfreizügigkeit und des freien Verkehrs von Kapital, Gütern und Dienstleistungen könnten bestehende Einschränkungen der grenzüberschreitenden Gesundheitsversorgung abgesenkt und damit die Wahlmöglichkeiten der Konsumenten zwischen unterschiedlichen Leistungserbringern, Behandlungsarten und (öffentlichen und privaten) Versicherungen erweitert werden. Schließlich ist für den Aufbau und den Erhalt von ‚capabilities' über den gesamten Lebenslauf eine stärkere Ausrichtung der Gesundheitsversorgung auf Vorsorge unabdingbar.

Alterssicherung – Die Bemühungen zur Verbesserung der Systeme der Alterssicherung zielen auf die Aufrechterhaltung der langfristigen finanziellen Nachhaltigkeit und eines angemessenen Einkommensersatzes im Alter. Durch den erhöhten Einsatz von privaten und kapitalgedeckten Systemen sollte ein ausgewogeneres Verhältnis zwischen verschiedenen Gestaltungsoptionen der Alterssicherung verwirklicht werden. Damit könnte auch die Nachhaltigkeit der staatlichen Systeme verbessert sowie der Umgang mit neuen flexibleren Lebenslaufmustern erleichtert werden. Größere Wahlfreiheit, Diversität und Flexibilität könnten durch die Individualisierung der staatlichen Systeme (individuelle Rentenkonten, beitragsdefinierte Umlagefinanzierung), durch den Übergang von leistungsdefinierten zu beitragsdefinierten Systemen oder die Entwicklung von privaten Systemen mit voller Kapitaldeckung, durch flexible Ruhestandslösungen und durch die Einführung von Teilrenten realisiert werden. Die letzten beiden Instrumente können auch der Aktivierung älterer Arbeitnehmer und der Verlängerung des Arbeitslebens dienen. In jedem Fall muss das tatsächliche Renteneintrittsalter durch die Einschränkung von Frühverrentungsoptionen erhöht werden. Im Übrigen muss die Altersvorsorge mit anderen Politikbereichen, die auch den Lebensstandard im Alter beeinflussen, abgestimmt werden: die Gesundheitsversorgung, die Langzeitpflege, die sozialen Dienste und die Wohnungspolitik. Zum Zweck der weiteren Absenkung von Mobilitätsbarrieren und der Verbesserung des internationalen Politiklernens sollte die OMK vorrangig auf die Erarbeitung von Indikatoren fokussiert und auf das Ziel einer wechselseitigen Abstimmung („streamlining") der verschiedenen, für die Sicherheit im Alter bedeutsamen politischen Bereiche, ausgerichtet werden. Die Koordinierungsmechanismen der sozialen Sicherungssysteme in der EU sollten weiter ausgebaut werden, insbesondere um die grenzüberschreitende Portabilität von betrieblichen Rentenansprüchen zu verbessern.

Familienpolitik – Die Vereinbarkeit von Arbeits- und Familienleben ist nicht nur für die Unterstützung der personalen Autonomie und die Erweiterung der individuellen Fähigkeiten fundamental, sondern auch und noch mehr für das Wohlergehen der Kinder. Dieses sozialpolitische Ziel kann vor allem dadurch erreicht werden, dass die Optionen für die Inanspruchnahme von Elternschaftsurlaub erweitert werden, dass die Rolle der Väter in der Kinderpflege gestärkt wird und das Angebot der Kindertagesstätten flexibler gestaltet wird. Durch eine verstärkte Berücksichtigung von Familien mit Kindern im Schulalter kann Familienpolitik auch dem zunehmenden Problem der frühen Schulabbrecher entgegenwirken. Dazu eignen sich verlängerte und flexiblere Lösungen für Teilzeitbeschäftigung und Stellenteilung, eine flexiblere Gestaltung von Schulzeiten und bessere Einrichtungen für die außerschulische Betreuung, die auch von privaten und zivilgesellschaftlichen (nicht-staatlichen) Organisationen erbracht werden kann. Die Rechte von Kindern müssen gestärkt werden, beispielsweise in Form eines Rechts auf eine würdige, angemessene Kindheit oder eines Rechts, von beiden Elternteilen erzogen zu werden; und sie müssen auch unterstützt werden, indem Kinderarmut bekämpft wird, Alleinerziehende entlastet werden und allen Kindern gleiche Bildungschancen garantiert werden. Sowohl die enge wechselseitige Abstimmung von Familien- und Beschäftigungspolitik als auch die Stärkung der Rechte der Kinder sind nur machbar, wenn sie auf einer verbindlichen unionsweiten Einigung beruhen, die ggf. entweder durch die Setzung von Rechten und/oder Mindeststandards oder durch die erweiterte Anwendung der OMK im Bereich der Familienpolitik durchgesetzt werden kann. Mit der Alterung der Gesellschaft steigt auch der Anteil alter Menschen, die von Pflege abhängig sind. Gleichzeitig steigt aber das Pflegeangebot nur unterproportional. Deshalb sollte Familienpolitik auch Männer und Frauen mittleren Alters unterstützen, die für die Pflege ihrer Eltern sorgen wollen. Dabei sollte die Verantwortung allerdings nicht vollständig auf die Familien und den Markt übertragen werden, sondern eine Mitverantwortung des Staates und der öffentlichen Institutionen bestehen bleiben.

Mindestsicherung und Armutsprävention – Im Bereich der Mindestsicherung und Armutsprävention verlangt der Übergang von einem kompensierenden zu einem ‚befähigenden' und aktivierenden Ansatz vor allem eine neue Bestimmungsweise von Armut: anstatt einer einkommensabhängigen Definition bedarf es eines mehrdimensionalen Konzepts. Darüber hinaus müssen neben statischen Indikatoren, wie bspw. dem geringen Einkommen zu einem gegebenen Zeitpunkt, auch Langzeitindikatoren betrachtet werden, die die Wohnungssituation, den Gesundheitszustand, die Möglichkeiten für Bildung und soziale und kulturelle Partizipation über die Zeit erfassen, um daraus ein umfassendes und dynamisches Bild der Strukturen sozialer Ausgrenzung zu gewinnen. Da Langzeitarbeitslose, allein erziehende Mütter und Kinder die Gruppen mit dem höchsten Armutsrisiko sind, muss der Schwerpunkt der Armutspolitik auf folgende Maßnahmen gelegt werden: die Vorbeugung gegen langfristige Arbeitslosigkeit, die schnelle Wiedereingliederung in den Arbeitsmarkt, die Vereinbarkeit von Arbeit und Pflege insbesondere für Alleinerziehende, die Erhöhung der weiblichen Beteiligung am Arbeitsmarkt und die Gewährleistung einer hinreichenden Mindestsicherung. Zu diesem Zweck und insbesondere hinsichtlich der Mindestsicherung können neben nationalen auch europäische Mindesteinkommen und/oder Armutsgrenzen festgelegt werden, die den spezifischen Bedingungen und Situationen in den einzelnen Mitgliedstaaten entsprechend angepasst und relativiert werden sollten. Die Maßnahmen zur Armutsprävention und sozialen Inklusion müssen mit einer aktiven Arbeitsmarktpolitik verbunden werden, insbesondere wenn der Anspruch auf soziale Fürsorge an die Arbeitsbereitschaft geknüpft wird. Dieser integrative Ansatz sollte auch bei der Anwendung der OMK auf diese eng verstrickten Politikbereiche verfolgt werden.

Beschäftigungs- und Arbeitsmarktpolitik – Eine ‚befähigende' Arbeitsmarktpolitik sollte vor allem zwei Ziele verfolgen: erstens die Aktivierung der Arbeitslosen und zweitens die Flexibilisierung des Arbeitsmarktes. Im Sinn des ‚flexicurity'-Ansatzes sollte die Beschäftigungspolitik generell darauf abzielen, Übergänge und Wechsel im Arbeitsmarkt dadurch zu erleichtern, dass die damit verbundenen Risiken reduziert werden. Diese Ziel kann auf mehrfache Weise verfolgt werden: indem die Investitionen in die Bildung von Humankapital mittels lebenslangem Lernen, Weiterbildung, Training, und Qualifikationsentwicklung erhöht werden, indem Aufstiegschancen erweitert und verbessert werden, indem dem Abstiegsrisiko und dem möglichen Verlust des Arbeitsplatzes aufgrund obsoleter Qualifikationen vorgebeugt wird. Rechtliche Veränderungen mögen zu einem weniger regulierten Arbeitsmarkt führen, mit geringerem Beschäftigungsschutz und schwächeren Gewerkschaften, doch in Verbindung mit Maßnahmen zur Förderung der ‚capabilities' der Bürger können sich die Beschäftigungschancen dadurch am Ende verbessern anstatt verschlechtern. Eine aktivierende Arbeitsmarktpolitik mit dem Ziel der Steigerung von Vermittelbarkeit und Flexibilität umfasst vielfältige Maßnahmen: die Dezentralisierung und Regionalisierung der Interventionen auf dem Arbeitsmarkt; ein System mit Steuerfreibeträgen auf Arbeitseinkommen anstelle von Lohnsubventionen; bessere und flexiblere Lösungen für den Übergang in den Ruhestand, für die Freistellung zu Studien- und Weiterbildungszwecken sowie für die Freistellung von Eltern für Pflege- und andere familiäre Verpflichtungen. Die Politikbereiche der Armutsprävention und Beschäftigungspolitik sind auf Europäischer Ebene von größter Relevanz, weil sie unentflechtbar mit der Wirtschaftpolitik der Union und der Mitgliedstaaten verwoben sind. Aus diesem Grund sollten Zielvorgaben für die Bereiche Beschäftigung und Armut, angemessene Mindestsicherungsstandards für Arbeitslose und Mindestlöhne für Arbeitnehmer auf Unionsebene bestimmt und koordiniert werden unter Berücksichtigung der nationalen und regionalen Unterschiede zwischen den Mitgliedstaaten.

Die wichtigsten Instrumente auf EU-Ebene für die Koordinierung und Orientierung nationaler Sozialpolitiken in Richtung einer ‚befähigenden' Wohlfahrtsgesellschaft sind: die Steigerung der Wahrnehmung seitens der Institutionen der Union, die engere Integrierung und Abstimmung der Politikfelder, die Strukturierung der Kompetenzen und Beziehungen zwischen den verschiedenen Steuerungsebenen und die Anwendung der OMK im Bereich der Sozialpolitik.

Steigerung der institutionellen Wahrnehmung und Anerkennung – Um die europäischen Sozialsysteme in Richtung des Paradigmas des ‚enabling welfare state' zu orientieren und zu koordinieren, muss das Bewusstsein für die Bedeutung und den Einfluss einer Europäischen Sozialpolitik auf die gesamte Politik der Union gestärkt werden. Dies betrifft vor allem die Wahrnehmung der sozialen Dimension der verschiedenen supranationalen Politiken. Es muss zur Kenntnis genommen werden, dass die politischen Entscheidungen und Maßnahmen der EU generell und unabhängig von den spezifischen Politikbereichen – Grundfreiheiten, Gemeinsamer Markt, Währungs- und Wirtschaftspolitik, Struktur- und Sozialpolitik – in jedem Fall soziale Bedeutung und Wirkung haben. Deshalb ist es vor allem notwendig, dass die Relevanz des sozialen Bereichs in den Institutionen der EU an höchster Stelle erkannt und anerkannt wird, insbesondere seitens der Kommission und des Ministerrates. Die Wahrnehmung der Wichtigkeit sozialer Themen kann durch die Formulierung von klaren, anspruchsvollen und vor allem quantifizierten Zielsetzungen gestärkt werden, insbesondere wenn diese Ziele öffentlich kommuniziert und diskutiert werden. Es scheint, dass das institutionelle Bewusstsein hauptsächlich durch einen klaren Kommunikationsprozess von oben nach unten erhöht werden kann, d.h. von den höchsten organisatorischen und hierarchischen Ebenen der Institutionen der EU über die politischen Abgeordneten auf allen Regelungsebenen – auch im nationalen Bereich – bis hin zur breiten Öffentlichkeit in der Gesellschaft. Darüber hinaus muss das Paradigma des ‚enabling welfare state' auf allen politischen Entscheidungsebenen in den Vordergrund gehoben und verankert werden. Dabei geht es nicht mehr um die soziale Dimension als solche, sondern um die Explikation der konkreten Funktion der Perspektive des ‚enabling welfare state' für die Ausarbeitung und Umsetzung der entsprechenden Reformen.

Integrierung der Sozialpolitik in verschiedenen Politikbereichen – Die Wechselwirkungen zwischen den verschiedenen Bereichen der Europäischen Wirtschafts- und Sozialpolitik erfordern mehr Dialog und Zusammenarbeit unter den Institutionen, in denen die Entscheidungsprozesse in den unterschiedlichen Bereichen stattfinden. Die vollständige Umsetzung der Grundfreiheiten, des Gemeinsamen Marktes, der EWU, des SWP, der Struktur- und Agrarpolitik, aber auch schon die schrittweise Überwindung der diesen Zielen entgegenstehenden Hindernisse beeinflussen in erheblichem Umfang die Ergebnisse der Europäischen und nationalen Anstrengungen im sozialen Bereich. Dies gilt umso mehr, je größer die Bedeutung von marktwirtschaftlichen Lösungen im Sozialbereich wird, insbesondere im Gesundheitswesen und in der Altersvorsorge. Da sich daraus konfligierende Ziele und Prioritäten ergeben, lassen sich übergreifende Zielsetzungen nur durch die gemeinsame Behandlung verschiedener Politikbereiche bestimmen und erreichen. Obwohl nicht alle Wechselwirkungen aufeinander abgestimmt, koordiniert und integriert werden können, besteht die Möglichkeit, die Feinabstimmung von Wirtschafts-, Beschäftigungs- und Sozialpolitik zu verbessern, indem Steuerungsprozesse umstrukturiert, zusammengeführt und gestrafft werden. Die mächtigsten Institutionen auf diesem Feld sind die Räte, vor allem der Ministerrat und der ECOFIN. Da die Entscheidungen der Räte einerseits die größte politische Wirkung haben und andererseits die Politik der Mitgliedstaaten widerspiegeln, lässt sich leicht erkennen, dass die Verknüpfung der Sozialpolitik mit den anderen politischen Handlungsfeldern auf dieser hierarchischen Ebene angesiedelt werden muss. Abstimmung und Integration zwischen den Bereichen können in vielen verschiedenen Formen stattfinden. Eine mögliche Option ist die gelegentliche oder permanente Erweiterung des ECOFIN oder anderer Räte, entweder durch gemeinsame Sitzungen oder durch die Teilnahme der Minister für Arbeit und Soziales im ECOFIN bzw. der Finanz- und Wirtschaftsminister im ESPHCA. Darüber hinaus könnten durch eine veränderte Ablaufplanung des Entscheidungsprozesses, welche die Abfolge und Zeiten von Entscheidungsprozessen in anderen Bereichen berücksichtigt und Verfahren und Programme besser verzahnt, die Möglichkeiten der gegenseitigen Kenntnisnahme und Abstimmung von Entscheidungen verbessert werden.

Definition der Kompetenzen und Beziehungen zwischen den Steuerungsebenen – Solange jedwede Form einer supranationalen Sozialpolitik als Verletzung des Subsidiaritätsprinzips und der Souveränität der Nationalstaaten in diesem Bereich angesehen wird, scheint eine europäische Sozialpolitik ein unrealistisches Szenario ohne große Erfolgsaussichten auf eine signifikante Koordinierungswirkung zu bleiben. Die Spannung zwischen der theoretischen Bedeutung des Prinzips einerseits, und den praktischen Anforderungen andererseits, führt scheinbar zu endlosen Debatten und Diskussionen über Kompetenzen und Zuständigkeiten. Da diese Konflikte nicht nur das

Verhältnis zwischen der supranationalen und der nationalen Ebene betreffen, sondern auch zwischen den verschiedenen Regierungsebenen im nationalen Bereich – einschließlich der regionalen und kommunalen Ebenen – bestehen, ist eine klare Zuweisung und Verteilung von Befugnissen und Zuständigkeiten auf die unterschiedlichen Ebenen notwendig. Dies gilt umso mehr, je komplexer die Beziehungen zwischen den Ebenen werden, bspw. durch die Interaktion zwischen der supranationalen und der regionalen Ebene in Form von regionalen und interregionalen Programmen der EU. Abgesehen von spezifischen Fällen kommt eine weitere Übertragung von Kompetenzen auf die europäische Ebene als Lösung der Auseinandersetzung über Zuständigkeiten nicht in Frage. Stattdessen muss das Zusammenspiel zwischen den Regelungsebenen verbessert werden, wobei das Hauptaugenmerk nicht auf die bereits streng regulierten Beziehungen zwischen der nationalen und den subnationalen Ebenen gelegt werden muss, sondern auf die Interaktion zwischen diesen und der Union.

Die Offene Methode der Koordinierung – Die OMK soll drei Zwecken dienen: der Vereinbarung von allgemeinen Zielsetzungen, dem wechselseitigen Austausch von Erfahrungen zwischen den Mitgliedstaaten und dem Leistungsvergleich der Ergebnisse. Obwohl die meisten europäischen Mitgliedstaaten in ihren Sozialsystemen ähnlichen Herausforderungen und Belastungen ausgesetzt sind, können die konkreten erforderlichen Maßnahmen aufgrund von spezifischen nationalen Gegebenheiten, institutionellen Rahmenbedingungen und historischen Vorgaben stark voneinander abweichen.

Die Einschränkung der OMK auf die einvernehmliche Festsetzung von Zielen ist dem Zweck angemessen, den Mitgliedstaaten einen möglichst breiten Spielraum für die Auswahl und Anpassung von spezifischen Instrumenten für die Erreichung dieser Ziele zu schaffen. Wenngleich im Fall von unzulänglichen Ergebnissen zugegebenermaßen keine formellen Sanktionsmöglichkeiten zur Verfügung stehen, können schlechte Resultate durch öffentliche Kampagnen aufgezeigt, angeprangert und missbilligt werden („naming, shaming and blaming"), deren Wirksamkeit umso größer ist, je stärker Leistungsvergleiche im Sozialbereich in der Öffentlichkeit, und das heißt auch seitens der Wähler und Politiker, wahrgenommen werden. Möglicherweise könnte dies zu einer Wechselwirkung mit positiver Rückkopplung führen. Ein stärkeres öffentliches Bewusstsein erhöht die Wirksamkeit von öffentlichen Kampagnen und internationalen Vergleichen, die deshalb vermehrt eingesetzt werden, wodurch wiederum die öffentliche Aufmerksamkeit für soziale Themen im internationalen Vergleich steigt, so dass am Ende die Verbindlichkeit von Zielsetzungen, die mittels der OMK vereinbart werden, zunimmt. Auf diesem Wege wäre es sogar möglich, kontinuierlich anspruchsvollere Ziele festzulegen. Im Übrigen ist zu erwarten, dass internationale Vergleiche umso transparenter werden, je mehr marktwirtschaftliche Elemente Eingang in die soziale Sicherung finden, wodurch der Druck auf die Länder mit den schlechtesten Ergebnissen weiter steigt. In Bezug auf das Politiklernen und den wechselseitigen Austausch von Erfahrungen gibt es bisher nur wenige Beispiele von transnationalem Lernen und Politiktransfer. Für die Umsetzung eines effektiven Lernprozesses über die OMK reicht es nicht, Länder durch Leistungsvergleiche unter Druck zu setzen. Darüber hinaus ist es wichtig, die Unterschiede und Besonderheiten nationaler Politiken in angemessener Weise in Betracht zu ziehen. Gegenseitigen Politiklernens darf nicht als Strategie eines simplen Politiktransfers missverstanden werden, durch den erfolgreiche Praktiken einfach von einem Land auf ein anderes übertragen werden sollen

Sowohl die bisherigen Erfahrungen mit der OMK, als auch die mittelfristigen Aussichten ihrer weiteren Anwendung zeigen, dass es noch einiger Bemühungen bedarf, damit die Methode als effektives Instrument für die hier geforderten Zwecke genutzt werden kann. In den Bereichen der sozialen Inklusion und der Beschäftigung konnten gemeinsame Ziele erreicht und einheitliche Indikatoren bestimmt werden. Auf dem Gebiet der Alterssicherung sind die vereinbarten Ziele noch sehr generell und die Lernprozess nicht über die Identifizierung von optimalen Verfahren und innovativen Ansätzen hinausgekommen. Im Gesundheitsbereich wird gegenwärtig für eine vereinfachte OMK argumentiert, die nur als Plattform für den Austausch von Informationen und die Festlegung von Vergleichsindikatoren dienen soll. Eine Anwendung der OMK in der Familienpolitik ist zurzeit nicht in Sicht.

Introduction

The Emergence of a 'Social Europe'

During the last decennia, European welfare states have become ingrained in the multifaceted institutional arena of the EU. Compared to 25 years ago, when the European Community was perceived as little more than a common market, astonishing changes have taken place. Hence, in the wake of the new millennium, such a limited view on what Europe should strive for no longer appears tenable. The member states of the EU are, slowly but surely, moving ahead in the building of new institutions that also involve concerted action in the social domain – a domain previously considered to rest on purely national authority. Today, politically delicate issues such as employment, social exclusion and pensions are moving beyond national borders and treated also at European level.

Yet there is considerable uncertainty about where this process is, and should be, leading to. At the seminal European Council of Lisbon in March 2000, the EU set itself the goal of becoming 'the most competitive and dynamic knowledge-based economy in the world, capable of realising sustainable economic growth with more and better jobs and greater social cohesion'. The decision was taken to strengthen the social dimension of Europe, primarily by rerouting policies in the direction of soft law and novel open methods of policy coordination and mutual learning (OMC). Five years later, however, the evaluation of this new policy route has rendered mixed results. While the Open Method of Coordination has definitely contributed to streamlining the European policy process and to offering more opportunities for learning through comparing 'best' social policy practices, it does not seem to have provided successful cases of the transferability of these practices across national borders. Moreover, it has largely failed as a device for motivating member states to modernise their social systems or recalibrate their welfare states.

The need for European welfare states to modernise and reconstruct their social systems – each according to their own path of development and taking stock of their historical, economic and socio-cultural roots – is incited by the economic, social and demographic pressures they face. For them to cope successfully with these challenges requires a thorough understanding of which forms of social protection are better adapted to the demands of the knowledge-based economy or better capable of covering newly emerging social risks. This also requires an awareness of the kind of political methods, procedures and policies that might effectively respond to new social, economic and political conditions at both national and European level.

This book attempts to study the potentialities and limitations of European social policy in relation to European macro-economic and monetary policies, as well as to national policies in these domains. The study reflects on the challenges and opportunities that will confront Europe in the next decade and seeks to answer the intriguing question of how European and national social policy might respond to these issues more effectively. The book argues that while there seems to be a sort of implicit normative consensus about the 'European social model', there is no 'one size fits all' perspective. There appears to be room for a multifaceted world in which distinct welfare regimes seek to maintain their own path-dependent ways of achieving a fair and just society with a high level of welfare for all. The book hinges on the notion of the 'enabling welfare state' and, in the empirical part, undertakes an appraisal of the various policies and reforms fitting that approach in a pertinent selection of countries. In doing so, it scrutinises four important welfare state sectors: health care, old-age security, family policy, and poverty prevention. Within each sector, the authors comparatively explore the policies and practices in countries thought to typify the various welfare regimes: Germany and the United Kingdom, Germany and Poland, Finland and Estonia, and Belgium and Denmark. These country studies shed light on how future social policies should evolve and on the ingredients required for relevant reforms.

'Enabling Social Europe'

The title of this study – 'Enabling Social Europe' – may be read in different ways that reflect the various lines of reasoning followed throughout the volume. It might be interpreted, first and foremost, as paying regard to the idea of an enabling 'European social model' as it has been enacted under the various national social welfare systems. It could moreover refer to the ways in which European welfare states are able to cope with present and future challenges. But it might also be construed in the sense of fostering the European Union as an important political factor in the social domain next to its predominant economic significance. This line of thought pertains to the question whether the EU should become a more powerful partner in shaping and creating social policy regulations, both at Union level and within the member states. Another interpretation of 'Enabling Social Europe' addresses the national or supra-national orientation of social policies in the EU. In this last meaning, the focus is on the paradigm of the 'enabling welfare state', conceived as an instrument for innovation and modernisation of social policy approaches in the European Union.

All the questions highlighted and raised by these different ways of understanding 'Enabling Social Europe' are closely interconnected and analysed accordingly throughout the book. The final part seeks to identify strategies or modes of action for defining concrete policy innovations at national and European level.

Approaches to European Social Policy

This book is the outcome of a unique form of collaboration between specialists working in different fields such as social policy, law, economics and philosophy. Consequently, it draws on different research traditions and renders insights derived from various disciplinary backgrounds. It combines sociological, economic, legal and philosophical analysis, and at the same time connects cross-national comparative analyses of quantitative data with an in-depth analysis of the development of European social policy. Our multidisciplinary perspective is embedded in a common conceptual framework of social policy. This approach can best be explained and summarised by focusing on the main lines of analysis pursued.

As indicated by the title, the primary line of investigation is based on the idea that social policy reforms in Europe should move towards the objective of an 'enabling welfare state'. The paradigm of the enabling welfare state thus transcends the notion of the active welfare state, going further in its vision of an activating, responsible and life-course-oriented welfare state that invests in people and enables them to work and take care of themselves. The book centres on this paradigm in defining new routes for both European and national social policy, thereby tuning both levels of authority in an improved way.

The paradigm of the enabling welfare state is underpinned by a second line of thought that attempts to provide a philosophical and normative foundation for the idea of the European social model. By reflecting on the implicit normative principles and values of social policy in Europe, it has been argued that an enabling approach to a social Europe serves broader social goals such as personal autonomy, social inclusion and distributive justice.

A third and last line of reasoning is related to the challenge of European diversity. The lack of unity between the national welfare states is striking, and the social and political salience of these differences makes it practically impossible to conceive Europe as a whole, let alone to propose one-size-fits-all options. Hence, most of our analyses on 'welfare regimes' rest on the principle of heterogeneity and use the classifications proposed by Esping-Andersen and others. This allows us to take account of the substantial differences in the cultural, economic and historical evolution of the national welfare states – differences that have led policy regimes to pursue disparate objectives and follow divergent routes in building their very own, distinct welfare state institutions.

The Structure of the Book

The book consists of four parts. Part one gives a general overview of social policy in Europe in the 21st century. It starts with a brief discussion of the emergence and development of social policy, followed by an outline of the current and future challenges it faces. In a next step, the ethical foundations of social policy in Europe are examined by reconstructing the basic ethical principles underlying the European social policy domain and by delineating the normative scope for social policies in Europe. Building upon this normative framework, the paradigm of the 'enabling welfare state' is explained in more detail. The presentation of main features of an 'enabling' approach for social policy in Europe further seeks to show how the challenges to current and future policies can be coped with more adequately.

Part two discusses the significance of the European Union's policies for national social policy from two angles. The one illuminates how European national social policies are or, in other words, how EU policies affect the framework for national policy-making. It examines the consequences of European Union enlargement for the size and direction of migration and worker mobility flows. It also deals with the consequences of economic integration within the Union for European and national social policy, and addresses the role Europe plays in the formulation of member state labour market and employment policies. The second angle looks at how national European social policy is with a view to the subsidiarity principle, which limits the scope of European interference in the social domain. According to this approach, the room for EU social policy action is determined by the degree of authority the member states are willing to hand over to the Union as a whole. This part also aims at throwing more light on the institutional and legal interrelations between national and supra-national levels of governance.

Following the general analyses expounded in the first two parts, part three presents the results of the comparative analysis of four significant case-studies of social policy – health care, old-age security, family policy, and poverty prevention. For each social policy area, the differing welfare regimes of two countries are compared (Germany and the United Kingdom, Germany and Poland, Finland and Estonia, and Belgium and Denmark). The focus in this empirical part is on the comparison of regimes and practices with a view to their potential for designing 'enabling policies' and their relevance for policy learning and coordination practices at EU level.

Part four draws upon the conclusions of the previous analyses to assess whether and how social policy in the EU can be oriented to the paradigm of the enabling welfare state. Finally, it recommends strategies of action and suggests instruments for the reform of social policies at both Union and member state level.

1 Social Policy in the 21st Century

1.1 Development and Challenges of the Social Policy

1.1.1 The Development of European Welfare Societies

This first section pursues two interrelated objectives. Firstly, it offers a synoptic reconstruction of the development of national welfare institutions in Europe from their origins in the last third of the nineteenth century through the period of accelerated growth after 1945 up to their restructuring in today's Europeanised and global environment. Secondly, it situates some key ideas and concepts in a socio-historical context, which will be used throughout the rest of the book.

The section is organised as follows. We start by describing the emergence of European welfare states in the second half of the nineteenth century. We then situate the emergence of national welfare institutions in a broader social and political context and discuss the different goals of social policy. Thirdly, we give an overview of the different factors driving the expansion of social welfare in post war Europe. Fourthly, we look at the limits to expansion because of different socio-economic pressures on welfare states since the mid-1970s and analyse some scenarios for adjustment. Fifthly, we describe the emergence of the European Union as an autonomous, supranational level of social regulation from the sixties onwards. We conclude this section by outlining some broad challenges for the further development of social policy in Europe.

1.1.1.1 The Emergence of National Welfare Systems in Europe (1834–1940)

Social policy has existed as long as there has been some kind of collective political action in address to a social risk. The attempt to address the basic risks of life (old age, disability, sickness, work injuries and unemployment – to name the most important) and to meet certain perceived basic needs thus pre-dates the modern welfare state. In pre-modern Europe, the most important organisations involved have been religious; the church, families and local communities were the major sources of what we call welfare or caring, though they obviously lacked the professional expertise that we associate with modern educational, health, and social support services (Crouch 1999: 367). During the course of the nineteenth and twentieth century, however, the management of social risks was to a large extent taken from the hands of families and locally anchored institutions (the church, guilds, nobility, etc.) and trusted to the state. In almost all countries of Western and Southern Europe some form of collective, nation-wide and compulsory social policy arrangements developed to protect people from misfortunes (De Swaan 1988).

The key element in the emergence of the European welfare state during the nineteenth century is the so-called 'social question' (see for instance, Rimlinger 1971; Flora and Heidenheimer 1981; Alber 1982; Donzelot 1984; Baldwin 1990; Wagner 1994). It basically refers to the emergence of a working class in the second half of the nineteenth century and to the new forms of misery and conflict involved (as compared to the 'pre-industrial' poor). In the nineteenth century massive capitalisation and the development of large-scale markets had initiated a dynamics that entailed a number of technological innovations, the growth of industry and the growth of cities as new economic-industrial centres. The Industrial Revolution and the concomitant migration flows from rural into urban areas inflated the ranks of the working classes. The life of the masses (as eminently described in Friedrich Engels classic study "The Conditions of the Working Class in England") was brutish and nasty. Workdays for young and old were long; working conditions appalling; wages, even at the best of times, were barely sufficient to survive. The workers were exposed to new insecurities (Rimlinger 1971; Flora and Heidenheimer 1981; De Swaan 1988; Wagner 1994). The dangers of grave accidents, of job losses, of unprovided old age all increased. The family as the traditional helping agent was less available not only because of increased geographic distances or smaller families but also because the relatives of the urban workers were also without means (in towns one could not even fall back on subsistence agricultural production).

The early nineteenth century state had little desire to regulate these new social problems or to place limits on what might be done in the name of ownership and the free market. Slowly, however, the direction of the wind began to change. From the 1820s onwards, states started laying their hands on social and economic life in ways, and to an extent, that would have been wholly beyond the imagination of previous political communities (Flora and Heidenheimer 1981; Van Creveld 1999: 202–224).

The growing involvement of the state had several reasons. As industrialisation caused vast numbers of have-nots to concentrate in the rapidly growing cities, elites felt more and more threatened and became "worried about the effect on the masses of the free markets of liberalism, which would uproot them, both releasing them from the bonds of social deference and throwing their lives into disarray" (Crouch 1999: 384). At the same time, the working classes, who started to organise and to capture political voice, were less and less willing to accept their working conditions and the glaringly unequal distribution of wealth within the new industrial society.

To handle the new risks and to reduce the mounting political unrest, most European states introduced a number of social policy innovations, such as labour laws, safety regulations, industrial accident insurance, and workers' sickness-insurance. (In the USA in contrast, measures equivalent to this first European wave were only introduced during the 1930s.) The first Factory Acts, prohibiting the employment of children under nine and limiting the working hours of persons under eighteen to twelve a day, were passed in Britain in 1834. An 1844 law prohibited women from being employed for more than twelve hours a day – this being the first of a very long list of statues which the modern state, claiming that women were weak and needed special protection, enacted in their favour. To enforce these laws, as well as the safety regulations gradually being enacted from the 1840s on, a system of inspection was established. During the early days it often met with resistance, not only on the part of employers who resented the intrusion but also on that of the workers themselves who did not want limits on the earning power of their youngest family members. Other countries followed Britain's lead, albeit reluctantly and often after a considerable interval (cf. Flora and Heidenheimer 1981; Alber 1982). For example, Germany got the twelve-hour day only after unification in 1871; France, where conditions were in some ways worse than anywhere else, even later.

With working conditions increasingly falling under its own control, the state started expanding its power into other spheres of public welfare. In Germany, for example, Chancellor Otto von Bismarck introduced insurance plans that would ease the workers' lot during periods of hardship. In the period 1881–1887 sickness and workers' accident insurance schemes were pushed through the Reichstag and became law. The state, the employers, and the employees were all obliged to contribute. Initially applying to factory workers only, the plan was later extended to other groups until, during the Weimar Republic, virtually all trades received coverage. Quickly taking up the German example, the Scandinavian countries established their own schemes and by 1914 several of them were in operation. By 1920 Sweden, Denmark, New Zealand, France, the Netherlands, Finland, and Belgium all possessed voluntary, state-run and state-subsidised unemployment insurance schemes. The first country to industrialise, Britain was remarkably slow to establish any kind of social security system; still, in 1908–11 ten years of argument were brought to a close. As in Germany, the reforms included a compulsory health and unemployment insurance system with contributions by employers, employees, and the state. On top of this came a maternity benefit and a universal, non-contributory, scheme for paying flat pensions to persons over sixty-five years of age with no other sources of income.

1.1.1.2 Theorising the Emergence of the Welfare State

The first wave of welfare state building in Europe came about for various reasons. Numerous studies in comparative history and historical sociology (cf. Alber 1982; Baldwin 1990; Flora 1986–1987; Flora and Heidenheimer 1981; Heclo 1974) have been drawing an elaborate picture of the determinants of the formation and development of welfare systems. The abundance of such research makes it impossible to give an outline of the relevant literature here. Hence, we will limit ourselves to situating the emergence of the European welfare state in a broader societal context and indicate shortly the goals which were pursued with it.

The Welfare State as an Attempt to Coordinate Economic and Social Policy

The development of the welfare state is an outcome of the processing of problems that derive from the mutual structural dependence of the market economy and the state. With the industrial revolution, European states could no longer neglect social problems, which resulted from the spread of private property and the institutionalisation of the market economy as a separate societal sphere.

The market, as it unfolded, could not take care of social problems on its own, but needed considerable amount of outside (state) regulation and legislation (Polanyi 1944). In this sense, the welfare state implies "the use of political power to supersede, supplement or modify operations of the economic system in order to achieve results which the economic system would not achieve on its own [...] guided by values other than those determined by open market forces" (Marshall 1975: 15). The welfare state can thus be considered as an attempt to supplement the economic logics and values with social reasoning and values. For example, in the course of the twentieth century welfare policies were implemented whereby the primary distribution of income by the market could be corrected by state-sponsored secondary distribution. In many cases, the primary distribution of wealth in society was and is itself channelled by state regulations such as systems of collective wage bargaining, minimum wages, and tax structures.

Compared to the pure liberal and socialist forms, the social welfare state form is less elegant, more tension-laden, and also more complex (Kaufmann 2003a; 2003b). It does neither start from the assumption that a market can be purely self-regulating, nor from the idea that the free market economy and private property need to be curtailed. The market and the state are not seen as opposites, but as complementary principles of control based on different logics. Both state and market complement one another in their efficient one-sidedness and provide checks for one another.[1]

The development of the market economy and the rise of the modern state are understood as processes of functional differentiation and growing institutional autonomy, whereby these two partial systems of society specialise in different problem areas and develop their own institutional fits and, with the

[1] This is also the starting point of the normative reflections we will develop in chapter 1.2.

help of their respective sciences, their own distinct 'logics' (Luhmann 1990; Kaufmann 2001 a). "The tension between the economic sciences on the one hand and the legal and social sciences on the other underscores precisely this real tension in modern societies, a tension that became constitutive with functional differentiation. This tension manifests itself in the rise of 'social problems' and cannot be fundamentally resolved. The tension between the dynamics of the economic system and the demands of the social welfare system is thus a constitutive feature of welfare states. It is a permanent challenge for politics to achieve synergies, again and again, among economic and social policies" (Kaufmann 2001 a: 28).

The Welfare State as a Project of Nation Building

The major objective of reform movements during the latter half of the nineteenth century was to establish some solidity and certainty into the social fabric. Some reformers were motivated by a genuine concern for the welfare of the people. Others, perhaps more numerous, by fear of the revolutionary consequences that might follow if nothing were done. Many reformers came from the bourgeois elites, and were anxious to safeguard social order and political stability. The latter seemed precarious, as the working class started to organise itself as a collective capable of defining and representing its own interests.

The basic idea of social policy was the socialisation of risk or, vice versa, the enhancement of certainty, for the workers in terms of securing their daily lives, and for the elites in terms of avoiding political unrest. This double nature of social policy has long been an issue of political debate (see, for instance, Baldwin 1990). Depending on the historical contexts and the perspectives of the observers, it could either be regarded as an achievement of the working class in struggle or as a paternalistic donation of the state to its subjects.

Whatever the exact nature of welfare policy, its outcome was a stabilisation of industrialised social order (Flora and Heidenheimer 1981; Leisering 2003). The introduction and extension of social security schemes arguably helped to integrate the nation, to secure the loyalty of the citizens, increase political legitimacy and unite a country by bridging social and territorial cleavages (between rich and poor regions, urban and rural areas). National social programmes created a network of relations of solidarity between citizens[2] and the central government throughout the country, thereby helping to define the boundaries of the national political community and enhancing the legitimacy of the state (Banting 1995). The foundation of modern social insurance by Bismarck, for example, was a major component of consolidat-

[2] Compulsory social insurance created important streams of redistribution between various groups of citizens. Note that we will use the concept of solidarity in chapter 1.2. in a less neutral, more normative sense.

ing the German nation that had been united as a political state only 10 years before social legislation began (unification in 1871, insurance legislation 1883–1889). Bismarck's social insurance was a means of integrating the new nation state and securing the support of the labouring classes (and at the same time blocking the emergence of the socialist or communist parties). The foundation of the British "welfare state" – eminently prepared by the Beveridge Report – emerged out of a feeling of national unity during and in the immediate aftermath of the World War II. In the North of Europe the introduction of various welfare policies reinforced the state's penetration of civil society and enhanced the latter's loyalty to the state and "nation" via materially substantial, organisationally efficient and symbolically strong "social sharing" flows (think of the Swedish notion of a 'Folkhemmet', i.e. the welfare state as the home of all people). And even in the USA, where the idea of state welfare is least rooted, the depression of the 1930s lead to Roosevelt's "New Deal" of 1935. The name indicates that the introduction of social insurance in that year was more than a new financial arrangement. It was a major component of a new social contract as the basis of a renewed national unity (Flora and Heidenheimer 1981).

The main instrument to ensure the pooling of risks on a nation-wide basis was the introduction of compulsory mutual insurance (Briggs 1961; Rimlinger 1971; Ewald 1986). De Swaan has argued that "the development of a public system of social insurance has been an administrative and political innovation of the first order, comparable in significance to the introduction of representative democracy" (De Swaan 1988: 149). Social insurance was indeed a real institutional breakthrough in the history of the European nation state. Prior to it, the management of social risks was predominantly in the hands of locally anchored institutions. These operated through occasional, residual and discretional interventions, considered as "dispensations" which society granted to persons often considered as undeserving. The actual delivery of assistance took highly differentiated organisational forms, on a very narrow territorial basis. Social insurance made a complete break with this traditional approach by providing standardised benefits, in an impartial and automatic form, based on precisely defined rights and obligations, according to highly specialised procedures and with a national scope: all citizens possessing certain requisites were subject to the new rules (Ferrera 2004: 96).

The Welfare State as a Project of Inclusion and Social Citizenship

The welfare state thus institutionalises a nation-wide solidarity through the pooling of risks (old age, disability, sickness, work injuries or unemployment) across the whole population – or large sections thereof. One of the core traits of the new technique – social insurance – was its compulsory nature. "It was precisely the obligatory inclusion of wide categories of workers that allowed the new institution to affirm itself as a powerful redistributive machine, capable of affecting the life chances of millions of citizens. Obligatory inclusion meant that risks could be shared across wide populations, with three big advantages: a less costly protection per insured, the possibility of charging 'contributions' (i.e. flat rate or proportional payments) rather than 'premiums' (i.e. payments differentiated on the basis of individual risk profiles, as in policies offered by private companies) and the possibility of granting special treatment (e.g. lower or credited contributions, or minimum benefits) to categories of disadvantaged members" (Ferrera 2004: 96). In contrast to private and/or voluntary insurance, compulsory social insurance could thus produce not only horizontal redistributions – flowing from the "non-damaged" to the "damaged" – but also vertical ones, from higher to lower incomes.

As Ferrera has convincingly argued, compulsory social insurance enabled nation-states to lay down the rules for membership of the nation-state and to govern exit and entry within the national territory. The possession of citizenship was seen as a condition for the enjoyment of the new social rights – a pre-requisite for partaking in national schemes of redistribution. In other words, compulsory social insurance and the involved social rights entailed an external and internal closure of membership spaces of the nation-state. For non-nationals, it was rather difficult to enter the solidarity spaces of other states, especially when it came to deriving benefits out of them. Under certain conditions, legal foreign workers could acquire citizenship and were admitted into the schemes and thus obliged to pay contributions. In general, however, non-nationals were put in conditions of systematic disadvantage in dealing with issues of contribution accumulation, transferability etc. (Cornelissen 1996). Nationals, on the other hand, were virtually "locked in" – being subject to the obligation to be members of public schemes.

The technique of compulsory inclusion is intimately connected with a more normative idea of the welfare state, understood as the institutionalised responsibility of the state for the inclusion and social participation of all its citizens. In the fully developed welfare state, the state takes responsibility for each and every one of its citizens. The British sociologist T.H. Marshall (1950) provided an influential early analysis that linked the welfare state to the emergence of social rights, implying an ethic of 'equal social worth'. Marshall designated social rights, as anchored in the United Nations General Declaration of Human and Civil Rights, to be a systematic supplement to civil and political rights accompanying citizenship and defined them as: "the whole range from the right to a modicum of economic welfare and security to share to the full in the social heritage and to live the life of a civilised being according to the standard prevailing in the society" (ibid.: 74).

The very purpose of the welfare state is to help the disadvantaged to enter mainstream society by countering processes of societal marginalisation. As such, the welfare state held out a promise of the enlargement, enrichment, and equalisation of people's life chances: "an equalisation between the more and less fortunate at all levels – between the healthy and the sick, the employed and the unemployed, the old and the active, the bachelor and the father of a large family" (Marshall 1950: 107). As political rights, social rights are not defensive rights against state intervention, but participatory rights; they refer to participation in social life. That makes these rights ambivalent for systematic economic policy since they can be guaranteed only by state intervention in economic and social relationships which the liberal credo would exclude.

Expanding on Marshall, Parsons and Luhmann have used the term inclusion in a sociological sense to underscore this characteristic of welfare-state responsibility. For Parsons, inclusion means the recognition of a person as a member of a social community, i.e. primarily a moral state of affairs with legal consequences. (Parsons 1977). In contrast, Luhmann defines the term functionally: "Each person must be able to gain access to all functional realms. Everyone must have legal status permitting him to start a family, take part in exercising or at least in supervising political power; everyone must be educated in schools, be able to obtain medical care if necessary, and to take part in economic activity. The principle of inclusion replaces that of solidarity based on belonging to one and only one group" (Luhmann 1980: 30). The requirement of inclusion is here clearly applied to societal modernisation, i.e. the abolition of feudal bonds. This made personal freedom possible on the one hand, but at the same time abolished the previously existing rights to protection and participation. Modern individuals no longer live in the closed and protective social environment of a manor, monastery or guild but need to have access to every function system. Everyone should be able to enjoy legal status and the protection of the law, everyone should be educated in schools, everyone should be able to acquire and spend money, and so on. Political and social rights are supposed to ensure that individuals get the opportunity to participate in the different realms of life (Luhmann 1965; Verschraegen 2002).

1.1.1.3 The Expansion of the Welfare State (1945–1973)

Gradually, and especially after World War II, social policy turned from a 'workers policy' into a general redistributive policy for the whole population in an individualised society. Influenced by the universalistic ideas of Beveridge, social policy became less and less characterised by class politics and evolved into general welfare politics, directed towards the inclusion and welfare of all citizens. Under impetus of Keynesian economics, the state was assigned the responsibility of ensuring continuous economic growth, economic stability and full employment.

To a very large extent, this post war expansion of welfare states was financed by high rates of economic growth, which allowed governments to expand their revenues without having to increase rates of taxation. The social pacts in many European countries redistributed the growing revenues and thus cemented the place of an extensive and comprehensive social policy for the whole population.

Though the form of social policy differed across countries and policy domains, it became a massive feature of European states in the second half of the twentieth century. In almost all European countries more than one-half of all government expenditures was (and is) devoted to social policy, as opposed to the economy, the military, law and order, infrastructure and other traditional functions of the state (Therborn 1995: 92-94). Four main factors have been driving this general movement towards an expansion of social policy after the World War II (Alber 1982; Rieger and Leibfried 2003: 76–83):

1. the extension of social policy to all societal groups;
2. the development of an income-guarantee;
3. the growing role of the welfare state as employer;
4. the shrinking duration of employment over the life course.

The Extension of Social Policy to All Societal Groups

At their origins compulsory schemes only covered employees, and typically only industrial employees earning up to a certain wage. Only Sweden introduced an old-age insurance from the beginning (1913) covering the whole population regardless of income or occupational status. Starting from the end of World War I all countries began a rapid process of coverage extension, which accelerated after World War II and resulted in a complete or at least near universalisation of welfare state benefit schemes and services. In Britain and Scandinavia this process developed with just a few big waves of inclusion. In the Continental countries the Bismarckian tradition prevailed, leading to a sequence of differentiated inclusions, typically flowing from industrial employees to agricultural workers, then to the self-employed and finally to other marginal or inactive categories.

In social insurance schemes, the income ceilings for participation in mandatory insurance were continually raised and social rights became integral to citizenship status (Marshall 1950). This also implied that social policy became increasingly divorced from employee status. Through various social services, tax deductions for educational expenses and home ownership, not only the working but also the middle and upper classes became clients of the welfare state, which thereby acquired a more democratic character. As result of this universalisation of social rights, nearly every citizen in the industrialized world was incorporated in one way or another into the welfare systems of secondary income distribution.

The Income Guarantee of the Welfare State and the Institutionalisation of Social Security

It was a new objective in the post-World War II development of social policy in Western Europe to maintain the level of income of the insured. Welfare institutions not only had to provide protection against poverty but also had to defend the social standing attained in competitive labour markets. Since the 1950s, systems of income protection together with high income replacement ratios, defended individually achieved living standards against income loss due to illness, unemployment, invalidity and old age – and also against the costs of long-term care (Flora 1986–1987). In all developed industrial countries, provisions such as indexing of pensions or an automatic cost-of-living adjustment were introduced into social policy. "Such measures served to strengthen the relationship between generations as well as between social groups by guaranteeing the equality of their material living conditions, their consumption prospects and their lifestyles independently of an individual's employment." (Rieger and Leibfried 2003: 76).

The new welfare objective of maintaining the level of income of the insured was part of the more general goal of 'social security'. Kaufmann relates this to the growing complexity of society and the ensuing acceleration of social change (Kaufmann 2001b). In modern, complex societies individuals increasingly have to plan their lives and take decisions in the long-term. The guaranteeing of a secure life span by the social insurance systems widens the temporal frame of action for the citizens, thereby integrating the life-course as a whole. By redistributing income over the life-cycle in relation to people's needs, basal – but not complete – life-continuity can be established. "Security and continuity are essential moments of democratic citizenship primarily because membership in the community is enduring and cannot tolerate any kind of rupture. In a private and market setting, however, income and employment security cannot be as far-reaching as is the security inherent in citizenship status. In democratic welfare states, therefore, the insecurity or a particular employment position is paired with a vaguely formulated, yet nonetheless politically consequential, 'right to work', in the sense of a public obligation to ensure adequate labour market conditions" (Rieger and Leibfried 2003: 77).

De-industrialisation and the Growing Role of the Welfare State as an Employer

From the early 1960s most European countries experienced a rapid decline in industrial employment. The associated job losses were partly compensated by an expansion of income- and status-preserving transfer payments, and by expanding government-financed provision of services (Iversen 2001). This expansion of welfare state services – public health care and educational institutions but also social work, nursing and other public facilities including labour exchange services and communal or other authorities providing social welfare – created not only a wide range of new, but also highly organised groups of employees who, virtually residing in nation-state reservation, were well protected from international competition. The expansion of the welfare state and the related social services also proved to be a powerful engine for the growing labour participation of both high- and low-skilled women.

The Shorter Working Life

The growth of the welfare state leads to an increase in the portion of life spent outside gainful employment, due to later entry into (education) and to earlier exit from the labour market (retirement). In almost all Western countries years spent in educational institutions have more or less doubled since World War II. At the same time, the expected duration of pension receipt has increased markedly: life expectancy at 60 has grown steadily in almost all European countries. Taken together, these two evolutions place heavy demands on the welfare state: the proportion of people who are to a large extent dependent upon social provisions has been growing constantly and this evolution is expected to go on. This also means that radical welfare state retrenchment is not to be expected: most people have simply become too dependent on the welfare state to want its dismantlement (Pierson 2001c: 411–413).

1.1.1.4 Limits to Expansion: Pressures on the Welfare State (1973–2005)

As many commentators have pointed out, after rapid development during the 'trente glorieuses', followed by the uncertainties of the oil crisis and the increasing significance of international economic competition, European welfare states in the 1980s and 1990s faced severe pressures. The context in which European social protection systems operate has, in practice, changed dramatically since the 1970s. Whereas the expansion of welfare states took place in a stable and favourable socio-economic context, today's economic and social environment is widely seen as limiting the scope of action of the welfare state, for example in determining the level of public spending and the incidence of taxation. Compared with the 1960s the level of unemployment has doubled or tripled in many countries, inequality has risen dramatically in others, and tighter fiscal constraints have reduced the capacity of governments to cope with these problems through expansion of the public sector.

Pressures on European welfare states derive from a wide range of sources. Based on a wide range of literature, we will identify four as of most importance (Pierson 2001a; Esping-Andersen 1999; Ferrera and Rhodes 2000; Iversen and Wren 1998; Kuhnle 2000; Scharpf and Schmidt 2000; Taylor-Gooby 2002; Sarfati and Bonoli 2002):

1. The first important trend is growing international competitive pressure, resulting from the acceleration of economic internationalisation (or globalisation) which is resulting in the ever closer integration of financial and product markets worldwide. The process is a consequence of several events and developments. These include the various multilateral trade liberalisation agreements concluded by the General Agreement on Tariffs and Trade and the World Trade Organisation; the internationalisation of communication technology; the gradual removal of national barriers to capital movements and greater exposure to low-wage competition from newly industrialising countries, especially in East Asia and Eastern Europe. Within Europe the increasing competitive pressure is to a large extent related to the removal of regional barriers to the free

movement of capital, firms, goods, and services by the completion of the European internal market. At the same time, the centralisation of monetary (Economic and Monetary Union – EMU) and – increasingly – fiscal decisions in Europe, for instance, has severely constrained the margins for manoeuvring domestic public budgets. The international competitive pressures resulting from economic globalisation and European integration restrict the states' capacity to regulate employment and to raise the tax revenues that are needed to finance welfare states. As Scharpf and Schmidt point out, governments pursue directions that damage international competitiveness at considerable peril. "Welfare states remain internationally viable only if their systems of taxation and regulation do not reduce the competitiveness of their economies in open product and capital markets – which implies [...] that redistribution must be achieved through public expenditures rather than through the regulation of employment relations, and that the costs of welfare have to be collected from the non-capital incomes and expenditures of the non-mobile population" (Scharpf & Schmidt 2000: 336). Although all countries are exposed to international competitive pressure, different welfare systems experience quite distinctive policy problems. Countries differ substantially in terms of the share of exposed and sheltered sectors, employment regulation, wage costs etc. and consequently will try different strategies of boosting international competitiveness. For instance, countries where wage costs are high are more or less forced to exploit all opportunities of increasing the efficiency of production and the quality of innovation.

2. The second important trend is the development towards a post-industrial economy, characterised by the employment of the vast majority of the population in the services sector. This evolution has important implications for the functioning of labour markets and social protection systems. Employment in services is very different from the type of industrial employment that was dominant during the post-war period. It is, for example, difficult to achieve productivity increases in many services, in view of the importance of human contact and service quality (see, for example, Baumol 1967; Iversen and Wren 1998). In such areas as education, childcare, health care and personal services, work cannot easily be performed more rapidly or more efficiently without a substantial loss of quality, and hence of value. This is, of course, less true in industrial occupations. On the whole, the rate of productivity growth in the advanced economies tends to decline as employment in the manufacturing sector, where productivity gains are easiest to achieve, diminishes, and labour is transferred to the service sector, where productivity improvements tend to be more gradual. This leads to declining growth rates just as population ageing and other factors increase cost-pressures. The trend towards a post-industrial economy

may also lead to more wage inequality. Productivity increases are particularly difficult to achieve at the bottom end of the earnings scale, in such areas as personal services, catering and cleaning. Wages in such occupations increasingly lag behind those in occupations where productivity gains are possible, generating the social problem of the working poor, a phenomenon that has come to the fore over the past decade and requires social intervention.

3. The third important trend, which is irreversible in the medium term at least, is the ageing of the population. In OECD countries, the proportion of the population aged 65 and over is expected to increase throughout the next three decades, from the current 15 to 17% to around 25% in most countries. The ageing of the population will have a substantial impact on pension expenditure, although it should be noted that the sustainability of pensions also depends on employment rates, which may increase over the next few decades. The impact of the ageing of the population will also be felt in terms of higher expenditure on health care and particularly long-term care.
4. The fourth and last important trend is the significant rise in the labour market participation of women. The entry of women (especially married women and mothers) into the labour force in rising numbers in countries where this has not previously been the case, increases demand for jobs and for social care services to cope with the needs traditionally met through women's unwaged labour.

These four trends will most probably continue to pose important challenges for the foreseeable future, contributing to a socioeconomic and political context aptly characterised as an era of permanent austerity (Pierson 2001 c). In the post war period, it was possible for welfare states to expand painlessly: productivity-led economic growth generated rising tax revenues, which could be used by governments to build new social programmes. In contrast, current social protection reforms are being adopted in a less favourable context, in which part of welfare expansion needs to be financed by increases in taxation, or by reductions in other areas of government spending. The combination of population ageing, de-industrialisation, rise of the service economy, growing international competitive pressure and fiscal constraints deriving from the EMU pose difficult problems and dilemmas. Iversen and Wren (1998) argue that welfare states confront a three-way choice, or 'tri-lemma', between budgetary restraint, income equality, and employment growth. While it is possible to pursue two of these goals simultaneously, it has so far proved impossible to achieve all three. Private service employment growth can be accomplished only at a cost of wage inequality. Therefore, if wage equality is a priority, employment growth can be generated only through employment in the public services sector – at a cost either of higher tax rates or of borrowing (both implying lack of budgetary restraint).

1.1.1.5 The Europeanisation of Social Policy

Over the last decennia, all the European welfare states have been recasting the basic policy mix upon which their national systems of social protection were built after 1945. As explained above, these domestic reforms were to a large extent constrained by the Single Market Programme, EMU and the European Stability and Growth Pact (SGP). Yet, the principle site of welfare reform remained the nation-state. The process of the Europeanisation of social policy can thus not be understood in terms of a vertical and upward transposition of competences from the level of the nation-state to institutions of the European Union, as was the case with the completion of the internal market and EMU.[3] The Europeanisation of social policy is best understood as a dynamic interaction process between national welfare regimes and the spill-over effects emerging from the dynamic of EU market integration (Pierson and Leibfried 1995; Scharpf 2002; Sakellaropoulos and Berghman 2004). European social policy is made simultaneously at two interacting levels, the supra-national and the national. When national policymakers together agree on supra-national economic and social policies, these in turn have a tendency to reconfigure national welfare regimes. Europeanisation in this respect has a double meaning in the sense that it reflects, on the one hand, the gradual emergence of the European Union as a distinct supra-national level of social regulation and to some extent redistribution (through the structural funds), and, on the other hand, a growing awareness of national policy makers of what is happening outside their domestic jurisdictions (spurred by the Open Method of Coordination – OMC). In this way, European integration processes are shaping a dynamic that challenges to some extent conventional categories of thinking in social policy, which are traditionally nationally based. European welfare states were to a large degree considered to be relatively self-contained nation states. Nowadays, there is a growing realisation of the fact that national policies affect one another through interdependence due to common location in the European market and legal framework. Through the OMC process this awareness of mutual interdependence is stimulated and cross-national policy learning enhanced.

Our understanding of Europeanisation recognises the gradual emergence of a distinct EU social profile, based on a combination of hard and soft law measures, but at the same time acknowledges the persistence of nationally entrenched welfare regimes.

A European Social Model?

The European Social Model has been characterised as a common commitment to social justice the recognition that social justice can contribute to economic efficiency and progress. Yet, we believe talking of one European

[3] See sections 2.2.4 and 2.3.

Social Model should be substantially qualified, given the huge differences between European welfare states. The common typologies stress the historical diversity of the European welfare systems, which might even be increased by decentralisation and flexibilisation processes and EU enlargement. From this comparative angle, it is a little bit illusory to speak of one European model. Nor would it be advisable, since it would only stand in the way of adequate reactions to increased international competition. To respond flexibly to the consequences of regional or sectoral economic shocks, diversity in working conditions and protection arrangements is called for. A European model in the sense of harmonised systems or common minimum standards is mostly not necessary, since there are often sufficient policy options for the separate member states to safeguard social protection via their own systems. More fundamentally, harmonisation of social policy is to a large extent foreclosed by the large differences in economic development between member states, which increased greatly after southern and eastern enlargement. Per-capita gross national product in purchasing power parities is about twice as high in Denmark as it is in Portugal and, excepting Slovenia, it is three to six times higher than in the central and eastern accession states. Thus, social transfers and public social services at a level that is considered appropriate in the Scandinavian countries could simply not be afforded by Greece, Portugal or the new CEE member countries. "In any case, however, citizens in all countries have come to base their life plans on the continuation of existing schemes of social protection and taxation and would, for that reason alone, resist major structural changes" (Scharpf 2002: 651). Voters in Britain could simply not accept the high levels of taxation that sustain the generous Danish welfare state and Danish families could not live with the low level of social and educational services provided in the UK. Any uniform European solution would more than likely mobilise fierce opposition in countries where they would require major structural changes.

The underspecified use of the concept of the 'European Social Model' in the singular thus understates the large degree of diversity in the European social landscape. All European welfare systems developed within a specific national framework. Consequently, quite substantial differences can be observed, not only in terms of the level of social redistribution, but also in the architecture of social protection. Looking at the different social welfare systems, one immediately discovers significant variance along several dimensions of policy design (methods of financing, eligibility and risk coverage, benefit structure and generosity, employment regulation, etc.).

Welfare Regimes and Legitimate Diversity in Europe

Most typologies – such as the welfare regime typology by Esping-Andersen – classify the different welfare states according to their market-correcting components, which developed according to the different religious, cultural and legal traditions, and the particular distribution of power in the society of each country (Esping-Andersen 1990; 1999; Scharpf and Schmidt 2000; Ferrera and Rhodes 2000). They differentiate the conservative welfare regimes typical of continental Europe, the social democratic regimes of Scandinavia, and the liberal ones of the USA and, with certain reservations, Canada and the UK. A characteristic southern European welfare regime, and a 'radical' model typical of Australia and New Zealand, has also been proposed. These welfare regimes can be distinguished by the relative importance they assign to the central welfare producers, state, market, and family; their different requirements for access to welfare services and payments, i.e. citizenship, need, employment, etc.; their levels of support; the degree to which they are able to maintain the social status of clients, and how much they pressure clients to join the labour force. European welfare regimes have different key factors, particular sectors that were and are central to the national sense of social well-being, e.g., Germany's pension system, the UK's national health system, and France's education system.

Another important aspect in the differentiation of European welfare states is the variety in the structure of taxation and the methods of financing. Traditionally, financing the main risks of life has been based on two fundamental principles: a system of voluntary individual insurance and a mandatory social welfare system. Taking a closer look at Europe, various derivates of these systems can be found; namely, the Anglo-Saxon (Beveridge) state centred system and the Continental Bismarckian model, which stresses social insurance and corporatist elements. Generally, the foundation for financing the Bismarckian social insurance model is payroll tax contributions to social insurance funds, whereas Beveridge systems are covered by general revenue, i.e. mainly taxes. In addition, both systems share a voluntary individual protection based on risk-oriented premiums.

Figure 1.1 gives an overview of basic financing options and also illustrates the scope of possible designs. It shows that risks can be either covered by a voluntary individual protection or by a mandatory social welfare system. An obligatory enrolment in private insurances could be seen as a mandatory welfare system as well as an obligatory enrolment in the social insurance system. At the level of financing, options range from out-of-pocket-payments and risk-oriented premiums all the way to contributions on the basis of wages (salaries) or general revenue. Most countries do not rely on only one of these options, but rather apply different parts of various systems.[4]

Given the important economical, institutional, political and cultural differences between the European welfare states, and the commitment of the

[4] A risk-oriented individual protection scheme is dedicated to the more market-oriented benefit principle, whereas payroll taxes are based on wages and often comprise different redistributive elements. Social insurance contributions can be described as a mix between the cost-oriented benefit principle and the ability-to-pay-principle.

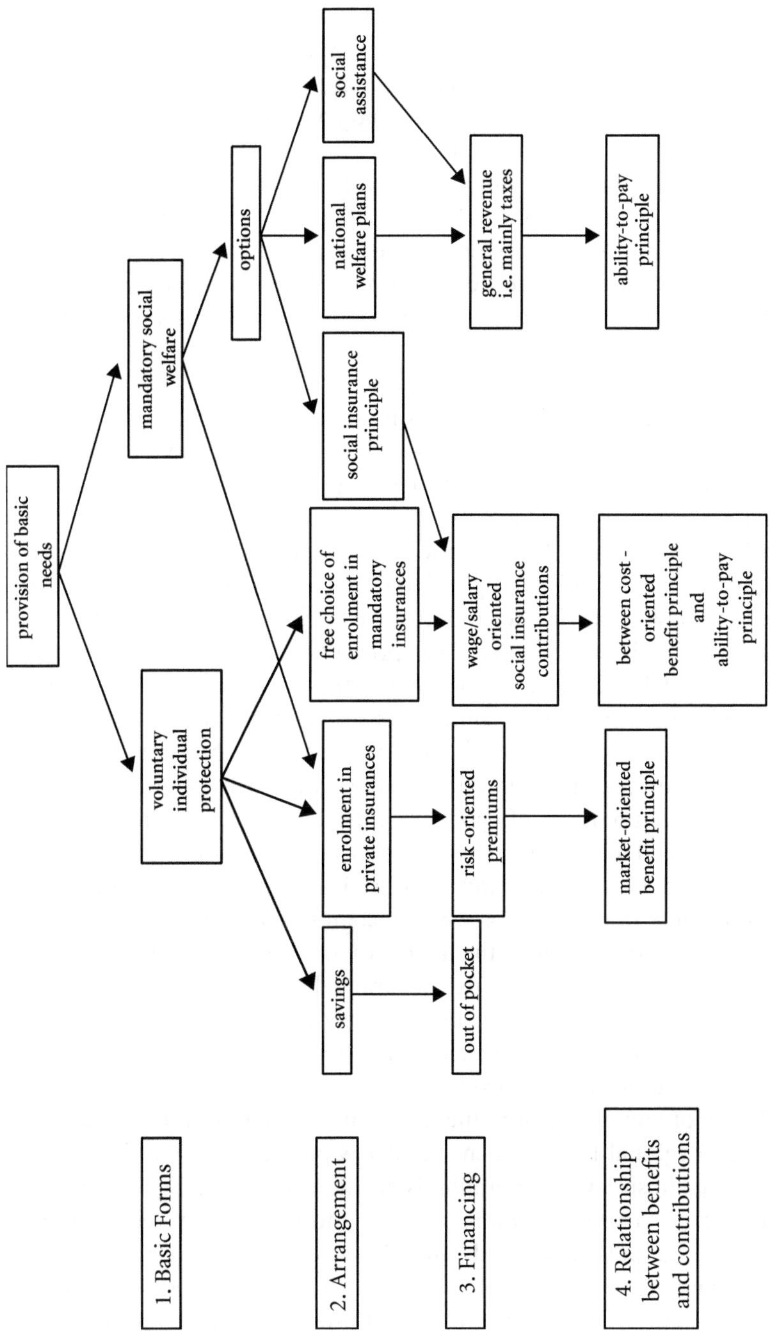

Figure 1.1: Provisions and financing of basic needs
Source: Zimmermann and Henke 2002

EU to maintain such legitimate diversity, it is very unlikely that we will witness the emergence of a fully-fledged European welfare state, replacing national welfare regimes in an enlarging European Union, in the long run. Because of the Europeanisation of social policy, however, we cannot simply go on studying 'national' welfare states. The analytical and political task has shifted its focus, from the domestic political settlements that underpin national welfare regimes, to the interdependencies between these domestic arrangements and the architecture of the European arena. As the different chapters of this volume make clear, anyone analysing national social policy in Europe will have to pay attention to the EU as an autonomous entity.

A Short History of European Social Policy

In the early days of the European Community, social policy received relatively little attention, and the Community's organs were provided with very limited powers in the social field. Social policy was, to a large extent, a means towards achieving other objectives. The structuring of the coal and steel industries, through the European Coal and Steel Community, involved social measures in aid of training and to finance adjustment. There was concern with removing barriers to labour mobility and ensuring that differences in the costs of social protection did not prevent competition in the supply of goods. In the 1970s the social dimension of the Community began to play a more important role. Many commentators attribute this in part to the debates which took place about the social aspect while the United Kingdom and others were deciding on accession to the Community. The issue was certainly raised by opponents of entry. Within the Community, the 1972 Paris Conference called for measures to reduce social and regional inequalities. In the social field the Commission produced a Social Action Programme, accepted by the Council in 1974, which recognised that the Community had an independent role to play in the formation of social policy and among the measures agreed were: first, to extend gradually social protection to categories of people not covered or inadequately provided-for under existing schemes, and second, to implement, in cooperation with member states, specific measures to combat poverty.

In terms of concrete action, the achievements were limited in scale and scope. The Regional Development Fund was put in place. The Social Fund was increased in size, with an emphasis on education, training and insertion into the labour market of young persons, and on regional redistribution: policy to combat poverty led to the first European Action Programme in July 1975 covering the period 1975–80, and a Second Action Programme for the period 1985–89. This was followed by a third programme stressing social exclusion and marginalisation. At the same time, the social dimension was receiving more attention generally in the Community. In 1989 the Commission put forward a draft of the 'Community Charter of Fundamental Social Rights' and this was adopted in modified form at the Strasbourg European

Council in December 1989 by 11 of the 12 Member States (with the United Kingdom dissenting). For our purposes here, the key paragraph is number 10, which states that: "Every worker of the European Community shall have a right to adequate social protection and shall, whatever his status and whatever the size of the undertaking in which he is employed, enjoy an adequate level of social security benefits. Persons who have been unable to enter or re-enter the labour market and have no means of subsistence must be able to receive sufficient resources and social assistance in keeping with their particular situation."

The opposition of the United Kingdom at the Maastricht European Council led to the Social Chapter being excluded from the final Treaty, but there was an attached Social Protocol in which the other members expressed their wish to continue along the path laid down in the 1989 social charter. With this compromise it was possible for the group of the remaining eleven countries to impose guidelines concerning labour conditions, consultation of employees, and equal opportunities for men and women by a qualified majority. With this, most market-correcting interventions remained dependent on unanimity within the Council (Streeck 1995). In a sense, the British opt-out marked the first step towards greater differentiation in European social policy development and reinforced the tendency toward a 'Europe of variable geometries'.

A second institutional innovation that sprang from the Social Protocol is the formalisation of the European social dialogue between the social partners at the community level. The Social Protocol allows the European social partners to sign collective agreements. Yet, until now, the results of the social dialogue have not entirely lived up to expectations (cf. Sakellaropoulos and Berghman 2004: 33).

The Role of the European Court of Justice

From the beginning onwards EU social policy initiatives were for the most part addressed at 'spill over' problems arising from European market integration. The sole purpose envisaged in the Treaty of Rome for social policy was to make a European wide labour market, in particular by enabling the cross-border mobility of workers. Very early, Community social policy abandoned the idea of harmonisation of social protection systems and limited itself to removing obstacles to mobility, especially of manual workers, among national – and still nationally governed – labour markets (Scharpf 2002). Leaving national systems basically as they were, EU policy concentrated on building interfaces between them (for example, the coordination law of social security), in particular by obliging countries to let EC 'foreign' workers enter freely to seek work and to eliminate any legal discrimination that impeded the free movement of labour as production factor across national borders (Streeck 1995). To this 'developing core of minimalist, market-making European social policy' the European Court of Justice (ECJ) later attached its own separate project by interpreting individual rights to equal treatment into the obligation of member states not to obstruct the free movement of labour. "Rather than equalising protection across national regimes, the Court removed discrimination within them, and in doing so it remained well within the limits of intergovernmental commitment to join market making" (Streeck 1995: 399).

In the literature much attention has been paid to the role of the ECJ as the main motor of integration and to its impressive capacity to make supranational law that overrides national law and binds national policy (e.g. Alter 2001). For our purposes, it will be sufficient to note that the imposition of market compatibility requirements partly by the Commission and the Council, but especially by the ECJ has led to an increasing erosion of the capacity of member states to regulate and control their social protection systems (see also section 2.2.4). For instance, in compliance with the freedom of movement, member states can no longer restrict welfare state access to their own citizens only. Workers of other EU countries must be automatically admitted too. In compliance with the active freedom of services, states must also grant access to foreign providers into their national welfare system (Ferrera 2004). According to Leibfried and Pierson (1995), all these changes have essentially transformed European welfare states from sovereign to semi-sovereign entities, irreversibly embedded in an institutional framework characterised by a certain pro-market bias and by a Court-led decision making process.

1.1.1.6 Adapting European Social Policy to a New Era

Over the last decades, European welfare states have faced severe pressures that require significant restructuring of existing welfare systems and a revision in the principles and philosophy of social protection. The circumstances which favoured the expansion of state welfare in the post-war 'golden age' – secure growth, full employment, moderate welfare needs and national politico-economic autonomy – have been reversed by both endogenous and exogenous transformations (rapid ageing, the shift to service employment, the rising pressures of globalisation and European economic and monetary integration). As suggested by Taylor-Gooby, a 'silver age' of social policy development is now clearly dawning in which "citizen welfare remains a major objective […], but is tempered by concerns about international competitiveness and cost-constraint" (Taylor-Gooby 2002: 599). As underlined by many commentators this new policy environment has not only made the status quo less sustainable and institutional reform more urgent, "but also seems to require a sort of quantum leap in terms of adaptability. European welfare states must not only re-adapt to the new context, they must become structurally more adaptable to environmental change" (Ferrera and Hemerijck 2003: 88–89). This process of re-adaptation necessarily takes place in a European, multi-level context where the policies of member states are embedded in and interact with the Community social policy architecture. In what follows, we shortly address the main challenges, which will be discussed in greater detail throughout the rest of the book.

Integrating Economic and Social Policy

European integration has created a constitutional asymmetry between policies promoting market efficiency and policies promoting social protection and equality. National welfare states are legally and economically constrained by European rules of economic integration, liberalisation and competition law, whereas efforts to adopt European social policies are impeded by the diversity of national welfare states, differing not only in levels of economic development, but also in their policy goals and institutional structures (Scharpf 2002). One of the main challenges is to put social and economic policies on more equal footing, but at the same time to re-adapt social policies to the new economic realities. In short, there is a need for a better integration of social and economic policies in the European policy architecture.

For instance, it is well understood by now that the economic affordability and long term sustainability of the various national welfare arrangements is dependent upon the labour market performance of member states. It is also clear that employment issues are strongly intertwined with social issues; it is only through a mix of economic and social policies that social issues such as poverty or social exclusion are efficiently tackled. Yet, today employment and labour market issues are mainly framed as part of the broad macro-economic policy issues and guidelines (see sections 2.2.2 and 2.2.3). At the same time, social policies in the member states not always successfully cope with the fiscal and economic restraints put on them. Improved integration and fine-tuning of employment and social-security policies is thus needed to make both systems more flexible and responsive to the needs of the new social and economic context in Europe. The challenge is to develop a conceptual model of policy making that recognises the interplay and reciprocal impact of policy areas in shaping social outcomes. Integrated economic and social policies are beneficial to the fulfilling of the ambitious goals the European Council agreed on at the Lisbon summit in the year 2000 and which indeed intend to achieve economic and social goals simultaneously.

Nowadays, all national European governments pay heed to the idea that economic and social policies reinforce each other. They all recognise that welfare states should be compatible with international competition and should become 'employment friendly' in reducing their cost (especially non-wage costs) and offering benefits that do not create disincentives (activation, making work pay). Yet, we feel that reforming the welfare state along the lines of activation will require more fundamental reforms in social protection, inasmuch as they involve not only modifying existing parameters and instruments of social policy, but also changing the overall logic of established social protection. In chapter 1.3 we therefore subscribe to the plea to redefine social policy as a dynamic investment in human skills and productive capacity, rather than as a means of protecting against interruptions to consumption. Social protection should be converted – to use the image of the World Bank – from a 'safety net into a springboard' (World Bank 2001). It should aim at empowering the individual by offering extended opportunities to participate in the labour market and in society in general. In addition to their traditional functions, social protection systems are therefore also expected to invest in human capabilities and to facilitate and encourage access to the labour market (for example, by providing incentive structures that are favourable to employment, offering lifelong education and training opportunities, particularly to unemployed or disadvantaged workers, and making it easier for parents to reconcile employment and family life). Some of these new demands are being taken up by current social protection reforms, which increasingly tend to emphasise the activation and employment promotion dimensions of social policy, but the efforts on the side of human and social capital formation remain insufficient.

Managing Diversity

The peculiarities and path dependencies of national social policies and employment policies, as well as structural and institutional conditions within member countries, have generated vast differences across countries and regions in terms of their economic, social and labour market performance. The actual growth and employment performance of European economies, as well as their overall level of economic development or their poverty rates, varies by country, and in particular, by region within countries. These differences have even been exacerbated by the different rounds of enlargement. Furthermore, the institutional structure of both social security arrangements and industrial/labour relations systems differ widely among EU member states. Countries differ not only in their average levels of total taxation and social spending, but also in the relative weights of various taxes and social security contributions on the revenue side, and of social transfers and social services on the expenditure side. Of even greater importance than these operational differences, however, are differences in take-for-granted normative assumptions regarding the demarcation line separating the functions the welfare state is expected to perform from those that ought to be left to the family or the market (Esping-Andersen 1990; 1999; Ferrera and Rhodes 2000). Hence, even those who agree that a 'positively' integrated 'Social Europe' must be created in order to compensate for the Common Market's corrosive effects upon national welfare states are unlikely to find it as easy to agree on any particular institutional blueprint according to which 'Social Europe' is built.

Yet, at the same time the mutual interdependence of national social and employment policy systems requires a greater amount of coordination than hitherto. The actual possibility of member states designing and implementing autonomous policies of social protection has been severely constrained by the EMU and the Single Market. Furthermore, to ensure effective economic and social policies in the new global economic environment, member states are to a large degree dependent upon each others' performance. A main challenge hence consists in aligning the need for greater coordination and cooperation on the one hand and the existing diversity of social protection arrangements and economic developments on the other hand. Faced by this dilemma, the Union has opted for a new governing mode, the Open Method of Coordination. OMC is based on a process of benchmarking and best practice exchange, as well as on the coordination of national policies towards the achievement of targets defined on a European level. Policy choices thus remain at the national level but are at the same time defined as matters of common concern. Through promoting common objectives and common indicators of achievement, and through comparative evaluations of national policy performances, one tries to improve and coordinate the different performances in member states. These efforts are certainly useful. Yet, it remains to be seen whether they will suffice to address the threats and challenges that the different EU welfare states currently face.

Building Social Citizenship in Europe

Historically, the introduction and extension of social protection schemes helped to define the boundaries of the national political community and to establish a sense of national citizenship; it enhanced the legitimacy of the state and diminished the importance of territorial and social cleavages. Nowadays, it is widely felt that in order to maintain popular support for both the deepening of European integration and the widening of its scope (enlargement), Europe must present itself to its citizens as a credible project of social security and protection, and certainly not as a threat to established social rights (Offe 2003). Recent events in various European countries have shown that even weakened labour movements are able to generate massive informal protests against European programmes threatening national welfare arrangements. Furthermore, insofar as the Union's member states derive domestic legitimacy from social policies, they are unlikely to cede complete control over these to the European Union.

The challenge thus consists in complementing the market-making negative integration through the abolition of tariffs and other hindrances of competition with some form of market-constraining positive integration, i.e. some form of social citizenship. It must be kept in mind however (as noted previously), that the European Union is not, and undoubtedly will not become, a federal welfare state like those of traditional nation-states. As Pierson and Leibfried rightly observe "this scenario was never plausible, since the EU arose in a different historical context and was layered on top of already deeply institutionalised and diverse social policy structures within each member state. Hemmed in by institutional and political constraints, the

European Union is incapable of the kind of positive, state-building initiatives of a Bismarck or a Beveridge" (Pierson and Leibfried 1995: 433). A full-fledged reconstruction of social citizenship at the European level hence appears rather unrealistic. The alternative social policy regime that is in the making will be of a different kind. Some ideas have started to circulate about the introduction of modest but symbolically significant social rights attached to EU citizenship as desirable corollaries to free movement rights, for example a universal minimum income guarantee (cf. Schmitter and Bauer 2001; Van Parijs and Vanderborght 2001; Ferrera 2004), but where this process is leading is uncertain.

1.1.2 Challenges for the Social Policy in the European Union

1.1.2.1 Introduction

In current European societies, several ongoing tendencies can be interpreted as challenges for rethinking the social policy, and additionally, as calls for interdisciplinary and comparative analyses concerning the social policy reforms and practical measures implemented in the European Union member states after the World War II. As was discussed in section 1.1.1, the social policy systems of the European societies have been built and developed as national projects and on varying regime bases, anchored on the modernisation processes of the industrial societies. In the contemporary situation of globalisation, the social policy has come on a new threshold of modernisation, which is going on as an international and interconnected process of different sectors and actors. In this section, the aim is to grasp some of the general social and economic trends, which challenge the developing of social policy.

The macro tendencies of economy, demography and social problems are factors which have influenced the social policy during the modern era, and those are also the tendencies which have been managed by social policy. As a theoretical frame, a circular development of social policy in an entity of complex reciprocal social processes can adequately be applied. In the current era of globalisation, the need for a socially sustainable development is challenging the social policy among the other policies which all are processed in the international context.

The political changes since the end of the 1980s, the collapse of socialist regimes as one of the most prominent changes, and the new division of global economy, the deepening formation of market regions in all continents included, are the axes which have to be taken into consideration while discussing social policy. International actors and stakeholders, such like multinational enterprises, international governmental agencies, non-governmental associations and movements, e.g. activities around the environmental issues and human rights, have increased their influence on the global level. Besides the market and international politics, the efforts of non-governmental organisations for building the global civil society have to be taken into consideration while rethinking the social policy, too. In the post socialist societies, the role of non-governmental organisations as actors of social policy is taking shape. The globalisation brings the international law as well as the international treaties and declarations on the agenda in a historically new way, too. After the World War II, in the (Western) European societies the human rights have had a strong dimension of social rights, which are also included in the constitutions of several European countries; usually the European society model was described as a social market economy. The new elements for the discussion are the post socialist countries which have had a societal development of their own.

The global economy is challenging the national economies, and this tendency has been highlighted in the post-socialist era. The economy can be understood as a prerequisite for the social policy, and vice versa: a dynamic knowledge-based economy is dependent on a well functioning social policy which enables the well-being of citizens, facilitates the managing of different phases of their life-course, and increases personal autonomy. The concept of capabilities as a compilation of approaches has started to appear in the theoretical discussions concerning the revitalising and renewal of social policy during the high-tech era (e.g. Sen 1985; for a review of the discussions see Robeyns 2004). However, the concrete empirical application of the mentioned capabilities approach is still scarce, which gives room for intellectual openings.

The development of the economy on a global level is a complex phenomenon which has different influences and impacts on the social policy at regional, national and supra-national level. The demographic changes are similar in all industrialised societies. Decreasing fertility, changes in the family structures and the ageing of the populations are the main tendencies, and those are also the issues highly discussed in the debates concerning the future of social policy. The regionalisation and regional differentiation reorganise the nation states and address new approaches for understanding the relations between local, national, supra-national and global levels. The social problems vary much in different societies; in the current European discussions, the social exclusion is used as an umbrella for the issues classically defined as social problems. As a particular phenomenon, the poverty is interpreted as the most prominent problem to be combated.

The new EU-countries, most of which are former socialist countries, are not only a challenge but a new factor in modifying the European Union and its future. The original characteristics of Europe, and the European Union, too, can be seen as a multidimensional entity of different histories, cultures, languages, policies and social practices. However, some similar societal tendencies and trends on the macro level are obvious in all European countries. For the social policy analyses, the recognition of those tendencies is of importance in contextualising and interpreting the phenomena of society. The European Social Model has been launched by the Commission as a strategic goal to be developed within the European Union (Palola 2004).

There are also internationally accepted regulative tools, human and social rights' declarations and conventions, which can be applied in developing the social policy of Europe.

1.1.2.2 Human and Social Rights as a Basis for Social Policy

Globalisation is primarily referred to when economic developments take a global effect, i.e. act across national boundaries. Among the actors of these developments are the multinational enterprises engaging in activities worldwide and being able to transfer them from one state to the other. The international economic order, for instance the World Trade Organisation, meets this structure, at least basically. As to the socio-political sector, corresponding global institutions and opportunities of action are missing yet. However, endeavours have been made since the end of the 19th century to create international regulations in the field of social policy. These endeavours had several objectives. On the one hand, the point was to bring the social standards in line in order to prevent distortions of competition through social dumping. On the other hand, by acknowledging social principles worldwide, the social protection of the underprivileged workers was to be improved, thereby contributing to social peace. With the increasing international mobility of the workers, a further motive arouse for setting up international standards for attempting to coordinate different national systems in order to avoid social disadvantages due to migration.

Universally, the above-mentioned efforts to create international social standards had results within the framework of the United Nations (UN) and the International Labour Organisation (ILO), and regionally within the framework of the Council of Europe, for example. As to the universal scope, the UN Declaration of Human Rights of 1948 contains personal rights of liberty and social rights which have each been differentiated further by a supplementary pact. The numerous ILO Conventions regulate social questions in still more detail. Thus the ILO Social Security (Minimum Standards) Convention No. 102 contains a comprehensive catalogue of social guarantees. By ratification the member states commit themselves to structuring their national orders according to the ratified conventions. As a rule, no direct legal claims arise out of that for the individual. Compliance with the assumed international obligations is rather supervised by a report system. A committee of experts of the ILO examines the national reports and ascertains deviations from the standards. Complaints and action procedures are also possible, but are hardly referred to.

Beyond these activities of the ILO, who strives towards the universal recognition of social standards[5], endeavours are also being made within

---

[5] On this, see the report presented by the ILO's Director General on the occasion of the 75th anniversary of the organisation: "Defending Values, Promoting Change: Social Justice in a Global Economy – An ILO-Agenda", Geneva 1994.

the bounds of trade policy to make the adherence to social standards compulsory by inserting so-called social clauses into trade agreements (Willers 1994). On the regional European level there are international law documents akin to those of the ILO and the UN, in particular the European Social Charta. The observance of the political and social rights standardised there, such as the right to work for example, is simply examined via reports. This differs from the Fundamental Freedoms and Human Rights declared in the European Convention on Human Rights.[6] In case of infringements of these rights there is a European Court of Human Rights that can be appealed to by the individual. Here the distinction becomes apparent between freedoms as rights to privacy from state intervention and social rights as rights of participation that cannot be absolutely guaranteed, since they are dependent on the state's capacity regarding the provision of social benefits. The division in both UN pacts is also based on this distinction. The differences have, however, been relativised by recent jurisdiction of the European Court of Human Rights who derived claims for participation against the respective state regarding social benefits from personal rights of liberty as the ownership, and the banning of discrimination (Schmidt 2003). In addition, relevant studies increasingly point out that the social rights are to be regarded as belonging to the fundamental human rights and that common functions exist in this respect (Iliopoulos-Strangas 2000; see also Bundesministerium für Arbeit und Sozialordnung 2000/2001).

A glance at the international law regulations shows that, from the universal as well as the regional point of view, there are numerous international rules also within the social field. The deficits, which become more and more apparent with the increasing economic globalisation, lie with the implementation of these standards. Insofar the transfer of appropriate powers from the national states to institutions at international level is not sufficient yet. The EU, who takes a unique position as a supra-national institution, is in a special situation: with regard to its constitution, the EU might develop activities also in the field of social policy, which institutions purely pertaining to international law cannot.

1.1.2.3 Changes of Economy and Labour Market

Currently, there occurs a need for new conceptualisations of global economy; the world-wide stock-market and the division of production according

[6] Of importance for the European social policy are the recent documents of the European Union: COM(1994)333 final, European Commission White Paper, European Social Policy: A Way Forward for the Union; COM(2003)305 final, Council Decision 2000/436/EC of 29 June 2000 setting up a Social Protection Committee; COM(2003)842 final, Communication from the Commission to the Council, the European Parliament, the European Economic and Social Committee and the Committee of the Regions.

to the developmental level of technology are challenging the regional, national and supra-national social policy, too. The following tendencies of macro economy and labour market can be recognised in the European societies:

- globalisation of market and division of production based on new technology versus fordist production of goods;
- development and extension of the new information and communication technology (ICT); knowledge production, and emergence of the entertainment and 'Erlebnis'-industry as a massive global phenomenon, 'macdonaldisation' of culture and social practices;
- emergence of network economy; internet-based marketing and communicative networking;
- new requirements for occupational qualifications; increasing demands for education; need of both generalists and specialists, and the demand of ICT-skills as a prerequisite for a successful entering into the labour market;
- concrete local changes of productive structures and organisations; regions with successful productive activities and regions in stagnation; as an outcome, the tendency of increasing gaps between regions, possible to be described as "silicon valley" regions and "looser regions" with the traditionalist agricultural orientation and with elderly inhabitants while the younger generation is leaving for high tech regions;
- imbalance of the high and/or long-term unemployment rates and, at the same time, lack of qualified labour force especially in the knowledge and innovation based production;
- unemployment and low employment rates connected with financial straits in social policy;

There is not yet much evidence of what the mentioned tendencies really mean for social policy. Also the question remains open what kind of social policy is needed if it still is thought in the classical frame of "industrial relations" which now are ICT-based industrial global relations. In a recent analysis concerning the knowledge economy of the Scandinavian countries, Benner (2003) has launched some results and conclusions worth to discuss. He summarises that the current societal development of Denmark, Finland, Norway and Sweden is connected with a strong path-dependency. The old coordinates of social policy – universalism, tax financing of social security, attempt to full employment through macro economic interventions by the state, social consensus – have been adjusted with the demands of new economy. However, the emphasis of public financing has been put on the investments of the knowledge-society infrastructure, e.g. on education, science, innovation centres, and the aim has been to strengthen the local ICT-based economy. This probably is the trend in several EU member states but systematic analyses of the developments are not yet available.

The globalisation[7] of production and market also puts the need for regulatory bodies on the agenda. The European Union can be interpreted also in this frame as a regulatory body in managing the Four Freedoms. Besides the nation states, the supra-national and local levels are structuring the market, labour market included. A complex system of markets emerges with consequences for social policy also in respect to its local dimension in the form of the concrete social services. In the social policy analyses, methodology for combining the local/regional, national and supra-national level is needed but there is a lack of regional and local statistical and other data, which could give opportunities for comparative analyses (Rauhala et al. 2000). The locus of social protection for the employees with persons dependent on them has to be considered. How to take care of children, of frail elderly, and of persons with physical and learning difficulties, and how to manage the long-term unemployment, are the questions to be answered. The social policy systems of the nation states established during the fordist industry seem not to function in a sustainable way in the IC-technology based economy and global market. The global economy and local everyday living of the citizens should be taken into the same context in order to develop an understanding of the adequate social policy.

As a consequence of the new economy, there also appear unexpected forms of unemployment. Of importance is to recognise that there is variation of unemployment according to life-course and personal capabilities; unemployment can be affiliated to and implicate different social situations. It seems to be increasingly difficult for the young people to enter the labour market even in the case they have education; the finding of the first full-day job has become difficult. According to EUROSTAT, around one in thirteen young people aged 15–24 was unemployed in the European Union in 2001 (European Commission, Directorate-General for Employment and Social

---

7 Social scientists have different conceptualisations with regard to the current globalisation. The historically oriented scholars, on the one hand, agree to the approach, which interprets the Age of Discovery/Exploration – as it is called in the Western history – as the beginning of the modern globalisation process. The next step in the process was the industrial revolution, which started in the 18th century or in 17th century, from the emergence of the modern natural sciences. During the 19th century and the first half of the 20th century, the fordist machine production drastically changed the world while making the global communication and trade possible through telephone, telegraph, radio, train, car, air planes, television etc. As a logical continuum for that development is the turning up to new digital technologies, to more advanced robotic systems, and to satellites, which extend the possibilities of global communication (e.g. email and internet).
Some analysts, on the other hand, interpret the appearance of the new technology since the second half of the 20th century as a profoundly new technological development in its own quality (as a second industrial revolution). To this qualitative character of the current era belongs the development of new social order, too. In the social sciences, a general description is often given by using the term "postmodern societies", or, while referring to the changed industry, post-fordist society.

Affairs 2003: 136). This represents close to 15% of the labour force of the mentioned age group, and young people under age 24 are more than twice as likely as people over 25 years to be unemployed. Additionally, among the young unemployed people there are relatively more women than men – Germany and the United Kingdom excluded, where the proportion of young unemployed men is higher than that of women.

In the EU-15, 3.3% of the labour force was unemployed for at least one year in 2001 (European Commission, Directorate-General for Employment and Social Affairs 2003: 138). Among the long-term unemployed, the young people form a special group, but in general, the group is heterogeneous. The persons who were educated during the 1960s and 1970s are qualified to work in the fordist production, and they often do not have the basic skills needed in using the computer technology. Many of middle-aged people (aged 50–64) have lost their occupational capacity at the end of the 1980s and during the 1990s while the enterprises were modernising their productive practices; this tendency has been prominent in the post socialist countries where the social order changed at the same time with the breakthrough of new technology.

EU-wide, 38.6% of the population aged 55–64 were employed in 2001; the highest figure (66.5%) among the older workers was found in Sweden in 2001 (European Commission, Directorate-General for Employment and Social Affairs 2003: 131). Because of the labour market changes, the long-term unemployed people are 'sitting in the waiting room of pension'. In several countries, middle-aged people are not considered as a potential for re-education, but – as in the case of the Scandinavian countries – the early retiring systems have been established in order to offer an exit out of the labour market to these people.

The immigrants with low education can have a vulnerable position in the labour market based on IC-technology which is applied also in cleaning, storing, retailing and transportation work – jobs or sectors where persons with low occupational education usually have entered. It seems evident that the EU-labour market competes for the best qualified and educated labour force, and only seasonally for workers with lower education. There is also a trend of the pendulum labour market in the border regions; people work in another country but have a home in the other. This kind of commuting is not classical immigration but a new life-style which is possible in a situation where there is a free access from country to country as is the case in the European Union.

From the viewpoint of social policy, the unemployment is one of the biggest challenges to be addressed, and a basic issue concerning the future of societies based on the waged employment. How is the social protection to be managed and developed in a situation where the structural long-term unemployment seems to be a stable phenomenon? The activation policies have been launched, but there are hesitations if it is enough in the cases where the basic occupational education and qualification of the young, middle-aged and immigrants are low or do not fit with the new demands of the knowledge-based labour market. The participation in the labour market is the leading factor of social order in the European societies, and that is why the tendency towards the "working poor" is ideologically a risk because it erodes the idea that by a full-time participation in the labour market the citizens can afford themselves the necessary everyday-life substances without any social subsidies. That has been the original idea of the industrialised European societies; additionally, by participating at the labour market citizens are also providing their own social insurance for the possible risks (sickness, accident, old age, temporary unemployment) which can occasionally or permanently interrupt working.

It is a big question how to continue with the aim of full-employment in the new situation where the production structures have changed, and where one tendency is that all citizens do not find a job in the restructured labour market. Very high standards are put on qualifications and, in the information production, on confidence and self-steered working, too. According to the conjunctures, one classical regulatory element, i.e. women remaining out of the labour market, has been lost because of the adopted dual breadwinner model where both women and men are educated and are expected to participate at the waged employment. New openings have been expected from the "third sector" (non-governmental organisations) as an employer but until now not any big success can be observed. It remains a fact that in the global economy, people still live their everyday life locally, and that is why local and regional solutions are needed. One challenge will be how to combine the supra-national and regional/local level in advancing the social and economic development, social policy included.

1.1.2.4 Changes of Demography

In the classical academic approach to social policy, the demographic development is regarded as one of the basic factors influencing social policy reforms and measures. The demographic trends belong to the processes, which can be forecasted and predicted for the next 30–50 years with a relatively high reliability, although the global migration increases the uncertainties in making prognoses. In comparison with the economic developments or political changes, the population trends based on fertility of the industrialised countries can be anticipated over a longer time span, i.e. over the next decades. Fertility, mortality, longevity and migration are the dimensions of demography, which all are of importance for the social policy prospects. At the same time, the age as such has increased its weight in social policy; the attention put on the children and the elderly has initiated debates of intergenerational relations and life-course based social policy (Leisering 2003).

In the European societies, the population is decreasing, and affiliated to that, a probable projection or hypothesis is that along the decrease of population the capacity/efficiency of production and the general prosperity are diminishing, too. The compensation of population decrease by immigration is a complicated issue; Europe has not been an attractive destination for highly educated people who according to political speeches are wished to come and give their contribution to the dynamics of developing the EU towards an innovative society.

Ageing

In all European and other industrialised societies, and in the developing countries, too, the population ageing is the dominating demographic tendency, and it is the major issue organising the current discussions and concerns of the future of social policy. The ageing of population means a qualitative change in the societal structures of the known cultures; there never have been so many old people as today. The ageing is a societal tendency, which makes the contemporary societies similar despite big differences in prosperity and other resources or characteristics (UNFPA 2002). The ageing is mostly discussed in the frame of increasing expenditure of pensions and health services. The developed social policy measures of old-age are taken under reconsideration in order to adjust the pension schemes and health costs with the rapidly increasing number of the elderly. The other issue is the everyday care of the frail elderly: one projection is a parental leave system for the middle-aged people who take care of their parents in the last phase of life.

From the analytical point of view, there appears a historical contradiction. During written history, the humankind has always pursued a long life for individuals, and the old age has been in many (not in all) societies respected as an achievement. Currently, the mentioned opportunity has become as one of the most predictable feature in the life-course of every person in the industrialised societies, and at the same time, the long life has begun to be a problem, even a threat for the societies, and as already mentioned, especially for the social policy in the form of increasing the pension costs and care expenditure. A possible projection is that in the future there will not be any possibilities to pay pensions comparable to those of today.

The increasing consciousness of ageing as a fact and as a structural change will have implications for the definitions of ageing. The ageing is an ideological construction, too, and it is assumable that many old people today suffer while hearing the political discussions in which the old age is interpreted as an expenditure-load. Not much is yet discussed of the opportunities which the ageing persons can give to society. On the other hand, ageing has made visible how important it is to take into consideration the demands of the socially sustainable development, e.g. to make decisions which concern the next generation as tax-payers and labour force in the care sector. In this context, the ageing is not interpreted only as a negative factor as such but as a challenge for rethinking the intergenerational solidarity and human responsibility as well.

It is a fact that the group of the oldest people aged over 80 (EUROSTAT defines the people aged 80 and over as the "very old"; European Commission, Directorate-General for Employment and Social Affairs 2003: 117) is increasing its proportion rapidly. Today, 45 % of those persons live alone, but in Sweden and Denmark more than 60 % of the oldest-old live alone. Among the oldest-old there are also persons with health problems and difficulties in organising the everyday life. It is expected that over the next fifteen years, the number of the oldest-old in the European Union will rise by almost 50 % (ibid.: 117). Today, the proportion of the elderly (aged 65 or over) is 16 % of the total EU-population or 24 % of the working age (15–64) population; the latter figure is expected to rise to 27 % by 2010. The great majority of the elder population will be women; every four persons over 80-years only one will be male.

In the analyses concerning the ageing as a structural change, something could be learned by comparing the current situation to the times when infant mortality began to decrease rapidly – not much longer than a good hundred years ago. The industrialising and urbanising societies initiated and implemented social innovations in order to adjust to the unexpected population change. Among those innovations were such like kindergarten, the free-of-charge elementary school system for the whole age-cohorts, the child welfare arrangements, the social insurance systems based on the number of children, tax systems adjusted to the small-family situation, family-housing policies, emergence of a retailing market based on the everyday needs of a nuclear family, and ideologically, the increasing understanding of the value of every single individual child (e.g. Myrdal and Myrdal 1935; Myrdal 1941; Hall 1952; Välimäki and Rauhala 2000).

An certain analogy can be recognised also with the situation when the child population made itself visible during the first massive globalisation at the end of the 19th century. Today, during the new technology era and the globalisation connected to it, the ageing population is prominently increasing its number everywhere, not only in the industrialised countries. From the point of view of sustainable development, active searching for social innovations in a situation of the population change is argued. The ageing population can catalyse, and it has already catalysed, new social and economic practices. The applying of signal technology in smart housing and home help services as well as the emerging branch of geriatric-medical technology are examples of that kind of development. The other field to be mentioned is the entertainment market for the older persons; the cultural activities and tourism specialised on the needs of the senior citizens is increasing its importance as a new market sector (Therborn 1995; UNFPA 2002).

Cohen and Hanagan (1991) have given empirical evidence on how children entered the stage of history, and how they started to be an important target group of social policy. Surprisingly or not, it was the decreasing number of children which made them social policy agents. The older people are making themselves visible by increasing their number in society. Golini (1999) has emphasised that the ageing of population has been an expected trend since decades, and the industrialised societies have anticipated the increasing number by developing pension schemes. In the beginning of the 21st century, the ageing in the highly developed industrialised societies does no more mean poverty and marginalisation as it often was the case fifty years ago. All industrialised societies have established pension systems which offer material maintenance to most of the retired persons. The current pension systems in the (Western) European countries have been developed during the last hundred years. It still remains open if it is possible to continue the chosen pension policies in the future (see chapter 3.3).

An issue to be taken on the agenda of the European Union are the social care services, which are not included in the health services. With help of social care services, the conducting of everyday living at home can be facilitated. In the Scandinavian countries, the social care services organised by the public authorities have been interpreted as the most peculiar feature of the Scandinavian welfare model (Sipilä 1997). From the point of view of ageing, it is worth investiganting how the need of health care services and/or institutional care could be prevented and/or postponed. One answer given by the Scandinavian analyses was that the social care services, especially the home help for the elderly can postpone the need of the medical in-patient-care services (Rauhala 1996). The financing of the home help services and care insurance schemes are opportunities to be developed further.

A crucial question is, if there will be enough persons who are willing to do home help and other social care work as an occupation. There are tendencies, which show that the younger generations are not interested to get education as care workers. The recruiting of staff for the social care services is going to be more dependent on the immigrant workers. The care work is undervalued, low-paid, and the workers are exposed to over-load and burnout (Rauhala 1991; Johansson 2001). It is also a probable perspective that the recipients of services in public and private settings will be customers who demand high quality and tailored services, and that also the relatives (adult children) are asking for quality of the services they pay for and/or organise for their parents. The services both in public and private care sectors will be dependent on the availability of care labour force. In the discussions on service provided by public, private and mixed sectors, the issue of ensuring sufficient educated personal will be one challenge for the future social policy (Pillinger 2001).

In 1999 the Commission has launched a special programme "Towards a Europe for all ages – promoting prosperity and intergenerational solidarity" (Commission of the Eruopean Communities 1999), in which one leading statement is formulated as follows: "[…] the very magnitude of the demographic changes at the turn of the 21st century provides the European Union with an opportunity and a need to change outmoded practices in relation to older persons. Both within labour markets and after retirement, there is the potential to facilitate the making of greater contributions from people in the second half of their lives. The capacities of older people represent a great reservoir of resources, which so far has been insufficiently recognised and mobilised. Appropriate health and care policies and services can prevent, postpone and minimise the dependency in old age. Furthermore, the demand for these services will open up new job opportunities" (ibid.: 21) There is much room for innovative openings on how to realise the pursued development while rethinking the ageing as a challenge for social policy concerning the old-age security, not only thought as pensions to be paid but as a new service economy, too (Leichsenring and Alaszewski 2004).

Migration and Multiethnic Development

Since 1989 the net migration has been the main component of the annual population change in the European Union, and in 2001 the migration represented 74% of the total population growth (European Commission, Directorate-General for Employment and Social Affairs, 2003: 120). The active policies of the European Union are to advance the free movement of persons within the EU-area; the diversity of cultural backgrounds of people is interpreted as a value and a factor for economic dynamics. On the European Union level the concrete social policy measures are developed for the mobility of labour force from one Union country to another. The other dimension of the migration is that Europe – as all rich regions of the world – is interpreted as an attractive destination by the persons in poor countries, and that is why the European Union like other rich regions of the world is developing regulatory measures for immigration. The third issue connected to migration is the international responsibility to be taken of the refugees, of people seeking for an asylum. With its manifold dimensions, migration is quite an ambiguous phenomenon.

The current migration and multiethnic development can be contextualised in the history of population changes in Europe during a longer period. Firstly, Europe has always been a multiethnic society with cultural and language minorities almost in all its countries; only Portugal and Iceland were linguistically homogeneous societies until the modern era. Secondly, Europe has lost population, especially to North America. Europe lost many people in World War II, too (Therborn 1995.) During the 1960s and 1970s, the labour market of several European countries needed workers from outside Europe, and at the same time, the migration within Europe began. Eastern Europe was excluded from the mentioned development, because of the Iron Curtain. The leading European industrialised countries – France, Germany, Sweden – have remarkable proportions of immigrants, and also Great Britain (which as The Commonwealth is an exception compared to other European countries).

In the multiethnic Europe, there always were tensions between ethnic groups, and the situation in Balkan still is obviously tensional. As a substructure, there exist tensions also in Belgium between the Flemish and Walloon people, in England between the English and Scottish, to mention some examples. Analytically, the ethnic tensions are one dimension in a context where the marginalising processes or even hostilities are developing. Often there are different factors which together increase prejudices towards some groups of people. In many countries, regional differences, urban-rural differences included, influence relations between population groups. Religion is increasing its meaning as a difference affecting not only positively the social relations by adding the cultural richness but as a distinction with negative implications, too. As an example of religious tension, the conflict since the 13th century in Northern Ireland can be mentioned. Today, the Moslems form the biggest religious minority in Europe by their share of 6–7 % of the European population.

In European policies, mutual understanding and respect are emphasised.[8] For social policy, the ethnic diversity means a challenge in providing social services for people with different languages and cultural backgrounds. Classically, social policy has been used as a balancing factor of class conflicts. Today, a challenge for social policy is to balance the ethnic relations and to advance mutual understanding and cooperation. In the contemporary social work education in several European countries, the ethnic relations are taken into consideration, and master-curricula for human rights have been developed in order to advance the awareness of multiethnic relations as a social fact of increasing importance.

In the global migration context, the current Europe is a receiving area. In 2000, there were 19 million non-nationals in the EU, of whom 13 million were non-EU nationals (European Commission, Directorate-General for Employment and Social Affairs 2003: 120). The two largest groups of immigrants in Europe in the beginning of the 21st century are Turkish people and the citizens from the former Yugoslavia; the majority of both groups live in Germany. Belgium, Ireland and Luxembourg – the countries where other EU nationals outnumber the non-EU nationals. Among the Scandinavian countries, Sweden has the largest share of immigrants among its population, and Finland the lowest.

The migration issues of the European Union are managed by the Treaty of Amsterdam: Title IV (Visas, asylum, immigration and other policies related to free movement of persons) includes principles, rules and proposals for the member countries. The Commission has launched a detailed programme of actions since 2000 in the "Scoreboard to review progress on the creation of an area of freedom, security and justice in the European Union" (Biannual updates available in www.migpolgroup.com/monitors). There is

8 The European Union is financing a monitoring tool of racism and xenophobia, The European Monitoring Centre on Racism and Xenophobia, EUMC (cf. www.eumceu.int).

no clear picture yet, how the European identity will be evolved in the European Union, especially in the former socialist countries. As border countries with very long histories of living close to and between different cultures, the new European Union countries have experiences of which the whole European Union could benefit in regard to the development of functioning relations with different ethnic groups (Lauristin and Heidmets 2002). On the other hand, there are challenges for the new member states to provide the rights of the minorities; e.g. in the Baltic countries (Estonia and Latvia), there are big minorities of Russian speaking people, who did not acquire the citizenship of those countries after the regained independence.

The European Union favours multi-cultural society as a source of the European identity and as a factor for effective production, similar to that of North America. In many European countries, the lack of labour force is interpreted as one of the main threats for the future development, and migration is expected to balance the decrease of the employees (Forsander 2002). A big challenge for the multiethnic development in the European countries is how to avoid the ethnic inequality, and especially, how to create social cohesion in a situation where the inequality is transferred to the next generation. For the European Union, the combating against the ethnic polarisation is one task to be conducted in the nearest future.

1.1.2.5 Regionalisation and Regional Differentiation

In all societies, not only in Europe but on all continents, there appears a tendency of regional differentiation. Many scholars interpret globalisation and localisation as reciprocal processes which all the time have influence on and are tightly connected with each other (Forsander 2002). Global and local levels are closer to each other than ever in the history of humankind, and much of that has to do with the online-information through television and internet but also with the global trade relations. Not surprisingly, the global social policy in a setting of regional differentiation has come on the international agenda, too (Deacon et al. 2003).

The main regional division is the one between the urban and rural areas, which in many countries means big economic, social and cultural differences (www.unfpa.org). Besides the ageing, the regionalisation seems to be the most prominent social trend in all societies (e.g. Loikkanen and Saari 2000; Entrena and Gómez-Mateos 2000; Graefe 2001). Strategies of both centralisation and decentralisation are applied in regulating the regional differences. There is no consensus on the question of how to divide competences between central and local/regional levels of government in different policy areas in respect to the purpose of enhancing a favourable development of the society. What is interpreted as a progressive development for which kind of regional development in different countries is a question depending on values. For example, in the Scandinavian countries, the centralisation is accepted as a strategy for building the communication and

ICT-infrastructure for advancing the knowledge and innovation based production (Benner 2003). The combination of centralisation and decentralisation perhaps explains the successful implementation of the activation policies in Denmark (Torfing 1999).

The new member countries of the European Union bring several dimensions into the discussions on regionalisation. In the old EU and in the Scandinavian societies, there has been a strong political will to support the agriculture based production or, in a wider context, the rural cultural and social development. In the budget of the EU, the structural funding for countryside areas has been on a remarkable level. In the new European Union countries, the agriculture as a production branch is still strong, but at the same time, the social and economic differences between rural and urban areas can be remarkably high; e.g. in the case of the Baltic countries, huge regional imbalances have been recognised (Rivža and Stokmane 2000). Besides the regional differences based on rural-urban distinctions, there are divisions within the countries according to the production areas; well-known classical examples of this kind of division are the Ruhr-area in Germany or Lombardy in Italy (Therborn 1995; Weihe-Lindeborg 2000). Currently a European region can be understood as an entity of different areas which often crosses country borders, too. Today, there are in the European countries ITC-areas where the high-tech and biotech enterprises accumulate the best know-how and expertise.

Another dimension of regionalisation are the differences within the European metropolitan areas. The term segregation of the big cities is used, and especially in France this tendency has been studied in the social geography science since the beginning of the 1970s (Helne 2001). A classical example are the china towns of big cities, but in the current situation there are several dimensions of urban segregation which can highly vary between the European capitals, too (Kortteinen and Vaattovaara 2001). Income levels and ethnic backgrounds are the main factors dividing metropolitan areas.

In the social policy literature, the analyses of regional developments have increased (e.g. Loikkanen and Saari 2000; Entrena and Gómez-Mateos 2000; Kainulainen et al. 2001; Graefe 2001; Weihe-Lindeborg 2000). The communication structures, economic relations, general social developments, demographic trends and innovative cultural capacities of the regions have been taken into the study settings. The ecological issues are emphasised, too, as a factor influencing on the welfare of inhabitants. The local organisation of the social services is on the study agenda, and according to that, the local identity is interpreted as an important factor in the decision-making on the provision of social services.

The question is how the local, regional, national and supra-national interests can be represented and reconciled with each other. In the European Union, one challenge for advancing the democracy and subsidiarity is the awareness of regional diversity. How should the competences of local, regional, national and supra-national level be reconciled in the issues of social policy? How are the rights of the EU-citizens ensured in the diversity of levels where social policy activities are practiced? It is obvious that effective measures for providing knowledge of regional developments in the European Union are needed in order to answer the mentioned questions.

1.1.2.6 Social Problems and Future Risks

Classically, the social problems include issues like dysfunctional social differences (strata and income), intergenerational and other social conflicts, psychosocial problems like drug-abuse and problems in conducting the everyday-life, poverty as a problem of maintenance, and, as a late-comer to the repertoire, the social climate in a form of decreasing solidarity and mutual help. As a tendency, there has been a shift from material poverty to multidimensional social exclusion, which is understood as a combination of different deprivations of an individual person or of a social group (Meeuwisse and Swärd 2002).

Besides the classical social problems, the future risks have entered into the discussions. Risks in this regard are the threats and unexpected crises and catastrophes affecting the whole populations, such as environmental accidents or nature catastrophes. On the other hand, the global terrorism has been taken as a serious threat for the societies. The European politicians have emphasised the human rights and justice based development of the international relations in order to advance the evolvement of democratic societies. In the global era, the organised international criminality can be interpreted as a risk increasing its meaning for local economies (e.g. drug trade).

In the European Union, poverty and social exclusion, all kinds of discrimination included, are considered to be the most serious social problems (Taylor-Gooby 2003; Heikkilä and Kuivalainen 2002). There is a vivid debate concerning the concepts of social inclusion and exclusion, and in that context, such topics like income, gender, generation and ethnic differences and polarisation trends of population are considered. In the debates and empirical studies, the deficiencies or factors hindering the fulfilling of life-chances, or the restrictions on the social participation are focused. It is asked how people can access educational and other institutions which ensure the skills necessary for participation in the labour market and social life in general, what kind of chances people have for learning, for developing their cognitive, emotional and moral capacities, and how they can acquire the cultural capital which is needed for organising life in the current global and complex society.

According to the statistics of the European Union, income differences and long-term unemployment are the main factors influencing poverty and polarisation of the inhabitants in Europe. The regional differences can be part of income differences and unemployment. The single-parent households are a risk group of poverty. The most vulnerable group to become poor are the long-term unemployed persons and their families. The "working poor" are also a group in the poverty statistics of the EU (European Commission, Directorate-General for Employment and Social Affairs 2003: 148–155).

Besides poverty, although connected with it, there is a continuing discussion of psychosocial problems which are increasing not only among the poor but among the stressed well-to-do persons, too, who are trying to combine work and family obligations, and who are often living socially isolated. The lack of social networks and solidarity are interpreted as indicators of a cold social climate. Among drug users, there are not only the people of lower social strata but also employed people as the debate on drug-testing indicates; in many countries, the enterprises already use drug-testing while recruiting employees. The social isolation and deprivation augmenting among the western societies induce the increasing loneliness and even alienation from social activities.

The context for discussing individual rights to privacy versus collective responsibility is not easy to create. In the welfare sociology, the changed self-identity has been discussed (Giddens 1991) but any empirical breakthrough has not happened in transferring the debates into empirical studies of social deprivation. Micro and macro levels are not connected which each other in the empirical study settings. Nevertheless, the managing of everyday-life seems to become a complicated task for people of the current era; hectic and stressful life-style indicates the problems in reconciling of the different fields of human life.

One of the prominent themes is the situation of children in the contemporary Western societies. There are debates on how to organise the everyday care of children according to the demands of work and family life. The labour market structures demand flexibility, and the everyday living has to be adjusted according the flexible working hours; for most adults with small and school-aged children it means puzzling of work, child care and family life without any stable rhythm. In the recent commentaries, the children's right to sustainable family relationships have been taken onto the agenda, with a risk for the debaters to become labelled as politically non-correct and conservative persons (Taipale 2004).[9] Nevertheless the well-being of children is one of the dominating issues in the current welfare studies, and big changes in the position of children, not only positive ones, have been reported (Therborn 2004).

Altogether, it seems that the social problems in the European Union are currently interpreted in the frame of social exclusion, an umbrella concept

9 In his criticism of the Scandinavian welfare model, David Popenoe (1994: 80) asked: "Why not take a moral position on the rights of children to have two parents during their formative years?" The critics as such could be reasonable but the Scandinavian welfare model can not be interpreted as a particular source for high divorce rates since these are high in all industrialised societies.

for factors hindering the participation at the labour market and other relevant social activities and networks. However, this frame has to be enlarged by the application of the capabilities approach. There is need for discussion of how to develop a conceptual frame in which both the individual level and different structural levels could be adequately studied (Robeyns 2004). Further, the relation between personal autonomy and activation measures is not an easy issue to be solved by social policy which anchors its principles in social *and* human rights. The sensitive boundary of privacy and public intervention has to be considered also in the social policy discussions.

1.1.2.7 Enlargement of the EU and New Countries as a Challenge for the European Social Policy

In May 2004, ten new countries acceded to the European Union, among those four former socialist countries (Czech Republic, Hungary, Poland, Slovak Republic), one former part of Yugoslavia (Slovenia), three countries with regained independence (Estonia, Latvia, Lithuania), and two Mediterranean countries (Cyprus and Malta). Through the enlargement the European Union increased its population by close to 20 %.[10] In the political statements of the European Union, the emphasis has been laid on the enlargement by the post socialist countries. Democracy, freedom, peace and prosperity have been accepted as the common European targets to be promoted and advanced. One dimension in the social fabric of the EU-25 is the development of the civil society, the enhancement of non-governmental initiatives intended to promote the local economy and social activities.

For the European Union, the enlargement is first and foremost one dimension of the Common Market, where the Four Freedoms, free movement of capital, workers, goods and services will be gradually realised. But the enlargement has also a strong political meaning, which is expressed in efforts of advancing of democracy and freedom in the post-iron-curtain era. In the political speeches and declarations, the common European values of democracy have been highlighted in a setting where the past socialist regimes are interpreted as a rupture of the democratic development in the Eastern European countries – a concept that refers more to political history than to geography (Pickles and Smith 1998).[11]

From the viewpoint of the European social policy, the free movement of workers and other persons in the form of migration has been widely discussed and will be analysed later in this book, too (see section 2.2.1). For social policy, the development of democratic political culture is of importance: political pluralism, participation of citizens, tolerance, trust and legit-

[10] For detailed information on the enlargement given by the European Commission see the website http://europa.eu.int/comm/enlargement/index.htm.

[11] Even after the enlargement of May 1st 2004, Helsinki geographically still remains the most Eastern capital city of the European Union countries.

imacy, clarity of political processes, level of individual responsibility and rules and customs of political behavior are mentioned as dimensions or indicators for the developmental level of political culture (Vihalemm et al. 1997: 201). Due to those indicators there occurs variation among the European Union countries, not only between old and new countries, but in general, too. However, if some ranking of differences between the new and old countries is asked for, the political culture is among the most prominent differences, which the low voting rates of the new countries indicate.

Social policy as an interest area of politics has been and will be developed in democratic processes where different interests are put forward, debated and contrasted. In all Western European countries after the World War II, social policy has been highly emphasised in doing national "nation building and/or restoring" politics, which is strongly built in the different social policy models, too (Esping-Andersen 1990). A similar development was not possible in the former socialist countries, but the social policy was practised as a part of paternalistic socialist order. However, there are not yet any analyses on how the personal networks of people in socialist countries were enabling the mutual solidarity and help: despite of a different political regime, the general human development indicators were and still are on a high level. Today, among the new EU-countries there are both societies with equality and democracy indicators similar to those of the old member states, and at the same time, countries with figures of a great difference.

There is evidence from the Baltic countries that the social policy was not interpreted as one among the first interest of the citizens after the regained independence. Instead of social security, the advancing of free market economy, free media, personal privacy and individualism as values have been accepted by the citizens as the issues to be conducted forward through politics (Lauristin et al. 1997; Vihalemm 2002). Lauristin and Vihalemm summarise by saying that "When material resources are scarce, the appropriate use of social and symbolic capital could help to overcome mistrust and despair" (2002: 59).

The new EU-member states are far from reaching the average gross domestic product (GDP) level of the EU-15 countries. GDP per capita of the new countries in purchasing power standards was between 39% (Latvia) and 76% (Cyprus) of the EU-15 average in 2003 (EUROSTAT 2004). As a result, income disparities between the EU member states have grown considerably with the enlargement. Additionally, most of the post socialist countries have had a favourable period of economic growth and successful privatisation processes. Since the mid-1990s, the figures of the growth of GDP have been in the new countries remarkably higher than in the old ones. In 2001, the annual growth rate of the EU-15 was 1.5% while the same figure was 4.0% in the new countries. The other side of the coin has been a high figure of (long-term) unemployment in several countries, both in the new and old ones. The structural problems of labour market are similar.

The new member states spend less on social protection than the old members. The EU-15 social protection expenditure was 27.5 % of the GDP in 2001, whereas, for instance, it was 18.1 % in Malta, 19.1 % in Slovakia, 19.9 % in Hungary, 25.6 % in Slovenia (European Commission 2004: 60). According to national data, Poland spent 23.9 % of GDP on social protection in 2003, and Estonia approx. 18 %. Nevertheless, the figures illustrate the fact that the new countries are much closer to the EU-15 in this respect than in the absolute GDP level. This can also be illustrated by the much higher position of the new countries in the human development index ranking than in the GDP level ranking. For example, Poland was classified at the 37th place in 2002 according to the HDI and at the 50th place according to the GDP per capita (UNDP 2004).

The structure of the social expenditure in the new countries indicates a passive character of social policy. Spending for transfers for inactive persons dominates, mainly for retirement and disability pensions; much less emphasis is put on active labour market policies, health care and family policy. Above all, the employment rates in the majority of the new countries stay behind the average in the EU-15. In many new EU-countries, deep reforms of social policy have been implemented after 1989. The reforms have changed not only the parameters, but the entire structure and logic of the social protection systems and are thus best described as structural or paradigmatic. The pension reforms are the best known example as will be analysed in the chapter 3.3, but deep changes are obvious also in the health care, education, and disability services. Until now, the new EU-member countries can be described as less affluent passive welfare states. Thus, crucial objectives to be mentioned are the economic growth and an activating social policy.

Kvist (2004) has asked in his recent analysis if the EU enlargement does start a race to the bottom, and his conclusion is that the internal challenges, like ageing of the populations, are of more importance for social policy development than the encounter of old and new EU-countries with different social policy systems. But he gives a comment according to which the social policy can be a competitive factor influencing on the attempts of countries "to attract workers, and, probably more important, to enhance their employment and fertility performance" (ibid.: 316). In other words, social policy can be one factor, which tunes competition for the best labour force among the European Union countries. In the new countries during their short history of social policy under market economy conditions, the extraordinary transformations of almost all economic, social and political institutions have facilitated the socio-political reforms too; political resistance to change was in this situation weaker and the political will for deep reforms bigger. If projections on the future of social policy are to be made, the new EU-countries with their reforms do not seem to support the hypothesis of a race to the bottom, but rather represent a clear case of an international policy-learning process.

In the former socialist countries, the current political life is highly professionalised – the politicians belong to both the new and old elites. In many survey studies a growing dissatisfaction with the decision-making machinery has been recognised. Citizen's alienation from and distrust of political institutions are obvious. This is also confirmed by the low voting rates (Munro 2001; Mishler and Rose 2001). Paradoxically, the implementation of parliamentary democracy has not brought about a stable political development in the post-socialist countries. There is room for different political interests in a wide range of political parties but, at the same time, also short-span populist movements are coming and going with their unrealistic promises. In a hectic political life, social policy as an issue of long-span reforms has not been taken on the agenda. The fragmentation of the societies during the transition period and the ideology of individualism have created an "unpolitical" atmosphere in which the citizens do not appreciate the political decision-making processes (Lauristin and Vihalemm 2002: 59).

While considering the future perspectives of social policy, much attention is nowadays put on the individual responsibility and on the civil society/third sector as a provider of social security. This is not a new discussion but one dimension of existing social subsidiarity patterns. In the post-socialist countries, there are some structural factors, which as such could facilitate the strengthening of this subsidiarity. The well-developed private networks can be mentioned as one important source of mutual aid and support, which were needed during the socialist regimes. However, the expected development of the former private networks towards non-governmental organisations has not happened. That is why the civil society development of the post-socialist countries is often interpreted as weak if compared to the development of the market sphere. Poland excluded, also the churches in the post-socialist countries have not got as much members and influence as was expected after the collapse of socialism. The individualistic orientation can erode the importance of networks. Younger generations do not understand what kind of structure the private networks have been for their parents and grand-parents living in a strictly controlled society. The family tights are still strong, and for example in the Baltic countries (Einasto 2002), the well-to-do adult persons support their pensioner parents with very low income with cash.

There is a lack of empirical data concerning the processes and institutions which the citizens of post-socialist countries interpret as agents of social security. In a situation where the renewal of the European social policy has been taken on the agenda in general, the mutual change of experiences of old and new countries could be positive not only for the single countries but for the European Union, too. The post socialist countries do not as such form an entity but a group of countries with varying social practices – a similar situation as the one of the old countries. However, some differences in the social policy can already be recognised; most of the new countries have launched private insurance-based pension systems, which are expected to improve the sustainability of pensions in the future. To a certain extend, it is argued that the transition in the field of social policy is still in progress in the post socialist countries which try to manage the high unemployment, increasing poverty and social polarisation. The same issues, trends and challenges for social policy are on the agenda of the old EU-countries too.

1.1.2.8 Concluding Remarks

Tendencies of the societal development in Europe have been discussed in order to form a general background for the analyses of this book. Economic and labour market changes, ageing, regionalisation, migration, social problems and enlargement of the European Union are the axes which influence the development of social policy in the future. Additionally, in the international and EU-community the accepted human and social rights are of importance in making projections for social policy. It is reasonable to speak of an intellectual matrix which consists of social tendencies and internationally accepted conventions according to which the social policy has to be analysed. In the draft of the Constitution of the European Union, Article III-116 is devoted to express the efforts of the European Union to advance economic, social and territorial cohesion and harmonious development. The sustainable financing is a challenge for the measures to be developed and implemented in order to gain the expected goals.

Recently the authors of a trans-disciplinary analysis of a sustainable health care system tried to find out how to develop a system which, at the same time and in its entity, satisfies high professional and ethical standards, is rational and controlled in its economy, and complies with the laws enacted in democratic societies (Gethmann et al. 2004). The sustainable social policy can be established on similar principles. Given the diversity of societies included in the European Union, the exertion of a common social policy will be demanding. Unavoidably the citizens of the European Union expect an improvement of their everyday-life in order to identify themselves as Union-citizens. It is a different question what role the social policy has in advancing the cohesion and social citizenship in the European Union. The regional and cultural diversification and differentiation, too, are remarkable features of the Union; local and regional economies can have greater impact on the social development than the economic performance of the Union. The new countries, finally, can be a challenge for the old EU, but are also an opportunity for new openings and ideas of social development. One challenge for the European Union is to advance the dialogues through which the diversity of the member states could benefit from the social and economic development of the whole Europe.

Palola (2004) has recently analysed the social policy discourse at the EU's highest strategic level, the European Commission. Former Commissioner for Employment and Social Affairs, Anna Diamantopoulou has given a serial of official speeches during 2000-2003,[12] which were taken into the analysis. Two conclusions crystallise how the Commission in its rhetoric advances the European social policy development: firstly, there exists a European Social Model, which can be preserved by radical changes, and secondly, a common discourse between the social and economic policies is needed in order to adjust the goals of the EU as a Common Market and as a social Europe. Diamantopoulou's speeches underline the open situation of the European Social Model; however, the speeches deal with the general level and emphasise a change whose character is not defined in details.

The general macro tendencies can be used in describing the main tendencies connecting the European societies. Many of those tendencies are similar to those recognised on the global level, too. While the European Union is thought to be a Common Market competing with other markets on the world-wide level, the current tendencies influencing on the well-functioning market and society are similar to those e.g. in the Asian and North American regions. New order of labour, knowledge-based economy, ageing of population, migration and multiethnic development, regional differentiation, new social risks, tensions between local, national and supra-national decision-making – to mention the main tendencies – challenge every industrial society. Human and social rights are accepted in the international community as principles to adhere to. From the point of view of the matrix developed in this chapter, it is argued to ask which could be the concrete proposals for a European social policy understood as an entity of statutory measures and as a social fabric creating social cohesion, too. In other words, we have to find out how the European social policy could provide the social citizenship needed in creating not only a competitive Common Market but also a dynamic region with a well-functioning market prolific for creating and cherishing the social resources too.

[12] Texts can be found on the website http://europa.eu.int/comm/dgs/employment_social/index_en.html

1.2 *Ethical Foundations of Social Policy: Personal Autonomy, Social Inclusion, Justice*

In this chapter, we intend to make explicit and systematise the general ethical principles and values, which are the essential basic features of social policy in Europe (the more concrete aims, means and instruments meant to give content to these abstract principles and values in the field of social policy will be discussed in the chapter 1.3). Historically the prime impulse for the unifying process in Europe has been the market, whereas social policy was put on the agenda only later. Nevertheless, we take it for granted that social policy cannot be reduced to an ideological epiphenomenon of economic interests and tendencies, but has to be taken as an independent sphere in need of ethical reflection and justification.

Before we can begin to unfold what we take the ethical framework of social policy to be, two clarifications – which are mainly methodological but have material consequences also – are in order:

1. As always in 'applied' ethics, it does not make sense to try to give a philosophical justification of the basic ethical principles and values. Since there is a never-ending disagreement among philosophers as to which and how many principles to take and how to justify them it is useful to start ethical reflection using mid-level or middle-range principles. The characterisation "mid-level" has two meanings which are not completely independent but have to be distinguished nevertheless.
   On the one hand these principles are "mid-level" regarding the level of justification. These mid-level principles can be found in nearly all ethical and meta-ethical approaches although often they are not the first or basic principles in these theories. These theories differ in the way they deduce or justify these mid-level principles and they take different principles as basic. But these mid-level principles are an overlapping part of the opposing ethical theories.[13] Starting from mid-level principles means to leave out the divergent ways of justifying them in the different ethical approaches and means to avoid the philosophical quarrel concerning the correct way of doing this. This is the *justificatory-sense* of "mid-level".
   On the other hand, the consensus constituted by these mid-level principles depends on two crucial features. Firstly, there is more than one such mid-level principle. And secondly, there is no strict hierarchy between them for ruling cases of conflict generally. This means that mid-level

[13] This idea has been made prominent in political philosophy by John Rawls' concept of an overlapping consensus (cf. Rawls 2001, Chapter 22) and is important in biomedical ethics due to the principlism developed by Beauchaump & Childress (2001); for a discussion of some of the meta-ethical premises behind this see Quante & Vieth (2002).

principles do not have an unrestricted scope of validity in the sense that they can be applied in all cases or have to be, in cases they can be applied, the dominant ethical principle to rule the case. This means for example that there might be ethical problems, which cannot be handled using a mid-level principle p at all, or that there might be ethical problems a mid-level principle q is relevant for but overruled by another such principle r. This is the *scope-sense* of "mid-level".

2. Neither social policy will start from scratch nor will the ethical reflection concerning the principles and values of social policy do so. Therefore, in our ethical approach, we will try to critically reconstruct which norms, values and principles frame the normative self-understanding of social policy in Europe. This means that we start from those principles and values which are implicitly or explicitly accepted (the reconstructive or hermeneutical aspect) but that we will try to make them more explicit and find out whether they are consistent and whether and how they are operative in social policy (on the level of aims, means and instruments). This is the critical dimension of our approach. Arguing in this way we understand our normative approach not in a revisionist[14] manner as if there were no social policy at all or as if the actual social policy were completely or in large part not acceptable from an ethical point of view. Starting from the normative framework, which is already operative in the reality of social policy, does not mean that the conceptual and normative framework we will develop here (and also in chapter 1.3) is only a defence of the status quo. It means rather that we try to keep in touch with what is going on in Europe and try to formulate an ethical framework which has the chance to be accepted 'in the field'.

Our two methodological commitments converge since the mid-level principles, which can be found in nearly all ethical theories, and the normative principles, which are operative in reality, overlap to a high degree (this is surely no surprise since ethical theories try to deal with our real lives). They deliver some constraints for the way we will go on in this chapter. Firstly, we will make explicit what the main principles and values of social policy are and how they have to be understood. Secondly, we will examine the idea of a just society lying behind this framework. Thirdly, the principle of justice, which is indispensable in the normative framework of social policy, is introduced as a third basic element of social policy. Having made explicit the overall normative framework of social policy up to this point, in the final two steps we will ask shortly in which sense there is a *European* social policy and how such a policy can be reconciled with universal moral claims.

---

[14] In philosophy a revisionist approach either ignores common sense beliefs and practices to a large extend or delivers arguments purported to show that our common beliefs are not false or ethically wrong; cf. Willaschek (1998) for more on this.

1.2.1 European Consensus: Personal Autonomy and Social Inclusion

In the official statements of various institutions of the EU dealing with social policy, two normative principles are dominant: autonomy of the individual person on the one hand and social inclusion of the autonomous individual on the other hand. Our thesis is that these two are the main features of the European consensus concerning social policy, which have been made explicit so far. "Consensus" is meant in a broad sense here covering implicit and explicit forms of general acceptance, so that in this book it is used synonymous with "common acknowledgement of the basic principles and values". As in the tradition of contractualism in political philosophy, it is useful not to reduce forms of consent or consensus on explicitly stated forms since by this way it would be impossible to describe and explain how such implicit and overlapping consensus evolves at all in an ongoing practice like social policy in Europe. Therefore, for our reconstructive and hermeneutical approach this use of the broad notion of consensus is useful (and should be kept in mind by the reader throughout the book).

Additionally we find two more concepts, which seem to be used as basic norms, too: subsidiarity and sustainability. As we will argue in the following these two should not be regarded as basic normative principles for social policy in Europe but as instruments. The motivation to refer to subsidiarity in social policy in Europe can best be understood if we take it as an instrumental advice, which is meant to guarantee that the relations between individuals, communities, the nation-states and the EU will be of the intended right kind (see 1.2.2.1 below). If we ask, which ethical grounds there are to introduce "sustainability", two reasons[15] come to mind: On the one hand, social policy should take into account the fact that human beings live their lives as persons who organise their live span in a biography and express their autonomy by developing a personality. Due to this fact the question of *intra*-personal long-term effects of social policy (e.g. provision for later stages in one's life) comes into view, i.e. effects during the individual's lifespan or life course (see 1.3.7). Therefore, given the natural (biological and cultural) facts about human beings the normative principle of personal autonomy naturally leads to sustainability. On the other hand, human persons of different age groups co-exist in a society and social policy normally will lead to distributions of burdens and means between these age groups. In this case, sustainability of *inter*-personal relations comes into play since *one* main justification for burdening the middle aged with funding the lives of the elderly would be that the whole system of distribution is stable and will pay the middle aged back in the future what they have to pay for others now. It is evi-

---

[15] This is not to say that the normative grounds for sustainability are limited by or can be reduced to these.

dent that in cases of interpersonal distribution the normative principles of autonomy and social inclusion are not enough to cover all ethical aspects of social policy. Therefore, the principle of justice (especially in the sense of interpersonal distributive justice) will be introduced as a third principle of social policy to deal with these aspects (see 1.2.3.1). Besides this, the demand for sustainability reaches even further if we take into account that moral claims of justice are universal in scope. This means that the impact of social policy on future generations has to be considered also (see 1.2.5 where we refer to this aspect as the *intergenerational* aspect of distributive justice in social policy).

In the following section of this paragraph we will unfold how "autonomy" and "social inclusion" as the two explicitly stated normative principles of social policy in Europe should be understood (and how, in our view, they are understood already).

In the context of social and political philosophy, it has become common knowledge that we can distinguish between negative and positive freedom.[16] Although this distinction is not without conceptual problems for our purposes it suffices to define *negative* freedom as the absence of external constrains or obstacles which hinder an agent to do what he wants. *Positive* freedom is not restricted to constrains or obstacles but focuses on the agent as the source of self-control, thereby reflecting on the conditions which have to be given so that individuals develop those capacities necessary for exercising their autonomy. In the ethical and political sphere negative freedom is understood in the sense that freedom consists in the absence of interference by another agent or social institutions: freedom in this sense consists of the space where an autonomous actor can act without being obstructed by others. Positive freedom, on the other hand, is understood in the sense that freedom can be developed and exercised only in social and political surroundings, which allow for the development of those capacities necessary for being autonomous. The possibility to lead one's life autonomously depends on a person's abilities (knowledge, skills, etc.). Some of these are gifts of nature (e.g. health or the absence of disabilities), but many of them are gifts of social conditions (e.g. education or the structure of the social system one lives in).

In the context of social and political philosophy, theories, which take autonomy of the individual as a core value, differ in their understanding of freedom. Those who primarily refer to negative freedom restrict the role of the state and welfare systems to a minimum since they are only interested in defending the space of freedom against interference (mainly against the state

[16] There are some other distinctions in use in the field of social policy, which are not covered by the distinction drawn above. For not complicating things more than necessary, for our purposes we restrict ourselves to the distinctions between positive and negative freedom made above.

and legal paternalism). Those who take into account the conditions of freedom also infer from these preconditions of freedom and autonomy that the community (mainly the state) has the duty to guarantee that individual's capacities are developed so that they can exercise their freedom and realise their autonomy.

As we see it, the consensus of social policy in Europe relies on the richer and more demanding notion of positive freedom. Therefore, the state is obliged not only to save individuals' negative freedom by not interfering (as far as possible). But in addition to this the state is obliged to provide a social setting wherein the persons can lead their lives autonomously. Furthermore, we think that it is possible to underpin the notion of positive freedom with a content, which focuses on autonomy and self-responsibility. In the capability-approach developed by Sen (1999) we find a theory of freedom which takes into account that freedom cannot be restricted to negative freedom but that interference of the state (or other social institutions) is primarily directed to promote and realise those capacities necessary for personal autonomy. An integral part of Sen's conception of freedom is that persons can choose between different forms of well-being, choosing different life-plans which express different rankings of values and capacities therein (Sen 1992). Since personal autonomy and individual life-plans are essential parts of this conception, the conception of positive freedom cannot be understood as the prescription of one particular conception of the good life. Therefore it is natural to understand the state's role here as an enabling and activating one (see chapter 1.3). On this understanding, which is presupposed and defended in our study, the main aim of the state's direct or indirect interventions is to enable persons to lead their lives autonomously in helping them to develop those capacities and guaranteeing them those social and material conditions necessary for taking over self-responsibility and encouraging them to actively participate in social and political life.

In this way, positive freedom means that the state primarily invests in the capacities of its members to make them more autonomous and give them space to design their lives according to their value decisions and individual needs. For sure, there are some constraints, which have to be respected here coming from the human condition. Human beings as biological and social beings cannot create their lives in a vacuum.

On the one hand, biological data (e.g. age, gender, sexual reproduction or health) deliver a framework in which every individual life-plan must be framed. Therefore, this basic grammar of possible biographies a human person might choose has to be taken into account. In the descriptive sense, this is done by the life-course approach where these temporal structures of a human person's existence are examined and modelled. In the normative sense, the biographical dimension of human persons has to be acknowledged within the notion of autonomy itself. Therefore the notion of autonomy cannot be reduced to autonomy of action or single decisions but has to be widened to the notion of personal autonomy covering extended life-plans and care for future chapters of one's biography (e.g. care for being old or becoming ill).[17] The topics we examine in the third part of this book (family, health and security in old age) cover essential aspects of every human life since they build the biological and anthropological framework within which personal autonomy can manifest itself in individual life-projects.

On the other hand, there are constraints to personal autonomy coming from the fact that human beings can exercise their autonomy and their individuality only within a community (at least in normal cases). Therefore, autonomy is impossible without social inclusion since recognition (in the normative dimension) and economic or material resources (in the descriptive dimension) are constitutive for developing and exercising autonomy. In complex modern societies social inclusion and recognition to a large part are related to labour, personal skills and economic power (therefore we consider poverty and labour throughout the third part of this book). Social inclusion requires active participation and recognition as much as material resources. Therefore, it must be an essential aim of every social policy to integrate human beings into society (objectively and subjectively). Since achievement is an essential criterion for recognition both for the one who is recognised and for the one who recognises, this aim cannot be reached by direct intervention (e.g. a right to work), because thereby the character of a personal achievement would be lost. For sure unemployment and poverty (at least if they last long in an individual's biography) endanger social inclusion and thereby personal autonomy. But since labour can be regarded as an instrumental value only (at least from the perspective of the normative framework developed here) the aim can only be to enable individuals and activate them to participate – welfare policy cannot be the surrogate for this mechanism of social inclusion, at least not for a longer period in a person's life. As we see it, personal autonomy and social inclusion are interrelated since the former can only be realised if the latter is given, but this leaves some conceptual space as to how to relate them normatively. In the next chapter (see 1.3), the conceptual framework developed so far will be filled in with the concepts of activation, enabling and life-course.

Before we come to the question of how the two basic principles are related to each other in social policy in Europe three concluding remarks are in order:

Firstly, focussing on activation, enabling and personal autonomy should not seduce one to forget that relief also is an essential aspect of personal autonomy. If there were no relief in some contexts of our lives personal

[17] For a more detailed analysis of the notion of personal autonomy and its relation to personal identity see Quante (2000) and Quante (2002, chapter 5).

autonomy in other contexts would be simply impossible for us because of overburdening. It should be left to individual choice at least to a large degree in which areas of one's life one wants to be relieved (e.g. in health care or in care for one's old age) and in which one wants to exercise one's autonomy primarily. Therefore social policy should take this need for relief into account as another integral part of the idea of autonomy (since, otherwise, it would become a kind of paternalistic coercion itself).

Secondly, focussing on activation and enabling should not imply that there are no other values relevant in social policy. For sure there are: if we take disabled persons who simply cannot develop those capacities necessary for exercising autonomy it is evident that a good (or just) society needs a richer conceptual framework than the one developed in our study. Since we accept this, we will not address the questions related to these problems here. But our arguments here should not be understood as the thesis that all questions e.g. in health care can be dealt with relying only on the notions of personal autonomy or the account of an activating or enabling welfare state.[18]

Thirdly, we think that there is one more core principle, which is necessary for every normative approach in social policy, which cannot be reduced to personal autonomy or social inclusion. We have the principle of (distributive) justice in mind, which will be introduced as a third pillar of European social policy (see 1.2.3 and 1.3.5).

1.2.2 Hidden Dissents? The Ethical Determination of the Relation between Individual and Community

As expected, the core elements of the consensus of social policy in Europe are abstract and leave a lot of conceptual space for philosophical interpretation on the one hand and for political concretisation in the sphere of social policy on the other hand. Therefore, the question whether some hidden dissents lurk underneath this common acknowledgement of principles and values is hard to answer. In the end it depends on the concrete interpretation of "autonomy" and "social inclusion" whether they are conflicting values or not. This is aggravated by the fact that some of these more specific understandings of the core principles will have implications for political philoso-

[18] There is a common misunderstanding that the enabling approach is suitable only for the strong and 'able' members of a society. But we have to distinguish two different standards which are in use here. On the one hand, we can use a general standard relying on anthropological, biological or social common facts. This general standard can be used in contexts where the principle of justice comes into play (e.g. compensation of social disadvantages in education). On the other hand, we need an individualistic standard relying on the potential a particular human being has. This standard of enabling comes into use for example if we apply the concept of enabling to disabled human beings. The idea then is to help them to realise their potentials in the most effective and autonomous way possible for them.

phy and, thus, for ethically acceptable social policy.[19] Conversely, there are also presuppositions of political and social philosophy, which already shape the concrete interpretations of "personal autonomy" and "social inclusion". Furthermore, it is of utmost importance to distinguish clearly different levels in the debate – we suggest to differentiate the following three (see 1.2.2.2):

- the level of ethical principles and values;
- the level of aims;
- the level of instruments.

Since there are many interrelations between the basic ethical principles and values, premises in social and political philosophy and the three levels just distinguished here, it is idle to try to develop through a deductive kind of argument *the* ethically justified social policy for Europe. We cannot simply 'apply' the principles and values we find in the diverse declarations concerning social policy in Europe. Therefore, we have to look at the real processes going on and refer to the more concrete normative arguments, which are defended and accepted in the debates concerning the different areas of social policy. Since this cannot be done without giving a detailed analysis of the concrete processes (which we give in part three of this book) in this section we have to confine ourselves to three tasks. In a first step, we will show that and in which sense the generally accepted principles operative in social policy in Europe are underdetermined. Then we will distinguish three levels of debate, and finally, we will defend the thesis that comparison of social policies of different European countries (in special areas presented in part three) has the structure of a twofold specification. Our arguments in this chapter are neither intended to justify the principles, values and specifications in a strict philosophical sense. Nor can we justify our theses in this section since doing this requires all the empirical data and descriptive tools we will present in the second and third part of this book. What we want to do now is to make explicit and clarify the conceptual space, which is open for a normatively justified European social policy within the overlapping consensus being manifest so far.

1.2.2.1 The Under-determination of Europe's Normative Consensus

At first, we have to make a general remark: in philosophy it is important to distinguish between the thesis of *in*-determination and the thesis of *under*-determination. To say that a sphere X, say the sphere of aesthetics, is in-

[19] Due to this, it is evident that the evaluative and normative framework reconstructed and developed here neither can entail a detailed catalogue of political measures nor is intended as giving such 'recipes'. The process of political realisation has to be left for the political process itself. Our hermeneutical and reconstructive approach here is meant to give orientation and philosophical exploration, not political advice.

determined is to say that there are no facts of the matter which could make our statements and theories true or false. To say that such a sphere is underdetermined is to say that all the facts available (in principle) will never be sufficient ground to decide which of rival theories is the true one (ruling out all rival accounts).

Our thesis is that the common acknowledgement concerning the basic values and principles for social policy in Europe is underdetermined, not indetermined. This means that the facts (declarations, social policies in Europe, etc.) are open to different interpretations, but these interpretations are not completely arbitrary or cannot be justified by reason in principle. There are normative facts on the one hand and the normative consensus concerning social policy in Europe clearly rules out as incompatible some suggestions what the right (or good) relation between individual and community should be taken to be (the importance of this fact is discussed in 1.2.4).

As we have seen above (in 1.2.1) one reason for this is that the basic principles and values are open to interpretation. The other reason, as we want to argue now, is this: there are at least two accounts of the ethically good or just relation between individual and community, which are both compatible with the given normative self-understanding of social policy in Europe, but can lead to incompatible social policies on the level of aims and instruments (although they do not have to have this consequence necessarily). Since these two accounts imply different rules of balancing the different values and give some values a different status (e.g. intrinsic versus instrumental value), they are incompatible as overall theories but might come to the same conclusions in dealing with concrete problems of social policy.

After having made this general remark we can distinguish four models. Although they are taken from debates in political philosophy where the focus primarily is on the relation between the autonomous individual and the state, in the definition of the four positions given now we do not refer to the state but to community more generally. On the one hand, we take the state to be one, indeed very important, instance of 'community', but we think that the issue at stake here is the right relation between autonomous persons and the community (or communities) they live in. On the other hand, the state is an important topic in social policy, too, since we can ask at which place (on which level) and by which means the state should be an actor in social policy (the same can be asked with respect to the EU as a supranational social institution). Nevertheless, this latter question should be distinguished from the more general ethical question how to define the right relation between individual and community (see section 1.2.2.2 for conceptual clarification of the latter question). But since the state always has been (and in some sense surely will stay to be) so important in political philosophy we explicitly mention it in the definitions given now.

- *Libertarianism*[20] holds that personal autonomy has absolute priority on the one hand and social inclusion (i.e. the integration of an individual in the community) has instrumental value only insofar as it is a necessary tool for realising autonomy (autonomy itself is defined as a non-relational state of the individual). Therefore, the state is an – most unavoidable – evil, which shall do nothing but guarantee the negative liberty of individuals.
- *Moderate Liberalism* also holds that personal autonomy has absolute priority, but also claims that autonomy cannot be restricted to negative liberty but has to cover positive liberty, too. Therefore, the community (especially the state) not only has to guarantee negative liberty but also has to deliver conditions in which individuals can develop and exercise those capacities which are necessary for leading autonomous lives. *Weak* moderate liberalism additionally holds that social inclusion is of instrumental value only for the development of personal autonomy, while *strong* moderate liberalism ascribes social inclusion the status of intrinsic value since social inclusion is taken to be a constitutive part of personal autonomy. But it is a distinguishing feature of moderate liberalism that the ethical claims of individuals always dominate ethical claims of social institutions (e.g. family, the state or other communities).
- *Liberal communitarianism* holds that personal autonomy and individual rights related to this are ethically important and at the core of our culture's normative self-image. Like strong moderate liberalism, it also holds that social inclusion is an intrinsic value and has to be taken as a constitutive part of personal autonomy. In opposition to moderate liberalism (in both versions), liberal communitarianism accepts that (i) social entities (e.g. social institutions like the family or the state, communities, traditions etc.) are bearers of intrinsic ethical value on the one hand *and* (ii) it leaves open the possibility that the ethical claims of these social entities might overrule the ethical claims based on personal autonomy in *single concrete cases* of conflict. Nevertheless, liberal communitarianism holds that prima facie personal autonomy is the overriding value so that the burden of proof is ascribed to those who claim that in single concrete cases personal autonomy should be overridden. (In some accounts which fall into the category "liberal communitarianism" it is argued that participating in the state is an essential element of full personal autonomy so that individuals are ascribed the duty to participate.)

[20] We are dealing here with political libertarianism, of course. In the free-will debate libertarianism stands for a philosophical position which holds firstly that free will and determines are incompatible, and secondly, that freedom of the will is possible for human beings; cf. Quante (2003, chapter 10 for details). Since in our context it is clear that the metaphysical debate concerning free will is irrelevant we omit the qualifier "political" for simplicity in the following. Therefore our use of "libertarianism" should always be understood as "political libertarianism".

- *Anti-liberal communitarianism* holds that ethical claims of social entities override the claims of personal autonomy and the related individual liberty rights. In the *strong* version, anti-liberal communitarianism denies the intrinsic value of personal autonomy; in the *weak* version, it acknowledges that personal autonomy is an intrinsic value. But weak anti-liberal communitarianism holds that claims of personal autonomy are generally overridden by claims based on the ethical value of social entities (e.g. social institutions like family or the state, communities, traditions etc.).

Both the basic principles, which have been identified and unfolded in this chapter so far, and the social policy, which in fact takes place in Europe, prove clearly that the normative consensus identifiable in social policy in Europe excludes anti-liberal communitarianism as well as libertarianism. Emphasising personal autonomy on the one hand and interpreting it in the light of Sen's capability approach on the other hand is clearly incompatible with anti-liberal communitarianism. But it is also, although this is sometimes missed, not compatible with libertarianism. This is due to two reasons: Firstly, personal autonomy cannot be reduced to negative freedom but is essentially dependent on conditions in which the capacities necessary for exercising autonomy can be developed. Secondly, to say that the state has to deliver these conditions in principle is not to say automatically that the state himself should appear as a directly intervening actor. We can also hold that the state's duty is to guarantee that the autonomy-promoting conditions are provided but that this can be done by delegating this task to those social institutions which can realise it most effectively – not only in terms of money but in terms of enhancing autonomy and increasing direct responsibility (Eigenverantwortung). Distinguishing the levels of values and principles, aims and instruments helps to avoid libertarianism's common fallacy that personal autonomy has to be restricted to negative freedom since the state should not be allowed to take an active role in social policy. This is why we can accept some of the libertarian arguments against a directly intervening state without restricting our normative framework and our normative understanding of the state's proper role in social policy in the libertarian way. As we will see later on (in chapter 1.3) the idea of an activating (or enabling) welfare state is the most plausible account for unfolding the normative consensus in social policy in Europe identified so far.

Subsidiarity – a Brief Digression

Roughly stated the principle of subsidiarity says that in social and political philosophy, firstly, the smaller unit (e.g. in terms of the relation between family and state) should be ascribed ethical priority and secondly the larger unit should help the smaller one to fulfil its duties. Since the smallest unit in social and political philosophy, at least in the normative dimension, is the individual person, the principle of subsidiarity also applies to the relation between individual and community (in the sense of all social entities).

The principle of subsidiarity, historically a central "Sozialprinzip" ("social principle") of the "Katholische Soziallehre" ("catholic social doctrine"), has gained prominence in diverse declarations. As we see it this comes as no surprise for at least three reasons:

1. It can be understood as a principle that serves to protect individuals and smaller social units against interventions, tutelage and restrictions of superior units or the state. In times where trust in the creativity, effectiveness and adequacy of such central social institutions vanishes, the principle of subsidiarity becomes attractive as claiming a negative right for non-interference.
2. The principle not only protects individuals against social institutions or the state, but also protects smaller social entities (e.g. families or regional communities) against the state (and, for sure, the national states against the EU!). Therefore, it is of equal interest for moderate liberalism and liberal communitarianism. Taken our diagnosis for granted this makes intelligible why the principle of subsidiarity is a good candidate for compromise and declaration.
3. But apart from these there is a third, more constructive, aspect. The principle of subsidiarity is not only a principle, which states a negative right, but at the same time it states a duty. The larger unit has the duty to help and – taken the basic idea, that the smaller unit should be able to fulfil its own tasks and duties by itself – this help can only be of the "enabling" or "activating" kind. Therefore, the principle of subsidiarity not only fits very well into the dialectical framework constituted by the basic principles and values commonly acknowledged but can also be integrated easily into the conceptual and normative framework we suggest here.

Thus, the prominent role ascribed to the principle of subsidiarity is not only evidence for the reconstructive and in this sense descriptive thesis defended in this section. It also can be integrated into our overall normative framework (on the abstract level) and into the account of an enabling or activating welfare state (on the level of social policy) introduced and defended below (see 1.3). But since the principle of subsidiarity is best understood as a rule which allows to guarantee (or protect) the basic principles and values it should not in itself be taken as an intrinsic value of its own and least of all as a basic principle of the same level as the three principles of autonomy, inclusion and justice. As far as we can see, its normative status should be that of an instrumental value and guideline which can help to overcome shortcomings in social policy on the one hand and encourage enabling and activating policies on the other hand. The ethical force of this principle comes from the underlying values, which are intended to be protected by it. Without them, the principle of subsidiarity, so we argue, has no ethical value (in this respect our interpretation must be understood as a correction of the status gained by the principle of subsidiarity at least at the surface of the declarations). Embedded in the underlying principles of personal autonomy and social inclusion the principle of subsidiarity should be taken as an integral part of the third principle important in social policy: justice. Taken this way subsidiarity is only an instrumental principle ruling the right relation between autonomy and inclusion within a just society.

Having distinguished four models of the right relation between individual and community (especially the state) and having shown that the prominent role the principle of subsidiarity has in the declarations concerning social policy in Europe can be understood within the conceptual and normative framework developed so far in this chapter, we want to conclude this section by stating which position can be taken as the normative platform of our arguments in this book. At first, we have to say that we will not decide definitively between strong moderate liberalism and liberal communitarianism, but that we take strong moderate liberalism as the normative starting point since it is accepted by all of us. Deciding between strong moderate liberalism and liberal communitarianism is not necessary for the purposes of this book since as we have seen strong moderate liberalism can accept social inclusion as an intrinsic value in social policy thereby capturing a central intuition of liberal communitarianism. The main difference between both options is marked by the fact that liberal communitarianism in principle allows for normative claims of social institutions (e.g. families or the state) to overrule claims of personal autonomy in case of conflict while moderate liberalism makes personal autonomy the generally overruling normative principle in these cases. But, as far as we see, this difference does not become effective in the material cases we discuss later (in part three of this book). Although there might be problems in social policy (e.g. concerning family or health), where strong moderate liberalism and liberal communitarianism are committed (or at least entitled) to conflicting normative answers for our purposes we can leave this open.

1.2.2.2 Diverging Tools, Aims or Values?

As mentioned above it is important not only to clarify the basic principles and values underlying social policy in Europe. It is also of utmost importance to distinguish different levels of argument. This will help to overcome talking past each other and will help to avoid fallacies (like the libertarian one mentioned above) and to make visible the divergent burdens of proof. In order to clarify the ongoing debates it is useful to distinguish three levels: the level of principles and values, the level of aims and the level of instruments. Each level requires arguments of a special sort and is in need of different expertise and justification.

In this chapter, we mainly deal with the *level of principles and values*. On this level the overall ethical framework for social policy in Europe has to be made explicit and – in case of diverging interpretations – to be justified. This is primarily the task of the philosopher dealing with conceptual analysis and analysis of ethical argument. If there are divergent normative claims regarding the basic principles, then philosophical analysis can make clear where this dissent comes from and which implications the different options may have. But the decision as to which normative route has to be taken cannot be delegated to the philosopher; in the end it is a political decision. Since we start from ethical consensus as manifest both explicitly in the declarations and implicitly in real ongoing social policy in Europe we do not need to justify the ethical framework of social policy in Europe. And since questions of balancing the divergent principles – as said before – should be answered in political processes, we restrict ourselves here and try to make visible only where the dissent might come from and might be solved.

The *level of aims* comes into view if we ask what the principles of personal autonomy and social inclusion might mean in the context of social policy. Which account of the role of the state for example and which structure regarding family, labour or health is best suited to realise the basic principles and values? This level, which will be addressed in the next chapter in a rather general way, will be dealt with in more detail in the second and third part of the book. Here we can only say that on this level conceptual and normative aspects combine with empirical aspects, since answering the question which structures of social systems best sustain autonomy or maximise social inclusion needs both: on the one hand, a clear understanding of the values to be sustained or maximised, and on the other hand, an empirical knowledge on how divergent systems work and which effects they have.

The *level of instruments*, for sure the most concrete and detailed level of examination, can best be understood if we choose the aspect of efficiency (including sustainability). This is primarily an empirical question and the expertise in answering it must be delegated to those who have the special knowledge in the different areas (therefore this level is mainly dealt with in the second and third part). Here we can say only two things: firstly, it should always be kept in mind whether we are discussing on the level of principles or on the level of instruments (especially if it comes to the questions as to what the duties of the state are regarding social policy and what is the right way of the state's intervention in social policy). Secondly, it should be clear that on the level of instruments the aspect of efficiency is always an ethically important one but that this does not mean that ethical reasoning can be reduced to the idea of maximisation. If the values and principles are given it is ethically important to find out which instrument is most efficient in realising them. But this is not to say that effectiveness is the sole criterion to decide which principles and values can or should be justified ethically.

1.2.3 Justice: the Third Normative Principle of Social Policy

In social policy no ethical framework can be adequate which does not include the principle of distributive justice (equality taken as an essential aspect of it). As long as resources of human societies are scare and core liberal values are respected there will be inequality of individuals on the one hand and the problem of inequality in the distribution of goods between individuals on the other hand. Social policy therefore always will have to redistribute goods and distribute benefits according to some distribution scheme which has to be established as just. Since the state (or other social institutions) needs resources to fulfil its duties in social policy there will always be the question of how to get them, from whom and according to what scheme of burdening. As long as there is equality, every normative approach to social policy will have the consequence that questions of distributive justice have to be answered and distribution schemes have to be justified. For sure, neither are we in the position to develop a theory of justice here nor can we defend one philosophical account as the best one. Therefore we restrict ourselves to some conceptual clarifications in the following subsection (1.2.3.1) and will deal with the question of how to conciliate the universal claims of justice on the one hand with the demand for context sensitive solutions in social policy on the other hand in section 1.2.3.2.

1.2.3.1 Justice: Some Conceptual Clarifications

We can roughly distinguish four constellations here: Firstly, we can ask how an individual should distribute his resources among the different phases of her or his life. These questions related to *intra*-personal justice can be dealt with mainly in the framework of personal autonomy (and the life-course approach), since here the distribution over the individual's lifespan (or life course) is under consideration. As long as we do not take the future stages within a person's life as separate future selves these questions do not touch the question of distributive justice directly.[21]

Secondly, there are the problems related to distribution and redistribution between coexisting individuals (take systems of health care, for example, where burdens are distributed unequally). Although Sen's capability approach does have some implications concerning distributive justice it is evident that these questions cannot be answered referring to the capability approach alone.

Thirdly, there are the problems of distribution and redistribution between coexisting age groups (children, middle-aged and old-aged persons), which will become more and more important in Europe because of the demographic developments (take old-age security systems, for example, in which the middle-aged are burdened to finance the pensions of the old-aged). Although some aspects of these problems are related to the life-course approach and to the principle of personal autonomy it is evident that these questions cannot be dealt with adequately without taking questions of distributive justice into account. Besides this, the principle of sus-

[21] Cf. Parfit (1973) for this line of thought.

tainability will become important in this context since stable and secure social systems will help to justify claims of justice between coexisting age groups.

Fourthly, there is the question of how to take into account the interests of future generations. Since today's decisions in social policy will have effects on future social systems questions of distributive justice have to be extended here, too. As in the case of coexisting age groups, the principle of sustainability can be of some help to answer these questions but it should be clear that this problem poses philosophical questions we cannot address here.

Equality and the compensation of natural or social disadvantages are essential aspects of social policy if we regard them under the aspect of a just society.[22] This means that the question of distribution or redistribution is unavoidable and, in a normative context, the problem of how to justify distributive justice is posed.[23] In principle, there are three options to justify burdens and demands, which are caused by distribution of goods and means in social policy.

Firstly, one can argue that such distribution and redistribution can be justified with reference to the informed self-interest of rational persons. Taken together the empirical structure of human beings, their interest in autonomy and their interest in social inclusion, it is rational for every rational person to accept distribution and redistribution which will help to fulfil his interests in the long run (i.e. during his own life).

Secondly, one can argue that justice (equality included) is a basic universal moral claim, which is as well justified as all other moral claims are. Whether we accept philosophical accounts intended to demonstrate the validity of such universal moral claims beyond rational self-interest or not: it is a fact that it is part of our modern and liberal normative self-understanding that equality and justice are accepted as central moral principles which are as basic as respect for personal autonomy.

Thirdly, one can argue that distribution and redistribution cannot be understood as justified by an universal moral claim of justice, but have to be justified relying on the principle or value of solidarity. Solidarity as a basic

---

22 Rawls himself had restricted the scope of compensation to social inequality but others have extended his approach and apply it to natural inequality, too. See Buchanan et al. (2000) for the most elaborated approach so far.

23 It is important to notice that justice cannot be reduced to equality at least not in the context of social policy, since here justice always includes a component of redistribution, which gives priority to the weak and poor members in society in some sense to be defined more precisely in an overall account dealing with justice, equality and priority; cf. Parfit (1991) for more in this. This means that diminution of the gap between the better and the worse off may not be aimed at if increasing this gap would benefit the latter more than diminishing it. Put another way: Even if one regards equality as an intrinsic value it should not be taken as the ultimate or sole value in social policy; this claim has been defended by John Rawls in his theory of justice.

ethical ideal is supererogatory in the moral sense.[24] But nevertheless, solidarity and altruism are essential features of human beings' overall psychological make-up and therefore part of their motivational structure and their own normative self-understanding. Additionally, solidarity is an important value which helps to build a just society and which is necessary to deliver conditions wherein personal autonomy can be realised.[25]

As we see it, reducing demands of justice to questions of informed self-interest is too weak a position either to reconstruct the common acknowledgement of the basic principles and values, which can be found in our societies, or to justify, which duties between coexisting persons and age groups should be accepted. It is for sure that if it can be shown that acting according to these duties also fulfils my informed self-interests, then this will help for motivation and acceptance. But such a coincidence should not be regarded as a constitutive element of the moral demand itself.

In the same vain solidarity as a supererogatory duty seems to be too weak to capture the sense of obligation we actually accept in our societies and should accept regarding distributive justice and equality. For sure altruism and solidarity are ethical values and ideals which are essential elements in the welfare systems of our societies (e.g. in the care for old-aged in families). Besides that altruism and a sense of solidarity are strong motives for persons to act. Therefore, social policy should always provide institutional space for this (e.g. in form of honorary engagements). But it would weaken the normative claims of social policy too much if solidarity as a supererogatory principle replaced the moral principle of justice completely.

Therefore, we take the principle of justice as a genuine moral principle, which can be supported by arguments relying on informed self-interest but cannot be reduced to it. The crucial question in social philosophy will be how and where to draw the line between the obligations backed by the principle of justice and those duties backed by solidarity. As we see it, this cannot be decided a priori but has to be left to the political process, to autonomous decisions of political subjects and to the experiences, our societies will make with the answers they give. All we can do (in the next section) is to relate the principle of justice to the principles of personal autonomy and social inclusion on the one hand and to argue that the universal claims of justice are compatible with context sensitive social policy in principle on the other hand.

---

[24] Actions which go beyond the demands of ethical duty and are morally praiseworthy are supererogatory. Supererogatory actions are not morally required by duty but are ethically encouraged by ethical ideals.

[25] Distinguishing between justice and solidarity in the way suggested here is revisionary since in the tradition of social policy both notions have been used nearly interchangeable (as described in 1.1.1).

1.2.3.2 Universal Normative Foundations and Twofold Contextual Specification

As we have argued so far equality is an essential element of justice and, within the framework of Sen's capability approach, the criterion for measuring whether a given state of affairs is just is 'equality of capabilities'. Since capabilities not only cover actions of persons to do things they value but also includes that individuals can reach valuable states of being, Sen's approach can integrate Dworkin's criterion 'equality of resource' (cf. Dworkin 2000). Both, Sen and Dworkin, start from Rawlsian theses, but both try to develop distribution schemes which are more individual-sensitive (cf. Kymlicka 1990: 76–85). While Sen is more interested in empirical sensitivity, which should allow for more specific political interventions to overcome injustice, Dworkin integrates elements into his ambition-sensitive distributive scheme which allow him to compensate the endowment-insensitivity of Rawls' approach. Thereby, like Sen, he can take the value of personal autonomy into account (i.e. that individual persons choose different life-plans expressing different rankings of value). So, Sen's and Dworkin's approaches allow for individual freedom and choice without giving up the ethical claim of distributive justice as a universal one (see 1.3.5 for more on this). Therefore, the idea of intrapersonal justice can be integrated into the principle of justice without restricting the universality claim (at least as long as the specific structure of the life-course and the individual life-plans chosen are taken both as basic).

But in the context of social policy in Europe this universality claim which is demanded from the ethical principle of justice has to be made compatible with two other things. Firstly, the principle of justice has to be specified to contexts of action. It would be a philosophical mistake and ethically not acceptable to conclude from the universality of justice that there must be one and the same distribution scheme operative in every context of social policy. This does not follow for two reasons: On the one hand, in the process of specification the special features of the goods, which have to be distributed, have to be taken into account (e.g. health). Additionally, the relation between personal autonomy and the sphere of social life we are dealing with (e.g. family) can make differences. On the other hand, there are more ethical principles and values at stake in social policy than only justice. This allows for context specific balancing between them so that demands of justice may be overridden by demands of social inclusion in some contexts (e.g. family) but not in others (e.g. old-age security). As we see it, context sensitive specification and context sensitive balancing of the principle of justice is and should be a general feature of every normative framework for social policy.[26]

---

[26] Specification is needed if a general rule is applied to a specific, particular case, while balancing comes into play if a plurality of principles (or values) has to be taken into consideration; cf. Richardson (1990; 2000) .

Secondly the application of the principle of justice has to take into account the national and cultural differences in Europe. The universality claim of equality, which is embedded in the principle of justice, can easily result in excessive demands that overburden both individuals and social institutions (especially the national states). The fact that the application of a principle causes excessive demands is ethically relevant; therefore, this alone can be taken as a justification for restricting the claim of equality in such a way that overburdening will be avoided. In moderate liberalism, and even more in liberal communitarianism, the stability, integrity and 'identity' of cultural and national social institutions (and traditions in social policy) can be taken into account as instrumental or even intrinsic values which have to be balanced against the demands of justice cast out in universally applicable distribution schemes. As in the case of context sensitive specification and balancing, a kind of national or cultural sensitive specification and balancing can be defended as long as the pluralism of principles and values is taken for granted in the normative framework of social policy.[27] This means that benefits of social policy in different areas cannot be compared by directly using the principle of justice. And it means that benefits of social policy in the same area cannot be compared by directly using the principle of justice if these benefits belong to different cultural systems or nations. In both cases, the principle of justice has to be specified and balanced in a context sensitive way. As long as the demand on universality is not taken to exclude this twofold specification and balancing, we can defend the universality claim of justice on the one hand and develop adequate and 'realistic' distribution schemes in social policy on the other hand.[28]

1.2.4 Social Policy in Europe or European Social Policy?

The questions whether there is and whether there should be a European social policy can have two different meanings. On the one hand, we can ask whether the EU is or should be an actor in social policy in Europe? In this sense, the descriptive task must be to identify contexts wherein European institutions and their activities do have influence on social policy in the member states. The normative task (hinted at by "should") then is whether having the EU as an actor in the field of social policy is rational, useful or effective for reaching the different aims social policy in Europe is after. This is to say that the normative claim here relies on instrumental rationality, not on ethical or moral considerations.

On the other hand, the transition from social policy in Europe to *European* social policy can be understood in a fundamentally different way: Does

[27] The increasing role of regions (or even smaller sub-units) within the social policy in Europe can be understood as an effect of this.

[28] For sure it will be a difficult task to find an ethically (and legally) acceptable balance between this and the pressure coming from the principle (and law) of anti-discrimination.

social policy in Europe have a unified evaluative and normative profile, which can be taken as the normative self-image of European social policy? Besides this descriptive aspect, we can ask whether Europe "should" develop such a normative self-image. In this context, "should" might rely on instrumental rationality only if we think that such a 'corporate identity' of the European member states and their people will increase the acceptance of the economical and political processes of integration of the single states into the EU. But we can also say that "should" can be related here to ethical or moral considerations, too. Making explicit the basic principles and values underlying European social policy by discussing and improving them in public debates is necessary since in doing this the autonomy of the European people and the principles of democracy are respected. Furthermore, the idea of active participation is taken seriously and then the result will at least be considered as criticism or at best as active identification of the European people with their European social policy. But for sure the result of such a process might be that there is no consensus at all, although one does not have to hold such a pessimistic view.[29] In any case, the value of pluralism therefore will have to be another integral part of Europe's normative self-image.[30]

For sure, the normative consensus existing de facto in Europe concerning social policy is, as we have argued in this chapter, underdetermined and leaves a lot of conceptual and political space for individual, regional, cultural or even national specifications which will hardly be unified. (Therefore, it comes as no surprise that the OMC has become so prominent in European social policy and its declarations.)[31] But as we have seen accepting this does not commit us to the thesis that there is no common acknowledgement of some basic principles and values at all. Personal autonomy, social inclusion and (distributive) justice are the main elements of European social policy, which deliver at least a rough shape of Europe's idea of a just society. Furthermore the under-determination concerning the ethically right (or good) relation between individual, community and state does not exclude that European social policy has a normative profile which is different from other conceptions of what a just (or good) society should be (in 1.2.2 we have identified at least two such conceptions which are not compatible with the European self-image). Although there is this identity-constituting feature, for sure the European consensus is oscillating between what we have called a

---

[29] Furthermore, dissent which has been made explicit can itself be regarded as a positive resource for developing better (social) policy since it can motivate to develop a plurality of strategies on the one hand and can motivate further development of existing policies or tools, too.

[30] Pluralism (as a value) has to be distinguished from relativism, since it is itself a moral principle; cf. Kekes (1996). Besides that, pluralism (as a normative principle) is backed by the normative principles of autonomy and justice.

[31] On the various functions of OMC in social policy in Europe see part two and three.

moderate liberalism on the one hand and a liberal communitarianism on the other hand. But nevertheless, strict libertarianism and anti-liberal communitarianism are two alternative models of a just society which are clearly ruled out by the generally accepted principles and values we can find, should defend and develop further in Europe.

Making explicit these differences by elaborating and justifying the normative option given in European social policy will not only help to strengthen the identification of the members with this normative self-image by active participation in forming it. This self-image will be made explicit for the members and for those who hold a different vision of a just society in such an articulation. Therefore, our thesis is that there already is a European social policy (in the ethical sense) and that Europe – at least in the long run – should actively try to make this normative self-image more explicit and elaborate. Evaluative and normative self-images are constituted by recognition – internally by their members, externally by those who do not share the normative view in question. This constitutive condition of every social policy can be fulfilled if Europe actively develops its own 'identity' in matters of social policy.

1.2.5 Globalisation, Universal Moral Claims and the Limits of European Social Policy

As we have argued personal autonomy, social inclusion and (distributive) justice can be regarded as the main principles and values of European social policy (in the sense we have elaborated in this chapter). We have shown that subsidiarity and sustainability can be understood as essential instruments to realise or guarantee these basic norms in the field of social policy. These principles are not only – as shown in our reconstructive and hermeneutical approach – part of European social policy in the sense of a normative vision implicitly or explicitly shared by all member states of the European Union as a kind of 'corporate identity'. They are (or are based on) moral norms and values (at least autonomy and justice) which cannot be restricted to social policy alone. It is a widely accepted feature of moral principles that they claim universal validity. For our context this means that speaking of European social policy – in the sense of a normative framework, not in the sense of Europe as an actor in social policy – has to be justified not only in the sense that its basic principles have to be justified but also in the sense that it is (or should be) restricted to Europe. For sure we cannot burden this chapter (and this book) with the task to give a philosophical foundation of the basic principles (European) social policy is based on. As we said in the opening sections of this chapter, we start from the fact that these principles and values are widely accepted and operative in the current social policy in Europe. Equally we cannot go into detailed examinations of how to justify the universal claims of morality or how to restrict them in scope. All we can do here is to distinguish several questions concerning this scope-problem and make visible the limits of our arguments.

In times of globalisation, it is obvious that there are a lot of relations between European countries and the rest of the world. Therefore, it should also be evident that such economic and political relations will have many influences not only on Europe but also on these other countries. If we understand social policy as an ethically acceptable and even demanded instrument to realise a good and just society based on e.g. free trade and if we take into account that these global relations constitute relations of responsibility, on the basis of morality's universal claims the question will arise why social policy should be restricted to Europe. We can take globalisation as *spatial* extension and the question then is why to draw and how to justify the border indicated by *European* social policy (in the normative sense). If we do not start from within the European market and the political union reached so far but from globalisation and universal demands of justice, then it is natural to claim that there should at least be a kind of European developmental policy for those countries, which are deeply infected, for instance, by terms of trades. For sure this help can be framed according to Europe's normative principles and should result in an enabling developmental policy (having the task to enable other countries to become good and just societies themselves). But the ethically acceptable limit will be defined by non-acceptable burdens for people in Europe (and not acceptable effects on our cultural identities) which might be the consequences of such help.

Besides this spatial extension which is caused by globalisation and which shows that distributive justice will cover the relation between all nations, the demand of moral universality will have a *temporal* aspect also. As already said the instrumental value of sustainability stands for this. Sustainability nowadays is well established within social policy in Europe and justified in the case of intrapersonal relations (via personal autonomy and life-course approach), in the case of interpersonal relations and relations between actually coexisting age groups (via social inclusion and the capability approach). Therefore, all we want to say here additionally is that sustainability together with the universality claim of moral principles like justice leads to extending the scope of sustainability so that future generations are included. Although we cannot discuss all the philosophical problems related to future generations, we think it evident that even actual political discussions concerning the sustainability of welfare states and systems of social security in Europe accept this extension. In this regard, social policy and environmental policy both show the power of the ethical claims.

Neither the spatial nor the temporal extension, which are both necessary because of globalisation and the consequences of our actions in the future on the one hand and the universal scope of moral principles on the other hand, should be taken as an argument against developing a European social policy (in the normative sense of a European social identity). But they should be taken as justifications for the thesis that taking the European perspective cannot be the end of the (ethical) story of social policy and justice.

1.3 Enabling Social Policy: Basic Goals and Main Tasks

1.3.1 Shifting Paradigms of Social Policy

The development of the European welfare states in the past 35 years might be characterised by three different stages. Up to the early 1980s, social policy in most countries has been characterised by a focus on the traditional tools of social protection: income replacement and income support. In the next stage up to the late 1990s social policy in these countries evolved into the so-called 'active' welfare state; benefits systems became less generous due to retrenchment policies, governments changed the balance between rights and duties, the linkages to work became stronger by way of active employment policies. In this stage governments tried to make working more attractive by 'making work pay' policies and by trying to raise the flexibility on the labour market (e.g. through permitting temporary contracts and promoting part-time work). This evolution has been mirrored in the debate about the 'crisis of the welfare state', which took place in virtually every European country. However, the way the 'active welfare state' is conceived by governments nowadays has changed due to the wake and the rise of the 'knowledge economy'. Social policy concern shifted, from a fear that employment and competitiveness were harmed if social costs and levels of regulation were too burdensome, to the goal of attracting mobile (financial and physical) capital by offering a high-skilled labour force, made available through active policies of investment in human capital (Room 2000). Policies of human investment and skill development were increasingly considered as active instruments for both, economic growth and innovation, and for rendering opportunities to all. Social protection policies had to be reconfigured and re-engineered in support of these shifts.

More specifically, the emphasis in welfare policy shifted from a compensatory logic to a preventive one. This preventive approach to social policy places less emphasis on providing income support to people out of work and more weight on fostering active participation in the labour force and enhancing the quality of human capital. Social welfare policies were increasingly designed to 'enable' more people at work and to make citizens responsible for their own conduct. This became manifest in practices and ideas about rendering citizens more 'free choice' to enable them to take up their own responsibility for managing their lives (personal autonomy), by increasing investments in the social and human capital of citizens (capabilities), and by taking the life course as a starting point for defining new routes to social policy. This has led to the introduction of arrangements in the domain of life-long-learning, training and schooling. Additionally, arrangements have been introduced which allow for a better tuning of working and private life (child rearing, caring). This third stadium of the development of the welfare has just been set in motion but it is clear that in some countries this already has inspired politicians to propose strong reforms. Also the European Commission in their policy documents speaks now about the 'active and dynamic' welfare state and makes a plea for an initiating and active role of the government in the creation of opportunities for social integration of every citizen. However, current policies in most member states serve as a warning against too optimistic voices about this presumed shift in real day-to-day policy practices. To date the dominant policy approach in most member states is much more oriented to the conventional poles of active labour market policies such as 'making work pay' policies and policies oriented at employment creation, training and education, and the promotion of flexible and part-time work. Within the Commission, the dominant view seems to evolve into assigning a more active role to the citizen and to governments at various implementation levels for improving people's opportunities for social integration.

This 'activating' or 'enabling' approach to welfare state policy, together with a life-course oriented approach, might define the contours of an innovative model of welfare state operation in the new era (Arts et al. 2004).

In the scant literature about the evolution of the welfare state, various conceptions and models of welfare state intervention have been developed and at the same time heavily debated. It ranges from the 'Night watch' state according to which state interventions in the social domain should be as lim-

Table 1.1: The role of the government, market and society in welfare regimes

Regime Role	Minimal state (night watch state)	Institutional model of the welfare state	Residual state	Enabling or activating welfare state (social investment society)
Government	**Passive** minimal	**Active** intervening	**Passive** residual	**Active**
Market	Market economy *(Liberal, market conform, Transition countries)*	Institutional market economy *(Institutional)*	Corporatist market economy *(Coordinated)*	The responsible market economy *(Societal responsible)*
Community	Church / charity / informal ties *(Informal)*	Formal social institutions *(Social-democratic / Bureaucratic)*	Organised social groups / civil society *(Corporatist/ Southern)*	Informal networks / community *(civil society) (Communal)*
Individual / family, associations, etc.	**Passive** (private interests)	**Passive** (collective interest, individual interest are subordinate)	**Active** (interest groups)	**Active** (the responsible citizen: individual and collective interest)

Source: Arts et al. 2004: 229

ited as possible and restricted to poverty relief and sustenance policies to the comprehensive, 'cradle to grave' or 'institutional' model of the welfare state, compensating the weak for a very broad palette of social risks. Such a model is sometimes also delineated as a 'productive welfare state' aimed at very detailed regulation of the rights and duties of the individual claimants and therewith in detail also regulating the relationship between the state and the individual. The welfare state models presented here set themselves markedly apart in their views on what responsibility should be assigned to the government or the public authorities, what should be left to the responsibility of the citizen himself and what should be left to the authority of the society, the free association of citizens or the civil society in safeguarding a decent life for all. As we argued earlier, the issue of the distribution of authority between the various social actors has always been and is still in the heart of the social policy debate. In table 1.1, we depict the distribution of authority between government, market and society in the various conceptions of welfare states as they are discussed in the literature.

In the subsidiary or residual welfare state the citizen is expected to be passive, but also the government is supposed to downplay its interventionist policies. The government should only intervene when the private initiative appear incapable of guaranteeing even a minimum level of social protection. Most continental welfare states have been developing in the past 100 years from a residual corporatist welfare state into the comprehensive institutional welfare states of the 20th century.

At the end of the 20th century the role of the government in this institutional model of the welfare state has been gradually eroded in favour of a less encompassing model appealing much more on the own responsibility of the individual citizen. This development might be characterised by the transition from a welfare state to a 'welfare society'. It is not the state, which is left solely responsible for providing adequate care and social protection. This responsibility is shared with the private initiative (the market), the society (civil organisations), and the individual and its informal networks (family, friends, neighbours). The relationship between state, market and civil society in a particular model might be called the 'democratic triangle' (Zijderveld 1997).

This step in the evolution of welfare states is characterised by the introduction of activating elements in the social protection systems on the way to the 'activating or enabling welfare state'. The 'activating welfare state' is aimed at formulating a different stance to the 'implicit contract' between the citizen and society. The role of the government is perceived as less protective and more directing or managerial, whereas the role of the citizen is believed to be more active, more focused on the citizen's own responsibility and his or her judgement and conduct.

In real day-to-day practice this leads to less governmental, less contractual and more individual, informal (social networks; non-governmental organisations; civil society) and market related arrangements. The responsibility will be shifted away from the government into non-governmental, private and informal social associations being part of the 'civil society'. Giddens (2000) uses the term, 'social investment' society to delineate such an approach. The way this debate in the 1990s came up reflects the transformation into modern complex societies of the late 20th century. Processes of economic privatisation and globalisation have shifted power distribution from the state towards 'capital' and markets. At the same time, NGOs and civil society organisations have taken on an increasingly important role in the field of welfare provision, and will continue to do so. By and large the social world has become more complex, fixed social configurations and collectively prescribed patterns of behaviour have been eroded, leaving individuals as active designers of their own lives. Does this mean a retreat to classical but 'minimum welfare states' (the 'Night watch' state) or does it reflect a new configuration of the modern welfare state in which the great feats of social engineering in the 20th century (the institutional model) are not withered away but are retained in a renewed and more modern form (the activating state)?

1.3.2 Goals in Different Welfare Regimes

Social welfare is a rather ambivalent concept, since it varies a lot across policy regimes in the way it is framed and operationalised in particular policy tools and instruments, notably in the context of the economic, social and political forces that determine the economic and social performance of the economy. The definition of social welfare differs according to the welfare goals the welfare regimes want to prioritise. The notion of a 'welfare regime' appears valuable in this respect to distinguish countries according to the way they have embraced a particular interpretation of the social welfare objectives (Esping-Andersen 1990; Goodin et al. 1999; Leibfried 1992). Social welfare might mean the sum of individual welfare, the concept of classical utilitarianism still underlying mainstream economics and notably embraced by liberal regimes. Social welfare can also entail stability of a social order that ensures continuity of labour income and a decent living standard for the male breadwinner like it is prioritised in the conservative or corporatist welfare regime. It can also mean basic or minimum income security for all citizens, individuals or families, seen as the basic unit of society like it is conceived in the social-democratic welfare states (Begg et al. 2001). As Begg, Muffels and Tsakloglou argue: "these different objectives of social welfare are not just arbitrary examples. They are deeply engrained in the way welfare states have become institutionalised over time and rooted in the norms, values and beliefs of citizens, both of which shaping the paths along which they will change in the future" (ibid.: p. 8).

Apart from how in these regimes social welfare is conceived as a particular combination of income, employment and social inclusion (the welfare triangle), they should be classified also along two other axes, which are derived from the rich literature on welfare states and which are also depicted in figure 1.2. First, they should be distinguished according to the role that has been designated to the market or the government for intervention to attain these welfare goals (the market-state axis measuring the level of de-commodification). The more the government intervenes in the market and hence, the less room there is for market operation, the higher the level of de-commodification. The second axis (the citizen-community axis) deals with the distribution of responsibility between the individual, the family and the community, in safeguarding each individual's or household's welfare. It pertains to the issue to what extent an active or a passive role has been assigned to the citizen and who carries the main responsibility for securing that social risks are satisfactorily covered: the citizen, the social network surrounding him (neighbours, friends and family) or the community at large (i.e. social partners and the civil society). Figure 1.2 is depicted in two-dimensional space, but there are actually three dimensions: the welfare triangle (the welfare policy mix of income, work and social participation, the market-state and the individual-community dimension, which brings the picture actually into three-dimensional space. On each element of the welfare triangle dimension (income, employment, social participation i.e. education, healthcare, housing) the score on the market-state axis and the citizen-community axis might be different indicating that dependent on the domain of social

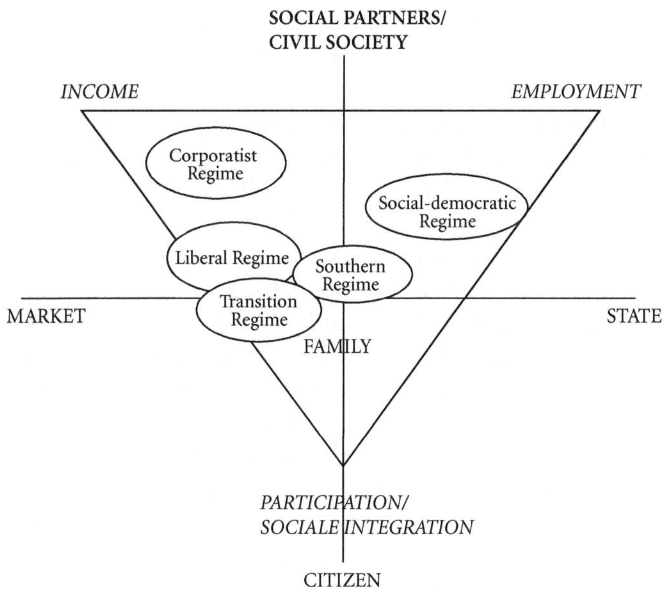

Figure 1.2: The location of welfare regimes within the welfare triangle

Source: Zimmermann and Henke 2002

policy, the regime-type classification might change. The general picture therefore masks the underlying variation that is likely to be larger when the focus shifts to other domains. The various domains of policy have adopted different logics rooted in different historical paths of institution building, different actors and different relationships between them based on different values, norms and behaviour. Thus, the entire set of policy mixes is, in theory, much larger and the location of the regime types is not as clear-cut as the picture suggests. However, in figure 1.2, an attempt is made to depict graphically the locations of the various regimes.

The location of each regime-type represents no single point but an area, signalling the large variation between countries pertaining to the same regime type. The indicated location of the regimes in the four quadrants reveals the focus of these regime types with a view to the particular welfare policy mix they aim to achieve and their location on the other two axes. It is rather obvious that the location of each regime is not fixed and might change over time. Note that in addition to the three regime types mentioned above we added a fourth Southern and a fifth post-socialist or transition country regime. The Southern countries set themselves markedly apart on the welfare mix dimension, featured by strict employment protection regulations and relatively poor levels of income and employment protection (Bonoli 1997; Ferrera 1996). The transition countries seem to constitute also a distinct type while due to their typical historical path of development they share some particular features like low employment protection levels, high unemployment rates and low levels of income support. They bear in these respects some resemblance with the liberal and Southern regimes. For that reason, we place the Eastern regime or the transition countries cluster close to the liberal and the Southern regime. The liberal Anglo-Saxon regime seems weak on guaranteeing generous income support whereas the corporatist regimes share high levels of income support but perform poorly with respect to the continued unbalance of their labour markets.

The location of the various welfare regimes on the market-state and citizen-community axes reflects in a way different conceptions or configurations of interdependence which are highlighted by the policy channels connecting the state, market, and the family. The liberal welfare states foster market and family dependency, but are intolerant to state dependency. Social-democratic welfare states are relatively intolerant to market and family dependency, but are to a certain extent relatively tolerant to state dependency. Corporatist welfare states are intolerant to market dependency and tolerant to state dependency. However, state dependency is accepted only insofar one's own social networks, the professional and civil organisations or the community to which one belongs (corporate actors such as social partners) fail to provide sufficient support. Eventually, Southern welfare states share much of the features of corporatist regimes, because they are tolerant to state and particularly to family dependency but also to dependency on the civil society when the state fails to support its citizens satisfactorily. Transition countries might just as the Southern countries be seen as 'soft' welfare states; they are intolerant to state and tolerant to market and family dependency although the market and the family is incapable to satisfy all the needs of the weak in society – a lacuna not filled up by the state (see table 1.1).

Dependencies on either the market, the state, the family or the community might on the one hand be seen as jeopardising the personal autonomy of citizens, but on the other hand might raise the autonomy of citizens by providing opportunities for improved integration in society particularly for the handicapped. It depends on the historical roots, moral values and cultural heritage of a country to what extent interference by these social institutions is perceived as threats or challenges for realising one's personal autonomy that is considered a basic condition for living a decent and worthy life. If in the counterfactual the same level of state interference would be introduced in liberal countries as in the Scandinavian countries, it would probably be judged by e.g. the British not as a valuable improvement of one's living conditions, although Scandinavians might be perfectly happy with it. There is obviously no single way or 'one-size-fits-all' approach to realise personal autonomy in the welfare state, but many different ways or conceptions, dependent on the social, economic and cultural conditions in society which are rooted in distinct historical pathways.

1.3.3 Welfare State Dependencies in Change and Personal Autonomy

From the perspective of welfare state regimes one might contend that welfare state policies are needed to protect the citizen from the adversities of the market like unemployment, low wages and weak employment protection. Welfare states are characterised according to Esping-Andersen (1990) by their 'de-commodification' policies or the degree by which the particular welfare state intervenes in the labour market to protect citizens from the adversities of the market, known as 'market failures'. De-commodification policies might then be understood as part of an attempt to insulate or protect citizens from these dependencies from the (labour) market. In the same manner we might speak of 'de-familialisation' policies, which mean the degree by which the state through income support or through providing services accepts responsibility for the care of small children and dependent elders. By doing so, the state interferes in the private market of care provision and, hence, takes public responsibility that would otherwise fall primarily and dominantly on wives and mothers. What does this mean for personal autonomy as an important aspect of welfare? One might adopt the view that the more someone is able to maintain a decent living on his own account without support of the state, the more autonomous the individual is. But for people who do not manage to extract sufficient resources from the market (or the family), welfare state support is beneficial and indispensable for safeguarding a decent living. If the guards for a decent life stemmed from the market or the family only, some of the weaker groups of people would suffer and would never be able to live a decent life. Welfare states play therefore an important role in balancing work, welfare and family dependency especially for particular groups. Through its policy programmes the welfare state provides a structure of choices between work, welfare and family. By offering a supplement to market and family provision when these fail to provide sufficient resources, it reduces the adverse effects of market and family dependency therewith raising personal autonomy. On the other hand it might also mean less personal autonomy for whom it concerns, since it leads to welfare state dependency: people being entrapped in prolonged states of income dependency without much perspective of escaping. Welfare states then have the effect of limiting the free choices of people instead of promoting them and this dependency also has been institutionally entrenched. Hence, the welfare state has also created new interdependencies and therewith bears the risk of jeopardising autonomy.

An important aspect of personal autonomy as a means of self-realisation and self-fulfilment is resource autonomy. (The lack of resources, particularly monetary resources, is traditionally considered to be the central impediment to self-development). However, resource autonomy is not exclusively dependent on monetary or even material resources alone. Education and other cultural resources that have been labelled as 'human capital' can also be considered as crucial for the autonomy and development of individuals. Health and education are also natural parts of any concept of resource autonomy. Welfare states have invested in the education of its citizens, the maintenance of their health, in order to maintain and improve their ability to participate in the labour market and to secure an income from employment. Furthermore, they have set up risk management systems that secure their citizens' resources after the incidence of certain risks such as unemployment or invalidity. By the redistribution of resources, welfare states have assured the resource autonomy of all segments of society. By providing security and stability, welfare states have guaranteed the continuity of this autonomy.

Thus, the welfare state's aim can be seen as the promotion, or better, the management of the autonomy of individuals and households, which can imply for some groups that the state intervenes but for others that this task is entirely left to the market or the family.

Structural changes in the economic and social contexts of the matured economies of Europe have affected the way these policy regimes pursued their distinct social welfare objectives through a particular design of the work-welfare policy mix as well as their performance to cope successfully with the strains put on them. For that reason the performances of these policy regimes have to be examined taken into account these changes which occurred at an increased pace and which stem from factors such as the process of individualisation, the increased international competition or globalisation, the growing impact of the ICT and the flexibilisation of labour markets. Concern about the shortcomings of welfare systems in either economic or social terms in coping with these changes has led to a variety of reforms aimed at modernising and reforming social protection and labour market policies. In particular, most EU Member States have sought to recast the link between work and welfare, have pursued more active labour market policies and have tried to make employment relationships and social protection more responsive to the new demands. These changes illustrate the need for a new paradigm of social welfare, that combines the elements sketched before and which constitute the basic aspect of the approach in this book: the capabilities approach, the life-course perspective and the activating or enabling welfare state which will be dealt with in the sequel.

1.3.4 Sen's Capabilities Approach

An activating perspective, based on personal autonomy and dependency, might be helpful for debates on welfare state reforms. Both workfare and activation policies increasingly take the form of individualising policies that emphasise individual agency and responsibility, while the development of human resources by investment in the social and human capital of people is an explicit aim of Scandinavian activation policies. Individual choice is both part of the rhetoric on privatisation as of some proposals for reforms in the Netherlands, Denmark and Sweden. What matters in this perspective is primarily the degree by which social policy contributes to the improvement of real freedom of choice. Social policy should aim at empowering the individual by offering extended opportunities to participate and integrate in society.

This view is often connected with Sen's capability theory, based upon the concept of human development (Sen 1999). According to Sen, what matters is not only the income or even the bundle of goods each person possesses, but also what they can actually do or be with them (ibid.). For Sen, it is important to focus on freedom to achieve rather than on resources. His theory is concerned with the actual freedoms of the people, and their 'capabilities' or actual possibilities to achieve certain things ("functionings" in Sen's terminology) (ibid.). Capabilities refer to opportunities or potential options for choice that are realistically available, but do not necessarily refer to real choices made. The broader someone's capabilities are the wider and the more effective their freedom to act is. In Sen's view, human development is, thus, a process of enlarging people's choices (Sen 1992).

Sen's capabilities approach actually boils down to the notion of 'freedoms' to get hold of what 'produces' welfare and contributes to well-being (Sen 1999). These freedoms are people's opportunities and capabilities to live the kind of life they have reason to value. For Sen, these freedoms are ends in themselves, not requiring any further justification on their instrumental effects on other outcomes such as economic growth. Sen's focus is neither on outcomes nor on people as passive recipients of these outcomes, but rather on individuals' acting and bringing about change, where the achievement can be judged in terms of their own values and objectives. The capabilities approach entails that social welfare policy goals have to be reformulated in terms of 'freedom to act' instead of 'freedom from want'. The challenge for policies is that instead of compensating the disintegrated for the consequences of lack of integration, it should aim at providing the individual with capabilities in a way that allows him/her to increase his/her future prospects (Muffels et al. 2002). This implies a shift from compensating income risks to providing opportunities to people. The aim of such policies is to enable people to achieve the functioning they have reason to value and to invest in their 'capabilities' to maintain or raise their human and social capital and to prevent their capabilities from becoming obsolete or redundant. It is exactly for this reason why according to Esping-Andersen et al. we need a *new* welfare state in Europe (Esping-Andersen et al. 2002). Instead of using the term 'inclusive society' he pays credit to Giddens' notion of a social investment society which has however a very similar meaning (Giddens 1990; 2000).

1.3.5 Dworkin's Equality of Resources

Sen's concept boils down to the notion of equality of capacities or capabilities rather than to equality of resources as a matter for egalitarian concern. However, Sen's view is questioned by Dworkin (Dworkin 2000) who claims that the equality of resources principle is a better principle for egalitarians to consider. What matters is not to make people equal in the overall capacity to achieve happiness, well-being, self-respect and similar desiderata whatever their tastes, dispositions, convictions, ambitions and attitudes might be, since this would just entail the false utilitarian goal of equal welfare or well-being. What matters instead is to make people equal in the resources they need to achieve these goals. In the view of Dworkin, it would be frightening to think of a government eager to bring about equality in people's capacities to achieve these complex goals of happiness, well-being and the like knowing that people vary in their ability to achieve these goals for countless other reasons for which we would not like the government to compensate for and to interfere (Dworkin 2000: 302). However, Dworkin presumes that this is not what Sen actually proposes. What Sen implies in his view is, that governments should strive to ensure that any differences in the degree, to which people are not equally capable of realising happiness or well-being and the other 'complex' achievements, should not be attributable to differences in the personal and impersonal resources (like education or infrastructure) they command. If it is attributable to differences in their personal choices and personality traits and the choices and personality traits of other people then it should not be of egalitarian concern. Dworkin then concludes that:

"If we do understand equality of capabilities in that way, it is not an alternative to equality of resources but only that same ideal set out in a different vocabulary" (Dworkin 2000: 303). Dworkin eventually agrees that people want resources in order to improve their 'capabilities' for 'functionings' they consider important for their lives. But, governments equipped with egalitarian concerns should deal with differences in personal and impersonal resources, not with differences in the happiness or well-being that people can achieve with these resources.

One device proposed by Dworkin to mitigate differences in personal resources is what he calls the "hypothetical insurance contract", according to which each person bearing the same social risks pays the same premium and receives the same benefit to cover e.g. the unemployment risk when it becomes manifest. Dworkin's equality of resources principle has some further and far-reaching implications for redistribution policies. It also implies that governments with egalitarian concern should try to mitigate intergenerational inequality by taxing, at a steeply progressive rate, personal resources which are not obtained through indigenous work or effort but by heritage. The government should use the means obtained in this way for the investment in improved education and training opportunities for those who are in need of it. From a life-course perspective these proposals sound interesting but they go beyond the purpose of this paragraph in which we seek to explain the development of goals and principles of social policy in Europe. From this perspective, the implications of Dworkin's concept are not very much different from Sen's approach, because it implies that social policies should aim at 'investing in the capabilities of people' by providing more opportunities (read resources) to people to manage their own lives.

1.3.6 The Relationship between Economic and Social Policies

Sen's approach also presumes that social and economic policies are closely interrelated because of which the perceived trade-off between economic and social goals might be avoided and social and economic goals be attained simultaneously. The ultimate goal of any modern democratic society is to enhance people's well-being. To arrive at this goal economic and social objectives have to align in order to be able to promote economic welfare and social integration. The inclusive or 'social investment society' (in Giddens' words) demands a reciprocal relationship between the three basic components of what produces welfare in a society: productive capabilities (inputs from labour, capital and technology), social capabilities (arising from norms, values, trust, institutions and social and human capital) and well-being (arising from consumption of goods and services, family ties, health, job security, community, freedoms and opportunities).

There are several linkages between these components. For example, social capability influences economic performance through cultural values, trust, interpersonal skills, social networks and institutions and their impact on risk-taking, incentives, innovation and economic participation. Economic performance, in turn, being part of productive capability, affects the social capability through its improvement of social cohesion by rendering opportunities for participation and learning and through increased public and private spending. There are also various positive and negative feedback loops between the concepts of productive and social capability, which tend to affect the outcomes in terms of well-being. Social capital, in the form of strong social ties within the community where one lives, fosters trust that may stimulate voluntary exchange and also reduces transaction costs. In turn, these developments smooth broader social interactions and, hence, increase the amount of human and social capital. There might also be some potential trade-offs: for example, a policy that boosts economic growth might be at the expense of some form of social capability. It may raise the tensions between winners and losers. A high level of social spending seems favourable from a social viewpoint because taxes and social revenues might be used to raise collective welfare. At the same time, however, high levels of social spending might reduce people's incentives thereby in turn hindering economic growth.

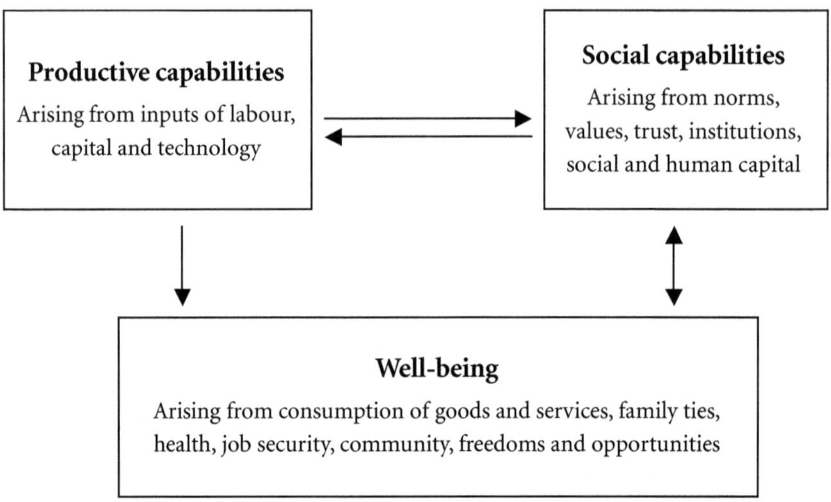

Figure 1.3: Capabilities and the 'inclusive economy'
Source: Ministry of Finance New-Zealand 2001

The basic idea is that the alleged trade-off between efficiency and equity is a simplification of the complex relationship between the economic and social performance of an economy and the wide range of interrelated factors that affect this relationship such as social and human capital and norms and values. Since there is a positive correlation of risk and return due to the market forces that tend to select activities with highest expected returns, the welfare state may enhance efficiency when it allows people to bear more risk.

Institutions enter the model in that they form part of the social capability domain affecting the economic performance of a society as well as affecting the overall outcomes in terms of the extent to which well-being is attained for all citizens in society. Important from the perspective of social policy is that the capabilities of the individual are directly and indirectly influenced by the possibilities (and restrictions) given by the institutions surrounding him, including the welfare state. Especially in modern societies, the welfare state plays a crucial role in determining the capabilities of its citizens and therewith raising their resource autonomy. A person with a limited set of capabilities will most likely have a too low level of resources to allow him to afford a minimum standard of living and to participate in society on a minimally acceptable basis.

1.3.7 The Life-course Perspective

An enabling perspective on welfare state reform requires a dynamic 'life-course' approach. Welfare state policies, which are concerned with fostering individual and household capabilities for self-realisation and self-fulfilment, can only be understood in a life course perspective as such policies require the tuning of welfare state interventions to the various stages of the life course (Vleminckx 2004). Esping-Andersen (2002), for example, considers a life course framework essential for the proper assessment of welfare state policies within the context of rapid societal changes. Such an approach might be able to connect the various parts of policy interventions, capturing the broad picture of reform, while at the same time being able to capture the dynamics of real lives of people.

The life course is very much institutionalised by the welfare state, even to such an extent that social policy might already be viewed upon as a 'life-course' oriented social policy. The welfare state is traditionally an important instrument for helping individuals and households to deal with the projects in life that are common to all of us: the chance to get a proper education, to find a job, to form a family, to have children, to combine family life with participation, to be economically secure in case of exposure to unemployment and illness, to get a pension at old age, to receive the necessary social services and care when the physical abilities are decreasing.

Thus, the welfare state's structuring of the life course has enabled individuals to plan their lives and its policies provide stability and security when something goes wrong. However, by recognising and institutionalising certain pathways and rejecting others it imposes constraints and restricts the life-course development of individuals. Modern welfare states institutionalised certain normatively dominant pathways, often associated with a "normal biography, making it more difficult to deviate. According to Kohli, the modern welfare state is built on the central institution of the modern life course: the labour market. Together they cause a so-called "tri-partition of the life course" (Kohli 1995). That is, the life course is composed of three distinct segments: an early part devoted to education and training, a middle part devoted to work, and a final part devoted to retirement. In the first stage, citizens are educated and trained in order to ensure that they are able to enter the (paid) labour market. In a second stage, they enter the labour market and thereby have the opportunity to gain a primary income. This income in turn enables them to exert command over resources that guarantee their social integration (Berghman 1997). In a third stage, the individual is allowed to permanently withdraw from the labour market and to receive a permanent retirement income, which assures social integration during old age. Transitions between these stages are governed by rules, regulations, and cultural norms (Kohli 1995).

However, the life course has become more complicated and the welfare states need to be adjusted to this new situation. Kohli asserts that the tri-partition of the life course will remain valid as long as our lives remain built around gainful employment as being one of society's core structures (Kohli 1994). Currently, we might observe some changes occurring and one of the important changes is that the life course has become much more diverse and involves more than three stages with the transition between the different stages becoming increasingly diffuse. In addition to the early education stage, we can for example distinguish a new, second stage identified as 'young adulthood', in which young people are not yet established in a stable job or in a long-term relationship and have limited responsibilities for care. During the third stage, most individuals establish both long-term professional careers and often have responsibilities for raising children. In a fourth stage, which can be coined "active aging" (SZW 2002), people have often withdrawn themselves from the labour market but their relative health allows them to participate in other activities. During a fifth stage, people become physically and psychologically dependent upon the care of others.

Although these stages are roughly associated with certain age groups, the borders between the various stages have become increasingly blurred. Furthermore, people seem more than before to combine various activities during the life course, which used to be compartmentalised. Young people now often combine education with gainful employment, while older workers more often interrupt their employment for education. Women more often than in the past return to the labour market after they have cared for their (young) children and, still quite rare, men occasionally interrupt their careers to care for children. Both men and women more often interrupt their careers in order to start another career, sometimes as self-employed workers or entrepreneurs. Thus, the life course has become increasingly flexible and is characterised by transitions of various kinds.

Important is that European welfare states adapt to this to this new situation and develop flexible strategies to deal with it. If individual 'life-course policies' change, then the institutions must also change. Greater variety and increased transitions necessitate more differentiated and more flexible forms of institutionalisation. Flexible retirement may serve as an example, allowing for greater flexibility of choice about when and how to retire. In order to allow for more individual choice in the management of the life course one could, for example, encourage people to retire partially through a system of so-called 'partial pensions', which allow workers to combine work and retirement and to withdraw gradually from work in the last stages of their working career.

More in general, this strategy of flexible life-course policies could be used to address problems of intergenerational distribution. The predicament of ageing, if unresolved, could provoke a 'generational clash'. The burden of pension and health care costs will rise strongly due to ageing in the near future. The revenues of good health care and pensions go to the oldest generations whereas the burden of payment is left to the youngest generations who have no guarantee whatsoever that once they grow old they will receive the same high standards of health treatment and pension benefits. How to distribute the costs and benefits as countries make the transition to ageing societies? How to ensure a fair distribution over the lifecycle and between older and younger generations as well as the general sustainability in the future? Goodin provides a good starting point, arguing that where the 'old policy' was aimed at spreading income across people, the 'new policy' should aim at a more balanced spread of money (income and wealth) and time during the life course. Time comes under severe pressure especially during the 'rush hour' of the life course, when people are at their middle ages and need to combine working, learning and caring activities at the same time. By acting in such a manner, modern policies do spread what is believed to be the final result of having money and time, self-respect and autonomy (Goodin 2003).

Esping-Andersen (2002) proposes a radical overhaul of the social security system and the distributional principles on which most welfare states are built. He argues in favour of a redistribution of resources and financial means to the youngest generations, particular to families with children. The underlying idea is that resources might be used much more effectively and efficiently if they would target to raising the 'capabilities' of the youngest generations rather than to endowing the 'vested' and older generations.

Others are not very supportive to this idea of creating such an 'age-selective' social security system. They make a plea for a more universal approach in which policies should aim at investing in the capabilities of all people during their entire life course. The balanced spreading of money and time over the life course asks for a recalibration of the tools and instruments of current welfare states. People should get more room in making the choices they value most for their lives and policies should aim at creating more options of free choice. We have already mentioned the example of flexible retirement.

More in general, one might think of a fiscal favourable treatment of savings devoted to build up a life-course fund to be used for being able at a later stage in life to combine working and caring duties or working and learning activities, e.g. by taking up a care or parental leave or a sabbatical leave (Bovenberg 2003). Such a fund might be created through public or private arrangements at firm, sector or even at national level. In the Netherlands such a 'life-course saving arrangement' has been enacted in early 2004, in addition to the existing social arrangements, but partly replacing them.

The existing schemes within the current social security and labour market systems in Europe are strongly exclusive: either one works or one receives a benefit but the idea of 'in-work' benefits has hardly come into practice. There is not much in between and very few – if any – opportunities exist to combine working, learning and caring duties. If there are, they appear either very unattractive or in practice hard to realise. However, the life-course approach requires the offering of opportunities to people for allowing them to plan and combine future activities. Also transitions between the various societal activities, such as working, caring, inactivity and learning or between various combinations of activities, such as working and caring or working and learning ought to be facilitated.

1.3.8 The Enabling Welfare State: Concluding Remarks

Reforming the welfare state along the lines of activation, investment in people's capabilities and a stronger focus on human and social capital formation during the life course might be profitable particularly in the longer run. The expansion of life-long educational opportunities and the upgrading of labour force skills can secure opportunities for all. Policies to promote investment in human capital, by shaping the skills that people bring to the labour market, determine to a considerable extent the prospective income flows that people can command and the protection of these incomes whose continuity and sufficiency it is the traditional task of social protection schemes to defend.

Yet, for labour market policies such an approach would imply a much more active way of interference with the labour market. Active (un)employment policies would imply to build bridges between out-of-work and in-work statuses to smooth the adaptation process to changing economic conditions and to make the system more flexible, responsive and less exclusive (Schmid and Schömann 2003). Improved integration and fine-tuning of employment and social security policies is then needed to make both systems more flexible and responsive to the needs of the rapidly changing societal contexts in an expanding Europe. If we consider family policies to be part of employment as well as social security policies it would imply that women gain more opportunities to choose the working-caring time combination they like most and which fit their household and life duties best. The opportunities provided should take into account the particular stage of the life course these women are in especially with a view to many women being in their rush hours of life when work, care and education have to be combined.

The challenge is to develop a conceptual model of policy-making that recognises the interplay and reciprocal impact of policy areas in shaping social outcomes. We believe that Sen's capabilities' approach can serve as a conceptual tool to arrive at some basic understanding of the principles and goals of the enabling welfare state. Though not very precisely defined, the model seem to hinge upon the idea that integrated economic and social policies are beneficial to the fulfilling of the ambitious goals all European countries as well as the European Union endorse and which are intended to connect and achieve economic and social ambitions and goals simultaneously (Muffels et al. 2002).

There is no reason to be too sceptical about the future because what has been achieved in the past is something to be proud of and not to be withered too easily away. It is certainly true that what has been achieved in the various countries is quite different also with a view to their success in shaping fair or just societies. But given these differences it is fair to say that the social performance of most countries is indeed a proof of the great feats of social engineering that these countries have achieved in the twentieth century. Having said that, it is also likely to become true that the national systems will remain rather different, that they do not easily converge but that there is no need for convergence when the social performances are steadily improving. It is to be taken for granted that whether and how we proceed on the road to a 'Social Europe', lastly depends on whether and how the countries are willing and capable of developing new and innovative routes for social policy that can resolve the challenges of the new era.

2 European Union and Social Policy

2.1 Preliminary Notes

In part one, the central issues – for the approach and the perspective followed in this book – of social policy in Europe at the beginning of the 21st century have been presented. In the first chapter, the historical backgrounds and the routes of development of the different social systems in the EU, as well as their most important and pressing challenges at present and in the near future have been illustrated. Although the EU has been considered as an important influential factor for the future trends and needs of development, it was not at the centre of the analysis. In the second chapter, the ethical foundations of the European Social Model – personal autonomy, social inclusion and distributive justice – and their specification with reference to the European context have been displayed. In the third chapter, the fundamentals of the paradigm of an enabling social policy, the capability approach and the life-course perspective have been exhibited and the basic goals and main tasks of a social policy oriented towards this paradigm have been described.

Although the object of the analysis was the European context – with the necessary reference to and embedding in the global situation, so far the European Union itself did not take centre stage. Thus, in this part the focus of the analysis will shift to the role of the European Union in the area of social policy. In two chapters firstly, the effects on the general framework for national social policies deriving from policies at Union level in other fields and, secondly, the function of the Union as an agent of social policy will be examined respectively.

The first chapter is divided in four sections. Firstly, the effects of mobility of workers and other persons within the EU on the national social systems and their coordination will be investigated. Free movement is not only granted as one of the Four Freedoms and thereby a particular trait of the Union, but it has historically also been one of the first and main causes for the need of supra-national coordination. In the future, it is expected to become even more important due to higher flows and new forms of mobility, especially between the former EU-15 and the new member states, requiring eventually broader and new efforts of coordination. Secondly, since the EU is already active in the area of employment policy, the effects of supra-national policies on the national and international labour market will be illustrated, thereby stressing the close inter-relatedness of measures of economic and social policies. Additionally, the requirements and possibilities of balancing and combining goals of security and flexibility in the employment sector will be discussed. Thirdly and further developing the issue of integrating policies in the economic and social domain, the consequences of the ongoing and widening economic integration on the national social systems will be described. The economic integration not only increasingly influences and limits the scope of action of the member states in the social area but has a significant impact on the competitiveness of their reciprocal positions and the exigencies of recalibration and coordination, too. Fourthly, the significance of the legal framework set by the European Union for national policies will be analysed in regard to the Fundamental Freedoms, the economic and competition law, the coordination rules and other provisions.

The second chapter deals with the European Union as an agent in the area of social policy, what the Union regardless of the principle of subsidiarity by many means is. In this respect, the objectives, contents, instruments, institutions and long-term trends of development of the social policy activities of the EU will be highlighted.

2.2 Effects of the EU on the Framework for National Social Policies

2.2.1 Mobility of Workers and Other Persons in the EU (Especially after the EU Enlargement)

2.2.1.1 Migration and European Social Policy

Migration is analysed in this book not as a separate subject but as an important context of European social policy. At the EU level, the free movement of workers (persons) has been acknowledged since the start of European integration as one of Four Fundamental Freedoms. This right has been firmly anchored in the EU (social) law. Unlike national social policies, the European social policy is also oriented, to a large degree, towards guaranteeing the implementation of the freedom of movement (for example, the coordination law of social security). There are also tight relations between migration and social policy at the level of member states. A better social protection in one country may attract migrants from other countries. The strength of this factor notwithstanding, migrants tend to represent higher needs for social protection. Migration can also change the balance between contributors to a social security system and its beneficiaries (on relations between migration and social security worldwide, see ISSA 1994). The issue of migration has played a special role in discussing European social policy in the context of the EU Enlargement of 2004, especially because of the large income gap between the old and the new member states.

People always have been moving between regions and countries. At present, international migration is mainly rooted in the expansion and consolidation of global markets (Massey and Taylor 2004). Migration is in principle a beneficial phenomenon (see section 2.2.2). According to economic theory, migration is an expression of mobility of the production factor labour. Migration flows follow in principle the wage differences between the sending and receiving regions. Thinking about migration, the single person takes into consideration both its benefits and costs (Barfuss 2002). The benefits include higher income and – closely related to it – better living conditions, better environmental standards, better life chances. The costs include the geographical distance, language and cultural differences, and the loss of social networks. An incentive to migration emerges when the individual rational comparison of benefits and costs shows a positive net result. In this way however, only a migration potential is being created. Only a part of those who would have an incentive to migration will actually decide to do this. Migration may of course also be mainly politically motivated, for example if wars or authoritarian regimes induce people to leave their country and look for an exile. In this text we only concentrate on economic migration which also plays a crucial role in the context of the EU enlargement.

The studies on migration differentiate between push and pull factors. The push factors which are located in the sending country include for example high unemployment, lack of affordable housing or bad environment. The pull factors derive from the receiving countries, and include mainly higher income, better working conditions or better career chances. The differentiation is not always quite clear: a higher wage in the receiving country works as a pull factor which means, however, that a lower wage in the sending country functions as a push factor (Fassmann 2002).

Recently, the importance of demography for the decision to migrate has been stressed (Fertig and Schmidt 2000). A relatively high share of younger people in the whole population is an important determinant of the migration potential, because younger people are much more easily willing to migrate.

Although migration is still mainly understood as a permanent move from one country to another, new forms of trans-national mobility have emerged: trans-national commuting, either regularly (for example as daily commuters) or irregularly (Fassmann 2002). These forms are also very important in relation to the latest EU enlargement in 2004.

2.2.1.2 Reality of Mobility in the EU

Mobility of persons is very low within the EU, especially when compared to mobility of goods, services and capital. Accordingly determinants and consequences of migration within the EU find relatively little interest in research (Fertig and Schmidt 2002). As a result of the successful European integration, former income differences between member states have decreased which has caused substantial reduction of international migration flows or backward flows to the home countries (Barfuss 2002). This is the main explanation for the low mobility within the EU at present. However, mobility is also hindered by language barriers.[32] Migration means very often a loss of social networks due to social and cultural differences. Higher labour participation of women can also be a barrier to migration of families, since in this case migration means that two new jobs have to be found for persons with different professional careers. The decision to migrate can also be negatively influenced by problems with finding a house or flat, according to one's expectations and financial possibilities (European Commission, Directorate General for Employment and Social Affairs 2002b: 16–17).

In 2000, the share of foreign citizens in the population of EU member states was about 5%. The number of non-EU citizens was twice as high as

[32] 47% of EU citizens declare to know only their mother tongue. A Eurobarometer shows that only 29% of EU citizens would be ready to live in an EU country, in which a language different from their native language is spoken (European Commission, Directorate General for Employment and Social Affairs 2002b: 16).

that of EU-citizens living in another EU country (see table 2.1). The numbers however, do not show the entire mobility within Europe. They neither include for example stays abroad for several months, nor "cross-border commuters", nor seasonal workers (Fertig and Schmidt 2002). Even if the scale of mobility in the EU is limited, public concern, especially after recent terrorist activities in Europe and some problems of multicultural societies (like recently in the Netherlands), has brought about a political emphasis on immigration restriction or prevention. "This is based on the assumption that the social and fiscal costs (and therefore also political costs) of immigration may outweigh its benefits" (Holzmann and Münz 2004: 10).

Table 2.1: Foreigners in EU member states in 2000

Country	Population (1000)	Foreigners in total (1000)	(%)	EU citizens (1000)	(%)	Non-EU citizens (1000)	(%)
Austria	8,103	754	*(9.3)*	99	*(1.2)*	654	*(8.1)*
Belgium	10,239	853	*(8.3)*	564	*(5.5)*	290	*(2.8)*
Denmark	5,314	256	*(4.8)*	53	*(1.0)*	203	*(3.8)*
Finland	5,171	88	*(1.7)*	16	*(0.3)*	71	*(1.4)*
France	58,521	3,263	*(5.6)*	1,195	*(2.0)*	2,068	*(3.5)*
Germany	82,163	7,344	*(8.9)*	1,859	*(2.3)*	5,485	*(6.7)*
Greece	10,487	161	*(1.5)*	45	*(0.4)*	116	*(1.1)*
Ireland	3,787	127	*(3.3)*	92	*(2.4)*	34	*(0.9)*
Italy	57,680	1,271	*(2.2)*	149	*(0.3)*	1,122	*(1.9)*
Luxembourg	424	148	*(34.9)*	131	*(31.0)*	16	*(3.8)*
Netherlands	15,864	652	*(4.1)*	196	*(1.2)*	456	*(2.9)*
Portugal	9,998	191	*(1.9)*	52	*(0.5)*	138	*(1.4)*
Spain	39,442	801	*(2.0)*	312	*(0.8)*	489	*(1.2)*
Sweden	8,861	487	*(5.5)*	177	*(2.0)*	310	*(3.5)*
United Kingdom	58,614	2,298	*(3.9)*	859	*(1.5)*	1,439	*(2.5)*
EU-15	374,667	18,692	*(5.0)*	5,801	*(1.5)*	12,892	*(3.4)*

Source: European Commission, Directorate General for Employment and Social Affairs 2002b: 115–116

In future, mobility within the EU may increase not only because of the enlargement, which will be discussed further below. Future migration flows may also be influenced by changes in the population structure and in behaviour patterns. Mobility will become a normal part of life for young people, especially students, highly educated people, and specialists with high qualifications. On the other hand however, in the long run, technological progress especially in telecommunication and transport, may reduce the importance of geographical migration as a tool of improvement of labour allocation (European Commission, Directorate General for Employment and Social Affairs 2002b: 17). According to economic theory (Heckscher-Ohlin Model), trade and foreign direct investment are substitutes for migration. This holds true for migration from Central and Eastern Europe to Western Europe. Increased trade and foreign direct investment have reduced migration flows considerably and are likely to have the same effects in future, too (Holzmann and Münz 2004: 28).

Residents from Central and Eastern European Countries (CEEC-10)[33] constitute a relatively small fraction of the total population of the EU-15. In 2001/2002, around one million citizens from the CEEC-10 resided in the EU-15. Their share of the total population of the EU-15 was around 0.25 %. The total labour supply of citizens from the CEEC-10 in the EU-15 is estimated at 430,000 full-time workers or around 0.25 % of the total labour force (Brücker et al. 2003: 3–4). Formal barriers, wage differences between CEEC and the EU as well as characteristics of EU labour markets caused a high level of illegal employment. The number of illegally employed from CEEC-10 in the EU-15 has been estimated to be almost twice as high as the legal employment in 1999 (European Commission 2001: 30). Migration from the CEEC is geographically concentrated. Around 70 % of the migrants from the CEEC-10 reside in Germany and Austria. Almost 600,000 or 61 % of all residents from CEEC-10 in the EU-15 live in Germany. Their share in the total population in Germany is 0.7 %. In Austria with 79,000 persons it was 1.1 % of the total population in 2001 (Brücker et al. 2003: 3–4).

The above figures concern residents, that is people who have left one country and live now in another. However, a different pattern of mobility between CEEC and EU-15 has also emerged. Some people (the so called "trans-national mobiles") are changing frequently between home in CEEC and workplace in the EU, on a daily, weekly or irregular basis, coming for example for several months. They are "between", "cross-border commuting" (European Commission, Directorate General for Economic and Financial Affairs 2001: 50). Studies on this type of mobility, also called "incomplete migration", have shown that this is the most important migration form between some CEEC and the EU (Okólski 2004; for Poland see: Jaźwińska and Okólski 2001). For example, it has been estimated that in the second half of the 1990s yearly some 500,000 short time migrants from Poland worked in the EU-15, mainly Germany. The migrants often work in lower segments

[33] CEEC: Central and Eastern European Countries. CEEC-8 include 8 new member states: Czech Republic, Estonia, Hungary, Latvia, Lithuania, Poland, Slovakia and Slovenia; CEEC-10 include CEEC-8 plus two candidate countries: Bulgaria and Romania.

Table 2.2: GDP per capita in purchasing power standards in acceding and candidate countries 2003 (EU-15=100)

Acceding Countries	
Cyprus	76
Slovenia	71
Malta	67
Czech Republic	63
Hungary	56
Slovakia	47
Estonia	44
Lithuania	42
Poland	42
Latvia	39
Candidate Countries	
Bulgaria	27
Romania	27
Turkey	25

Source: EUROSTAT 2004a: 1

of labour markets like household services, house construction and repairing, seasonal work in agriculture, simple work in restaurants or hotels (Okólski 2000: 153–158).

2.2.1.3 EU-Enlargement and Labour Migration

Before and after the latest enlargement, due to transitional rules discussed further below, the access of workers from CEEC to labour markets of most of the EU-15 has remained very restricted. Unlike goods and services, which could move almost without restrictions even in the pre-accession times, during which CEEC had already been economically integrated in the EU-15, the mobility of workers will be – after the transitional rules – "probably the most significant dimension of economic integration to change after accession" (European Commission 2001b: 51).

Potential migration after the enlargement in 2004 was a source of concern in the EU right from the beginning, even more than in the case of the Southern enlargement. This concern has mainly been motivated by large income differences between the old and the new member states (see table 2.2) as well as the geographical proximity (European Commission, Directorate General for Economic and Financial Affairs 2001: 9). Studies on potential migration, however, do not support the concerns for the whole EU.

Migration Potential (How Many?)

Many studies have been made on possible migration flows from CEEC into the EU-15. The migration potential has been assessed using two methods: micro analytical studies try to assess the potential by asking the people directly through interviews or surveys, while the macro analytical approach is based on econometric models. In studies concerning the latest EU enlargement very often data have been used from the migration flows following the Southern enlargement in the 1980s (Greece, Spain and Portugal). In model estimations the relative convergence rate, i.e. the difference between GDP growth rates in new and old EU member states, is a crucial element, because for the decision to migrate not only the actual difference but also the tendency is important. People may stay if they see that the situation is improving. Estimates of migration potentials should of course be treated very carefully as they rely on many assumptions which may not realise (Sinn et al. 2001: xxviii). Despite different methods, the results of the estimates are surprisingly similar (Straubhaar 2001; see table 2.3).

"The long-run migration potential from the candidate countries would be in the order of 1 % of the present EU population" (European Commission 2001: 3). "The overall impact on the EU labour market should be limited, both on the negative and positive side" (ibid.: 8). "Fears of a massive wave of immigration proved unfounded at the time of past EU enlargements, and in this respect are also without foundation for the forthcoming eastward enlargement in a few years' time" (Hönekopp and Werner 2000: 7).

Table 2.3: Estimates of potential migration into the EU-15 from CEEC under conditions of free movement

	CEEC- 8		CEEC-10	
	Stock	Flow/year over first 10 years	Stock	Flow/year over first 10 years
Brücker and Boeri (2000) *(only workers!)*	860,000 (after 10 years)	70,000 declining to 30,000	1.4 million (after 10 years)	120,000 declining to 50,000
Brücker and Boeri (2000) *(all migrants!)*	1.8 million (after 10 years)	200,000 declining to 85,000	2.9 million (after 10 years)	335,000 declining to 145,000
Sinn et al. (2001) (a)	2.7 million (after 15 years)	240,000 declining to 125,000	4.2 million (after 15 years)	380,000 declining to 200,000
Salt et al. (1999) (b)	2.25 million (after 15 years)	140,000		

(a) Czech Republic, Hungary, Poland, Romania, Slovakia
(b) Baltic states excluded
Source: European Commission 2001: 34

"Die Vorhersagen, die geborstene Dämme und neue Völkerwanderungen prognostizieren, sind und waren sachlich nicht gerechtfertigt, übertrieben und vielfach politisch motiviert" (Fassmann 2002: 85). The effects will be spread very unequally over the old member states. The labour migration will be further concentrated on Germany and Austria which will absorb some 80 % of the overall migration. 'Cross-border work' can also exert some pressure on the labour market in border regions, but can be costly for the new member states, too.[34]

Structure of Migration (Who and Why?)

The structure of the potential migration from the new member states will probably resemble the structure of the migration from these countries so far. The majority of migrants from CEEC have been: men, young, with relatively high formal qualifications (Fassmann 2002; Fassmann and Münz 2003). An 'incomplete migration', inter alia in form of seasonal employment like in agriculture, will be chosen by less qualified persons (for Poland see Okólski 2000).

Table 2.4: Push and pull factors of potential migrants (general preference) from Czech Republic, Hungary, Poland and Slovakia

Potential migrants (Answers in %)	Home country				Total
	Czech Republic	Slovakia	Poland	Hungary	
Higher wages	96.8	96.8	96.4	92.6	96.2
Better work conditions	74.5	83.6	75.6	81.2	79.7
Interest, adventure	87.6	71.2	79.5	80.1	79.0
Better career chances	47.5	41.6	53.1	74.4	52.0
Further qualifications	54.1	39.1	42.0	65.8	48.9
Unemployment	35.5	44.1	26.6	21.0	33.8
Political situation	22.7	38.9	27.3	34.0	31.1
Bad environmental situation	21.5	26.9	27.5	40.2	30.8
Family abroad	27.2	26.4	29.8	28.2	27.7
Ethnic minority	11.4	7.7	7.1	3.4	7.7

Source: Fassmann 2002: 79

[34] "Cross-border work can be costly to the country of residence, which may not receive income tax revenue from the worker but has to finance social expenditure and local infrastructure (including perhaps subsidised transport) for the benefit of the worker's family. Conversely, the employing country enjoys corresponding financial advantages" (European Commission 2001: 12).

Although after the EU accession many persons in the new member states, especially in Poland, will certainly leave the agrarian sector because of its low productivity, the impact on labour migration should not be large.[35] Surveys on motivation to potential migration[36] show clearly that pull factors of western labour markets motivate people from CEEC to move to the West (see table 2.4). As already mentioned, separate factors are partly interrelated. The results show also that unemployment does not play a crucial role as a motive to migrate, it can however increase the role of pull factors. The situation on the labour market is difficult in many new member states and unemployment is certainly one of the highest costs of transformation. This results to a high degree from factors which are specific for transformation countries, like the reconstruction of economy and privatisation. It is however, also related to factors, which CEEC and Western European labour markets have in common, especially the low flexibility.[37] Low flexibility on the labour market works in the receiving countries as a pull factor. Even with high unemployment, there is still work in many old EU member states, which their citizens do not want to do.

Demography also influences migration decisions. The demographic situation has changed dramatically in the CEEC after transformation. Natural population increase has become negative in almost all CEEC-10 and the fertility rate has fallen under the level of EU-15. This will lead to an ageing of population also in those countries, which now still have a relatively young population. Life expectancy is much lower than in EU-15 too, although in the same period a clear improvement has taken place. Thus, although at present a high pressure is still coming from the younger population on the labour market and partly also on emigration in some countries (for example Poland), in future the population and consequently the labour and migration potential of these countries will decrease.[38] Fertig and Schmidt (2000) attribute the small migration potential from the CEEC to the relatively small share of younger people.

It is often argued that higher social benefits in receiving countries also belong to pull factors of migration and that they lead to further increases of financial burdens of welfare states receiving migration. This is for example

---

35 Most people leaving agriculture will become pensioners or find a job in other sectors. Those who will become unemployed are older and less qualified persons, with barriers and lower motivation to migration (European Commission, Directorate General for Economic and Financial Affairs 2001: 61).

36 In table 2.4 and the following text, results of a survey are presented, which was conducted in 1996 by Fassmann among some 4,300 persons in the Czech Republic, Hungary, Poland and Slovakia. The respondents answered questions in a personal interview, which was conducted by national Gallup-institutes and lasted on average 45 minutes (Fassmann 2002: 66).

37 “Overall, one can conclude that the labour markets of the CEEC are predominantly Western European” (Belke and Hebler 2000: 230).

38 Between 2000 and 2025 population aged 15–65 will decrease by 3.6 % in Slovakia, 8.6 % in Poland and 25.8 % in Estonia (Fassmann and Münz 2003).

illustrated by the fact that immigrants living in Germany have in sum been net recipients of state benefits (Sinn et al. 2001). However, whether this will hold also in the case of migration from CEEC into the old EU depends mainly on the structure of migration. Empirical evidence leads to the conclusion that social insurance systems have very little influence on decisions to migrate. The high dependence on social assistance among the foreigners living in Germany may hence be a specific trait of past migration patterns and should not be projected on future migration flows (Bauer 2002).

Duration of Migration (for How Long?)

Both the experiences of migration from CEEC into the EU so far and surveys on potential migration after the enlargement have shown clearly that temporary migration between both regions will dominate. According to the citied survey, in the Czech Republic, Hungary, Poland and Slovakia one quarter of the persons thinking about migration declared that they would stay abroad no longer than one year. Another third wanted to go for one to two years and only about 8 % of those generally considering migration would do it for ever (Fassmann 2002: 81).

The strategy of commuting instead of migrating or "working in the West and living in the East" is the most desired option. Many arguments are raised that this will remain so also after the enlargement (Fassmann 2002):

- open borders,
- income differences,
- geographical proximity,
- ethnical networks,
- adaptation ability of "trans-national mobiles".

The consequences from this behavior differ from that of traditional emigration. From an economic perspective, it "helps the acceding countries to overcome structural problems on labour markets and helps the EU-15 to close gaps on their labour markets. Temporary movement is in effect a strategy to stay "at home" (Fassmann 2002: 87–88).

Transitional Regulations Concerning Mobility

Due to political sensitiveness of the issue, the free movement of workers belonged to the most difficult topics in the negotiations on EU accession. In the accession agreement with CEEC-8 (all accession countries apart from Cyprus and Malta) a transitional period up to seven years for the introduction of free movement has been agreed upon. The former EU member states can take national measures to regulate access to their labour markets by citizens of the new member states. At first, every country can fix a transitional period of two years, in the following another three and than eventually again two years are feasible. This regulation also allows separate member states to react flexibly to specific circumstances.

On the one hand, this restriction may be seen as an objectively (economically) unnecessary postponement of important market adjustments. On the other hand however, the transitional period has been a politically necessary measure for enhancing the public acceptance of the EU enlargement by the "old" member states. As shown by some opinion polls, without such a regulation, citizens of some of the old EU member states would not have accepted the enlargement. A possible advantage of the regulation for the new member states may be seen in some restrictive effects on the emigration of highly qualified people ('brain drain').[39]

CEEC: from Emigration to Immigration Countries

In the context of the EU enlargement, only the movement from CEEC into the EU-15 is analysed. In reality however, a large immigration is already going on into the new member states. Seven out of ten (Cyprus, Czech Republic, Estonia, Hungary, Malta, Slovakia, and Slovenia) had a positive migration balance in 2002, i.e. more people moved into than out of these countries (European Commission, Directorate General for Employment and Social Affairs 2004: 43). And in the remaining countries the negative balance decreases due to increasing immigration. They will soon become net immigration countries, too. Even with emigration of host residents and high unemployment, there is for example in Poland a large, mostly irregular immigration and commuting, especially from countries located further East (Okólski 2000). The EU accession, growing income and the expected demographic development, will make this tendency even stronger in future.

Concluding Remarks

The central argument of the EU enlargement is that integration of new countries into the EU will help to decrease the income gap between the old and the new members through faster economic growth, foreign investments and the EU structural funds, all of which will decrease the pressure to emigrate (Barfuss 2002). In this sense, the enlargement is "the best anti-migration policy" (Straubhaar 2001). "The present EU member states, especially Germany and the countries in Southern Europe, will rather not be able to cover their future demand for immigrants through the future free movement of persons within the enlarged EU. In the middle term we will not be even able to hope for migration from Central and Eastern Europe, so many are now afraid of" (Fassmann and Münz 2003: 32).

---

[39] It should be noted however, that even before the enlargement highly qualified labour from CEEC, for example physicians, were welcome in many EU countries and this continues despite the general restrictions.

2.2.1.4 Migration and Future of European Social Policy

Immigration can be one method to improve the financial situation of pay-as-you go social security schemes, because every pension system, whatever its financial sources and whatever its design, is a mechanism for organising claims on future production. Thus, output is crucial and immigrating labour is one way to increase output (Barr 2002). EU member states with aging populations and large pension systems will thus be partly condemned to immigration. Migration is also raised as an argument for reforming pension systems in Europe (see chapter 3.3). Mobility of persons in conjunction with freedom of services has also risen some important issues for European social policy, especially for health care systems (see 2.2.4 and 3.2). In the context of the latest enlargement, proposals have been made in some countries to postpone full access to social security benefits for migrant workers from the new member states in order to protect the old member states from migration due to higher social standards. However, as long as migration flows between the two groups of countries remain very limited, such an important change of EU rules on social security seems rather unlikely. Aging populations and segmenting markets create a persistent demand for immigrant workers in Europe. As these trends are likely to intensify in the near future, migration policy will become a key political issue in the EU in the twenty-first century (Massey and Taylor 2004; Holzmann and Münz 2004).

2.2.2 Employment Policy and Labour Market

In chapter one (see 1.2 and 1.3), we stated that personal autonomy, social inclusion and distributive justice are believed to be the basic principles and aims on which European social policy is grounded. If we accept this statement, it goes without saying that it is indispensable for European social policy to address the role of employment policy and labour markets. Work offers people autonomy and integration opportunities, partly associated with the income attached to working, which affords people to be engaged in social activities, and partly associated with the involvement in social networks through work. Equal access to work is therefore an important policy goal at the national and European level to realise these basic principles and aims. Moreover, it is beyond doubt that the better the quality of work people are engaged in is (in terms of working conditions and labour relations), the better the principles are safeguarded and the better the aim of social integration or social inclusion is to be achieved. In this section we therefore discuss the role of the labour market and of employment policy aimed at increasing the opportunities for people to be engaged in work of high quality and to fulfil their preferences for working a particular number of hours during the week in a job of their choice.

This section is organised as follows. We start by briefly sketching the European Employment Strategy (EES). Then we present some figures about the extent by which the labour markets in the various countries are capable of attaining the goals the member states of the European Union have agreed on at the Lisbon summit in 2000 and in 2003. In this part, we take already account of the different outcomes by regime type because one of our basic premises was the role policy regimes might play for explaining the different labour market outcomes across countries. Then we discuss the performance of the national labour markets within the various regime types with a view to the EES guidelines and the pressures faced by the member states to attain these goals. We will discuss the current and future role of the OMC. In the final part, we formulate some conclusions and recommendations about the way to proceed in the future in order to deal with the major dilemmas and constraints European social policy faces with respect to employment and labour market issues.

2.2.2.1 The European Employment Strategy (EES): Flexibility and Employment Security

The overarching objectives of the EES were from its start at the European Jobs Summit in Luxembourg in 1997 on, directed to attaining full employment and maintaining competitiveness of the European economy. It consisted of four pillars: employability, entrepreneurship, adaptability (flexibility) and gender equality. The objectives were to be achieved in a process that was called the Open Method of Coordination, known as OMC. It represents a new form of policy integration in which the role of the European bodies is to set the framework and objectives, and to orchestrate the monitoring and review of the action plans, while leaving member states free to decide on detailed policies and their implementation. In figure 2.1 we depict the OMC

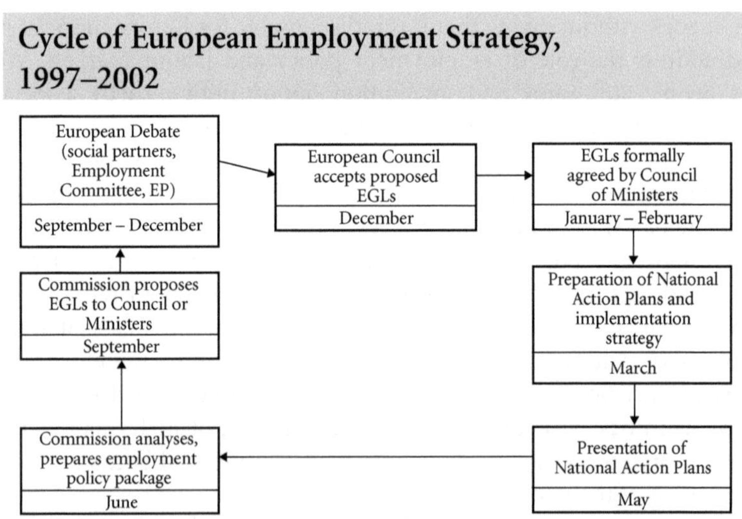

Figure 2.1: The (annual) cycle of the EES between 1997 and 2002
Source: Jørgensen 2004

process as it has been functioning for the EES in the period between 1997 and 2002.

The annual cycle starts in September with the debate across the European actors and the member states about the guidelines as proposed by the Commission to the various governments. Part of this European debate is the consultation of the social partners, the Employer's and Worker's Unions at the European level, the so-called 'social dialogue'. This results eventually in acceptance of the proposed and amended Employment Guidelines (EGL) at the end of the year by the European Council. In the first two months of the following year the EGL are formally agreed on by the Council of Ministers. During spring, the national governments are engaged in preparing the National Action Plans (NAPs) for Employment and the formulation of a strategy for implementation of the proposals. The plans are presented in May to the Commission who then starts to analyse the plans and the policy packages involved also with a view to the attainment of the European wide agreed principles and targets for the Employment policy. In August, the Commission then launches its report on the evaluation of the attained achievements and on the formulation of recommendations how to improve the achievement of the commonly agreed goals. Until 2003 this process has led to the acceptance of 20 guidelines. In 2001 at the Lisbon summit, the leaders agreed on setting quantitative targets on the employment levels for different population groups: 70 % for the total labour force by 2010, 60 % for women and 50 % for older workers of 55 to 65 years of age. At the same occasion the close relationship between these employment objectives, the macro economic policy goals and the social policy objectives were accentuated.

After the evaluation of the EES in 2002 the objectives of the EES have been rephrased in both the 2003 EGL and the Broad Economic Policy Guidelines (BEPG): to create the conditions for full employment, to improve the quality of work and productivity and to foster social cohesion and inclusive labour markets. Also in the 2003 guidelines, a number of quantitative targets have been formulated on activation measures for the reintegration of unemployed people and the long-term unemployed in particular, on lifelong learning, on postponing early retirement, on education of young people, on the availability of child-care to allow women to participate in work and on early school-leavers. In more detail the agreed targets were the following:

- Every unemployed person has to be offered a new start within 6 months of unemployment in the case of young people and 12 months in the case of adults.
- By 2010, 25 % of the long-term unemployed have to participate in an activating measure, with the aim of achieving the average of the three most advanced member states.
- By 2010, at least 85 % of 22 year olds in the European Union should have completed upper secondary education.

- The European Union average level of participation in life-long learning should be at least 12.5 % of the adult working-age population.
- By 2010, at EU level an increase of the effective average exit age from the labour market from 60 to 65 should be achieved.
- By 2010, childcare should be provided to at least 90 % of the children between 3 years old and the mandatory school age and to at least 33 % of children under 3 years of age.
- By 2010, an EU average rate should be attained of no more than 10 % early school leavers.
- All job vacancies advertised by national employment services should be accessible and available for consultation to anyone in the EU by 2005.

The tools to achieve these goals were to promote more flexible work organisations, and to facilitate labour mobility, both geographical and occupational, while taking into account the need for job security (European Commission 2003b). That the European Guide Lines 2003 promote mobility as a separate goal distinct from promoting adaptability might be associated with the overarching objective of the EES to improve the conditions for attaining full employment. A higher labour mobility would imply that people better respond to existing wage differentials and that they move to countries yielding the highest return on their investment in skill formation and education. This would foster competition on the labour market, it would improve the efficiency of the labour market and its allocation and it would lead to increases in productivity growth. The more Europe becomes integrated in economic terms, the more likely this will lead to rising factor mobility of labour and capital and increasing mobility flows of (high-skilled) employees (immigrants, commuters) and employers on the labour market. Migration is therefore actually a beneficial event, it might reduce regional disparities in unemployment levels and it might boost the flexibility of the labour market by creating the conditions for adapting the labour volume more quickly to changing socio-economic conditions (Dekker et al. 2003).

Within the framework of the knowledge economy one might expect a rising flow of employees, particularly the highly skilled people, moving from one region into another. But, due to the enlargement of the Union to 25 member states, there will be increasing numbers of fairly low skilled people from the East seeking for better employment opportunities in the West, which might affect these labour flows. (See section 2.2.1 on migration; for an overview of the performance of the accessing countries in dealing with the unbalance of their labour markets, we refer to the next passages where we review parts of the evidence.)

One might also think of a growing number of pensioners, due to the ageing issue, moving more or less permanently into a foreign country of their choice and the consumers of health care services looking for services in another region where they are of high(er) quality and available at low(er) prices (see chapter 3.2). What the likely impacts of these mobility flows on national policies will be, depends largely on the sheer size of their numbers, which in their turn are strongly affected by the strictness of the national and supra-national migration policies as well as the differences in institutional support for migrant workers. At the EU-level there is in principle free movement of employees in Europe but this seems not to have resulted in large migration flows as the current evidence shows (see 2.2.1). The larger these flows will turn out to be, the more they will affect the national labour markets in terms of their employment rates, the distribution and level of their wages, the level of employment protection to various categories of workers and the distribution of jobs across the various population groups. Apart from their likely effects on the operation and (distributional) outcomes of the national labour markets, they will also affect the OMC process itself and the extent by which it will be effective in attaining improved fine-tuning of policy interventions and improved coordination.

Labour Market Flexibility and Work Security

The strive for more labour market flexibility (part-time work, flexible working-times) in the EES went along with the pursuit of safeguarding income and work security. The EES therefore, seems to hinge on the view that the more governments aim at the pursuit of economic and social goals simultaneously the better the performance of their economies will be in economic (flexibility, productivity, competitiveness) and social terms (income and employment security, social inclusion, and quality of work).

The 2001 European Employment Guideline 13 under the Adaptability pillar explicitly addressed both, flexibilisation and security goals as it invited "to negotiate and implement at all appropriate levels agreements to modernise the organisation of work, including flexible working arrangements, with the aim of making undertakings productive and competitive, achieving the required balance between flexibility and security, and increasing the quality of jobs." In the Council decision on the revision of the EGL it is stated that providing the right balance between flexibility and security will help support the competitiveness of firms, increase quality and productivity at work and facilitate the adaptation of firms and workers to economic change (Wilthagen and Van Velzen 2004). Also in a recent OECD report a lance is broken for a combined strategy of raising flexibility taking account of individual's need for job security and a good work-life balance (OECD Employment Outlook 2004).

Employment Taskforce

In March 2003 the Commission launched the Employment Taskforce headed by ex prime-minister Wim Kok of the Netherlands to propose practical reform measures for attaining the employment targets which turned out to be unmet in 2003 and for implementing the revised EES as well as to achieve its objectives and targets. The first report was issued in November 2003 and the second report "Facing the Challenge" one year later in November 2004. The Kok I report was called "Jobs, jobs, jobs: creating more employment in Europe". The main ingredients of the taskforce's recipe were the following:

- increasing adaptability of workers and enterprises;
- attracting more people to the labour market;
- investing more and more effectively in human capital;
- ensuring effective implementation of reforms through better governance.

The notion of adaptability and flexibility is stressed in the report by claiming to "anticipate, trigger and absorb change if more jobs are to be created and filled". It also stresses the importance of striving for more flexibility but with safeguarding appropriate levels of security. Further to this, the report states that "flexibility is not only in the interest of employers, it also serves the workers' interests to combine work with care and education or to allow them to lead their preferred lifestyles". The report draws particularly attention to the 'best practices' in Denmark and the Netherlands for creating a balance between flexibility and security. With a view to the specific policy arrangements in these two countries, the report stresses the need to remove "obstacles to temporary work agencies to allowing them to be effective and attractive intermediaries" and to "offer improved job opportunities and high employment standards". The report further accentuates the creation of jobs and the need to respond effectively to economic downturns and increased competition: "to attach more people to the labour market so that they can achieve sustainable integration in jobs". With respect to the need to increase investments in human capital, the report states that "the productivity of enterprises and the overall competitiveness of our economy are directly dependent on building and maintaining a well-educated, skilled and adaptable workforce to embrace change". In this respect the report favours the role of lifelong learning and of 'active ageing' to involve the older workers and to downturn early retirement practices. Another characteristic of the report is its attention for what is called the 'Partnership for Change' to mobilise the social partners at the national and even sub-national tiers of government around the Lisbon targets as has also been stressed at the last Spring Council and the Tripartite Social Summit in 2004 (cf. Vandenbroucke 2004).

The November 2004 report of the high level group: "Facing the challenge" had a wider scope. It is not solely aimed at the employment policy but embraces the entire set of Lisbon targets including the social agenda. The report once again underlines the major highlights of the first report: adaptability and employability, labour mobility, skill formation and human capital investment, and social support (modern and efficient social policies) to equip people and firms to adapt to change. Health care is mentioned as a separate important domain in this respect to combating disease but also to sustain a productive workforce and to promote economic growth. Hence, the notion of social security or the welfare state as a productive factor is again underlined.

Even though the report took notice of the number of jobs created since 2001 and the reduction in the levels of unemployment and long-term unemployment, it should not be swept under the carpet that the reason to establish the taskforce in the first place was the underperformance of national policies with respect to the Lisbon targets. Again, the report stresses the need for the right balance between flexibility and security. The report pays heed to a new paradigm, according to which the preserving of jobs for life is not the main concern but to build up people's ability to remain and progress in the labour market. This very much underlines the aim of promoting 'employability' but also mirrors the 'capability' approach and the activation of people in a labour market in which people have to invest in their 'human capital' so as to acquire the skills necessary to be able to successfully adapt to change. The focus on the investment in human capital formation as being the main ingredient of the recipe to transform labour market policies into activation, signals the importance attached to the role of skill formation in the 'knowledge economy' and the establishment of a high skilled labour force to maintain Europe's competitiveness and innovative power.

The Proposals for a Change of the EES

However, in the debate on the EES, changes were proposed for the near future, which resembles the critical assessment of the strategy in 2003 by some actors in the field. Many see the EES as a technocratic device rather than as an open and transparent process of relevance for policies and policy learning experiences and practices. The cycle process at European level seems not well-tuned and badly integrated into the national policy making process and NAPs are considered a European exercise which has to be fulfilled but not involving domestic strategic considerations and operations. Or as another critical author conveys it: "Elite actors are socialised to new procedures and identities, while retaining their former ones, systems of actors are transformed by the introduction of new resources but policies seem to doggedly stay on their national tracks" (Rogowski 2004). Also in the two reports of the taskforce on the EES and the High Level Group on the Lisbon targets some doubts are raised about the process because of which the need was expressed to improve the 'governance' system. At the press conference at the occasion of the release of the Kok II report in November 2004, it was accentuated by the chairman responding to questions whether the 2010 targets has lost all credibility, that "we do not have the luxury anymore of just exchanging pleasantries". Some criticasters state that the OMC process has evolved from a discursive operation to learn from each others best practices and to implement change, into a sort of 'beauty contest' showing one's successes but omitting one's failures. The idea of 'naming, blaming and shaming' and the use of indicators and soft sanctions that the process seem to be build on, appear not to enforce compliance with plans and targets (Jorgensen 2004). The focus seems to lie much more on inputs than on outcomes and impacts. But instead of using indicators, a system of output evaluation would imply a shift towards a system of 'diagnostic monitoring'.

To date, there are few examples of policy learning and cross-national policy transfer for which reason there seems to be underutilisation of the learning potential. Also a recent review article of Casey and Gold (2005) shows that learning effects are rather small. They report that though there is some evidence at EU level of policy convergence through the adoption of certain labour market targets, there is little evidence neither of systematic learning nor of significant efforts at emulation. General impediments to organisational learning and perceptions of cultural and institutional differences between countries constitute possible explanations for the absence of significant learning effects. For these reason changes are proposed to make the system of governance more of a success whereas others ask for stronger policy instruments to enforce national government to apply to the agreed targets by implementing reforms and changes (Jorgensen 2004).

Without going into depth here it should also be mentioned that the one-year cycles seem not very efficient due to the crowded European and national agenda, but also due to the time needed to implement reforms and to adapt to the new and more ambitious targets. The process of governance of the Lisbon targets needs a longer time frame to achieve its targets for which reason a prolonged cycle of three years is proposed. The cycle would then be tuned with the process of the Economic reforms in the framework of the Cardiff process on the BEPG and the Internal Market Strategy. This would strengthen the endeavour of the Commission to attain more integration between the various policy domains which, yet, operate in a rather isolated and separated way. Another change that has been proposed concerns the decision making process itself and the role of the social dialogue. The Spring Summit of the European Council would become the main decision making forum to be proceeded by a Tripartite Social Partner Summit to commit the social partners to the process and its results. This streamlining of the decision making process should improve the decision making process itself as well as the coordination and integration of the various policy domains that the Commission is aiming at.

Targets too Ambitious?

At the same time a more fundamental debate has been commenced under the leadership of the late 2004 appointed new chairman of the Commission, Barroso. This upcoming debate questions the usefulness of the ambitious target setting in the employment area, since it might be detrimental or contra productive to the primary goals of the Union to arrive at more coordination in the economic and monetary domain. If the targets seem to turn out as being too optimistic or too ambitious and year after year the achievements stay behind, one may cast doubt about how realistic and attainable the targets are, especially for countries that perform far below the European average. This would particularly entail the problems the new accession countries face in meeting the high standards set by the 15 others. Wouldn't it be much more encouraging for these member states if, instead of having to report failures year after year, successes can be booked, even though these appear to be only small steps on the road to become, in the end, one of the best performing economic regions in the world? The setting of a more realistic agenda might be beneficial for quite a few member states lacking much behind. On the other hand, one might argue that the setting of a more realistic agenda, resembles a lack of ambition to set high standards and to attain them in due course. Lack of ambition might in this view turn out to lead to acceptance of the 'second rank' position of the European economy in the world, which might induce a spiralling downwards process without much chances for recovery later on. They especially voice a warning against the competitive pressures coming from the Eastern transition countries (cheap and productive labour) and the Asian economies, particularly China, as being a low wage country with a highly skilled and productive labour force. They believe that massive investments in 'human capital' formation, in innovative technologies and production processes as well as in demolition of Europe's labour market rigidity and bureaucracy is needed to be able to maintain Europe's competitiveness and attained high living standards.

2.2.2.2 Flexibility and Security: the Role of Welfare Regimes

The member-states widely vary in the way they have build up their labour market institutions. Partly at least, these differences reflect the typical features of the various welfare or employment regimes which exist in Europe. As we already showed in chapter 1.3, regimes differ in the way they are capable of attaining prevailing economic and social goals simultaneously. Their success is dependent on how they manage to attain a high level of labour market flexibility and at the same time to guarantee adequate levels of social protection and work and income security (Headey et al. 1997). This is known as the 'flexicurity' thesis (Kongshoi Madsen 2004; Wilthagen 1998; Wilthagen and Tros 2004). The pursuit of promoting a more flexible labour market without distorting income and employment security seems to be part of the EES as sketched out in the preceding sub-section (2.2.2.1). The idea is to implement a mutually reinforcing relationship between the flexibilisation of employment relationships and the protection afforded by safeguarding social and work security. The regimes that are identified in the literature differ substantially in how they combine employment regulation and welfare benefits to deliver social protection (Esping-Andersen 1990; Gallie and Paugam 2000; Goodin et al. 1999). Though these regimes are not static entities, they might shift over time due to economic or political developments, there is little reason to expect these widely different systems to converge easily across the enlarged Europe of the 25. If we try to place the various regimes according to the literature (see chapter 1.3) on the flexibility-security nexus, the following dimensions might be distinguished (see figure 2.2):

- level of employment protection regulation;
- level and duration of unemployment benefits;
- active or activating labour market policies (employment creation, training and 'employability' policies).

The Balance between Flexibility and Security

On the labour market, minimum wage regulations, strong employment protection rules and tight legislation with respect to allowing flexible working time practices might lead to rigidities or inflexibilities hampering economic productivity, employment and economic growth rates. Also the impact of strong unionisation or what is called a high trade union density and union coverage might jeopardise the 'adaptability' of workers and firms to change whereby deadweight or efficiency losses have to be accepted. Apart from these rigidities, a study by Nickell (1997) shows the likely adverse impact of the Unemployment Insurance Benefit system on the 'adaptability' to change, which pertains to the disincentive effects of the level of the unemployment benefit or the replacement rate (ratio of income out of employment to that of unemployment) and the duration or length of receiving benefits. Eventually, the content and scope of 'activation' policies (employment creation, creating education, training or life-long learning opportunities for people and policies to enhance worker's 'employability') seem also to exert a significant effect on the performance of national labour markets.

According to standard economic views, there is a kind of trade-off between flexibility and security. A high work security due to its disincentive effects will endanger mobility and flexibility, whereas a high flexibility might endanger the work and income stability of the weaker groups such as the low skilled. However, according to the 'flexicurity' approach, we might also think of a positive relationship, according to which flexibility is needed to safeguard a high level of work security.

The existing evidence (Muffels and Fouarge 2004; 2002) suggests that the Social-Democratic and Scandinavian Regimes (particularly Denmark) perform best in attaining a high level of labour market flexibility and at the same time providing much security in terms of income and employment security. Although generalisations are risky, the Southern European, traditionalist regime seems to deliver the worst combination as regards the flexibility-security nexus.

The transition countries or the East-European countries are represented as a separate cluster although the within regime variation is rather large. The evidence in table 2.5 suggests that these transition countries share a common high level of flexibility due to the absence of strict employment protection regulations. The evidence also indicates that the level of income and work security achieved in the transition countries is rather weak and comparable to the level of the Southern regimes. They seem to share an immature social security system and strong reliance on family support. For that reason, we placed the Eastern, former Socialist, regime or the transition countries cluster close to the liberal and the Southern regime type. Furthermore, the evidence shows that the Anglo-Saxon regime is weak on guaranteeing generous levels of social security while the corporatist regime does not perform particularly well as regards flexibility (see figure 2.2).

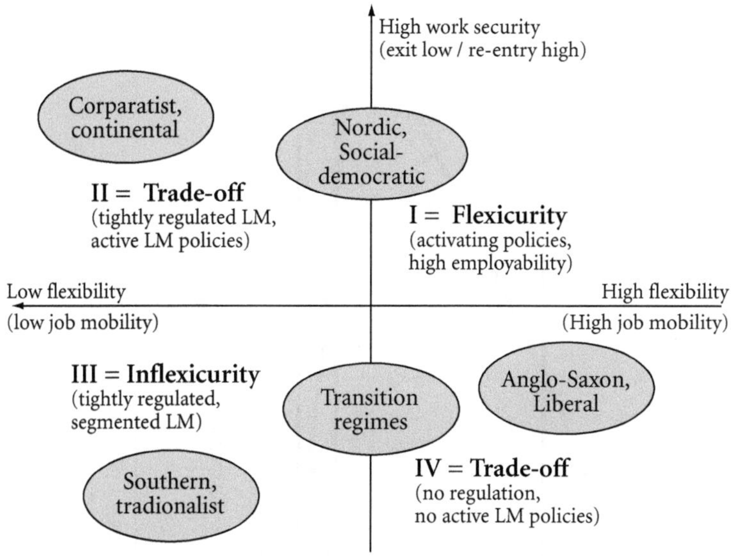

Figure 2.2: The location of welfare state regimes on the flexibility-security nexus (LM = labour market)

Some more evidence is given in the next figure in which we depict the evolution of some indicators for work security, job mobility and contract mobility for the years 1994/1995–2000 (see figure 2.3).[40] Job mobility is defined here as occupational mobility, the mobility between jobs of different occupational class level. Contract mobility is the mobility between permanent jobs, flexible or temporary jobs and self-employment during the observation period. Work security is calculated as the weighted average of the entry chances into permanent jobs plus the staying chances in work, minus the exit chances out of employment (permanent jobs, flexible jobs, self-employment) into non-work during the period and minus the chances of remaining out of work (Muffels and Luijkx 2005). The straight lines show the European average (for 14 countries) with respect to the various measures for job mobility, contract mobility and work security.

[40] The figures are based on evidence obtained from the European Community Household Panel on 14 countries covering the years 1994–2000. Sweden is excluded due to lack of information. We calculated average mobility rates over the seven years of observation.

The results also exhibit the large differences across the various regime types. Job mobility is lowest in the corporatist regimes and work security is lowest in the Southern regimes, just as we expected. Notice the high levels of job mobility in liberal regimes, but particularly the high levels of work security in these regimes, which are nearly as good as the ones for the social-democratic countries.

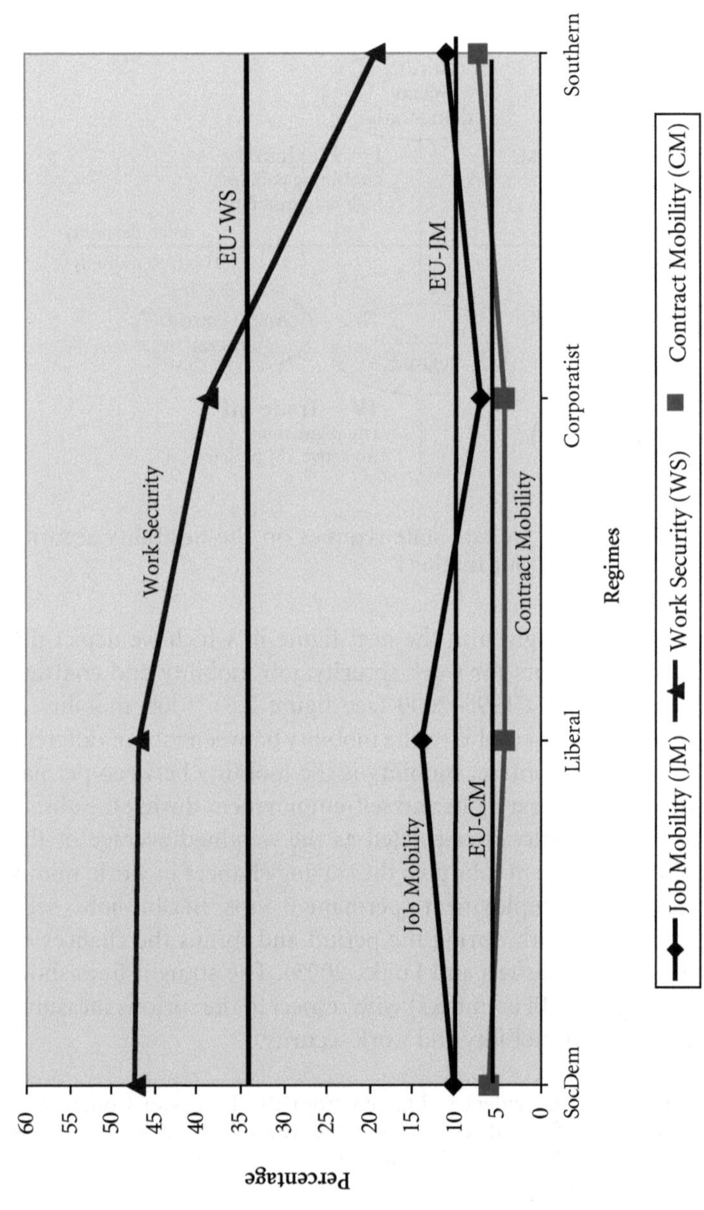

Figure 2.3: Work security, job mobility and contract mobility by regime type
Source: Muffels and Luijkx 2005 (based on ECHP data for 1994–2000)

OECD Mobility Index

In the Employment Outlook of 2004 the OECD (OECD 2004) has developed a so-called mobility-index that is aimed at measuring the overall level of labour market mobility, but as such is also considered to be an indicator for the level of flexibility in the labour market. The measure is also based on the year-by-year transition matrix of labour market statuses as calculated from the available evidence of the European Community Household Panel.[41] The outcomes indicate that highest mobility is attained in the liberal regime of the UK and next highest in the social-democratic countries, Denmark, the Netherlands and Finland. The lowest rates are again observed in the Southern countries, Spain, Italy, Greece and Portugal. The corporatist countries are performing hardly better than the Southern countries except for France and Belgium whose performance is the worst of all countries in terms of labour market mobility rates. Ireland's mobility rate comes closer to the corporatist rates than to the UK one. These results confirm our earlier findings with respect to flexibility.

Another interesting result is that when we make a simple cross-tabulation of the level of labour market mobility and the employment rate for each country, we find a close and linear relationship between the two. The countries with the highest mobility rates also exhibit the highest employment rate and the countries with the lowest mobility rates the lowest employment rate. Portugal is the only exception with much higher employment rates than the mobility rates would predict. The score of Denmark is rather impressive with a very high employment rate and a very high mobility rate but also the UK results are quite impressive. If we consider the employment rate as an indicator for the level of attained work security the findings confirm what we found earlier about the flexibility-security nexus phrased as the 'flexicurity' thesis, using more refined indicators for job mobility and work security as in figure 2.3.

Indices for Strictness of Employment Protection Rules

The OECD has elaborated indices for calculating employment protection levels ranging from 0 (no protection) to 5 (strict protection) for the 1990s (see table 2.5). Although the measures are controversial, they are used in several publications of the EU e.g. European Commission 2003b). The indicators deal with overall strictness of employment protection and of particular

[41] The mobility index is based on the trace of the transition matrix of labour market statuses. It is a measure for the extent by which the actual mobility rate into the destination state is dependent or independent on the origin state. The measure is one in case of independency or perfect mobility and approaches zero in the case of dependency or perfect immobility. Alternative measures based on the determinant or the eigenvalues provide largely similar results.

Table 2.5: OECD indicators for strictness of employment protection legislation by country and employment regime

Regime and Country	Overall (un-weighted) average regimes	Overall (un-weighted) average countries	Ranking regime-types	Ranking countries
Liberal	1.0		1	
United Kingdom		0.9		1
Ireland		1.1		2
Corporatist	2,5		4	
Austria		2.3		8
Belgium		2.5		9
Germany		2.6		10
France		2.8		11
Social-democratic	2.1		3	
Denmark		1.5		3
Finland		2.1		6
Netherlands		2.2		7
Sweden		2.6		10
Southern	3.5		5	
Spain		3.1		12
Italy		3.4		13
Greece		3.6		14
Portugal		3.7		15
Eastern Europe				
(transition countries)	1.9		2	
Hungary		1.7		4
Poland		2.0		5
Czech Republic		2.1		6

Source: European Commission 2003b: 35

provisions for regular employment, temporary employment and collective dismissals. The results reveal a large variation across countries with indices ranging from 0,9 for the UK to 3,7 for Portugal. We averaged the country values by welfare regime type and the results indicate that the liberal and Eastern European countries exhibit the highest level of flexibility and the corporatist and Southern countries the lowest. The social-democratic countries are in between. Denmark seems to be a special case given its relative high level of flexibility compared to the other countries within this regime type (Kongshoi Madsen 2004). Sweden in particular, seems to have rather strict employment protection laws that are at the same level of strictness as in France and Germany.

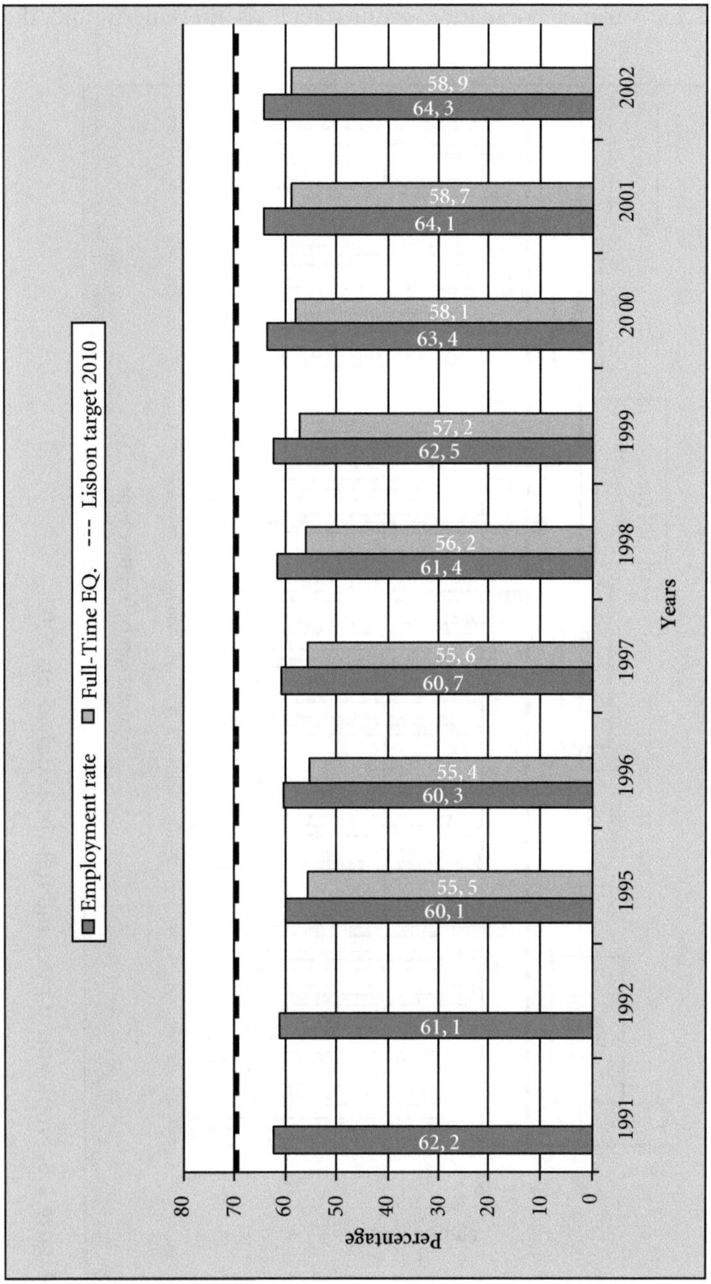

Figure 2.4: Employment rates and full-time equivalents 2002 (% pop. aged 15-64)
Source: European Commission 2003b: 35 (based on Labour Force Survey, Eurostat)

2.2.2.3 Welfare Regimes, EES and Labour Market Outcomes

That the Lisbon targets are not easily to be achieved in all of the 25 member states might not be taken as a surprise, but the differences across the countries emerge to be quite large. In the following figures we depict some of these differences with respect to employment rates for the professional population and for various population groups such as the elderly and the

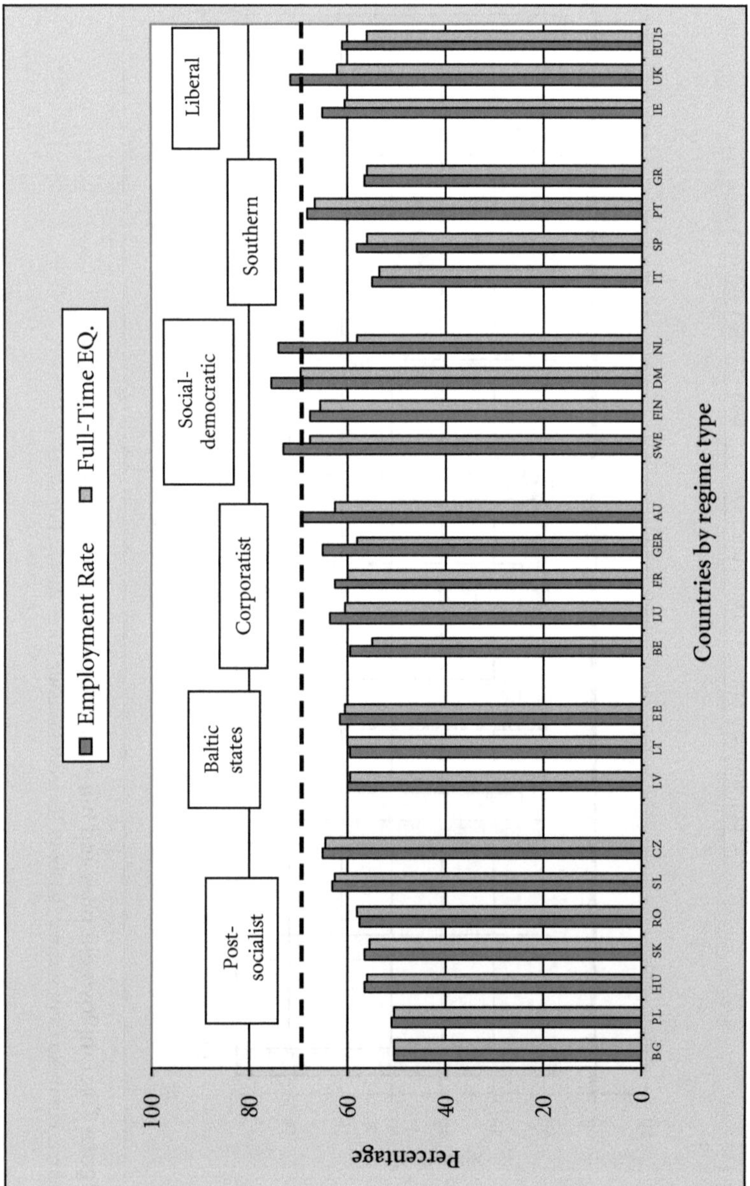

Figure 2.5: Employment rates and full-time equivalents 2002
Source: European Commission 2003b: 30-31 (based on Labour Force Survey, Eurostat)

females, since the employment targets vary by these categories. We have clustered the countries according to regime type (weighted averages of the various countries, weighted with the population of 16 to 64 years of age).

The employment rates by regime type for 2002 indeed show that the lowest values are achieved in the transition countries (55,5 %), next lowest in the Southern regimes (57,7 %), slightly higher in the Baltic states (60,5 %) and the corporatist regimes (64,3 %), but highest in the liberal (71,3 %) and particularly the social-democratic regimes (73,5 %). Though sharing a socialist economic system in the recent past, the Baltic States are generally performing better with a view to their labour market achievements than the transition countries in the East. The Lisbon targets for the total employment rate for 2010 have been achieved for a few social-democratic and one liberal

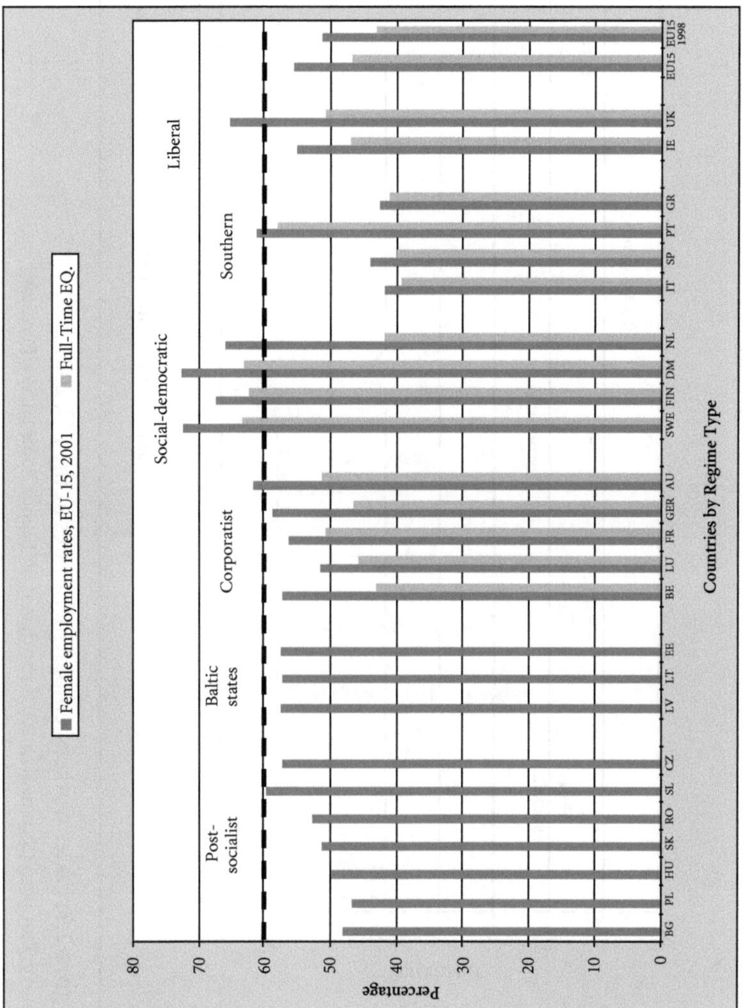

Figure 2.6: Female employment rates in the EU, 2001

Source: European Commission 2003b: 31 (based on Labour Force Survey, Eurostat)

country only: Denmark, the Netherlands, Sweden and the UK (see figure 2.5). But also the 2005 target of 67% would have been achieved only by a very few countries more such as Austria and Portugal. The transition countries, the Baltic States and most of the Southern countries are far away from these targets.

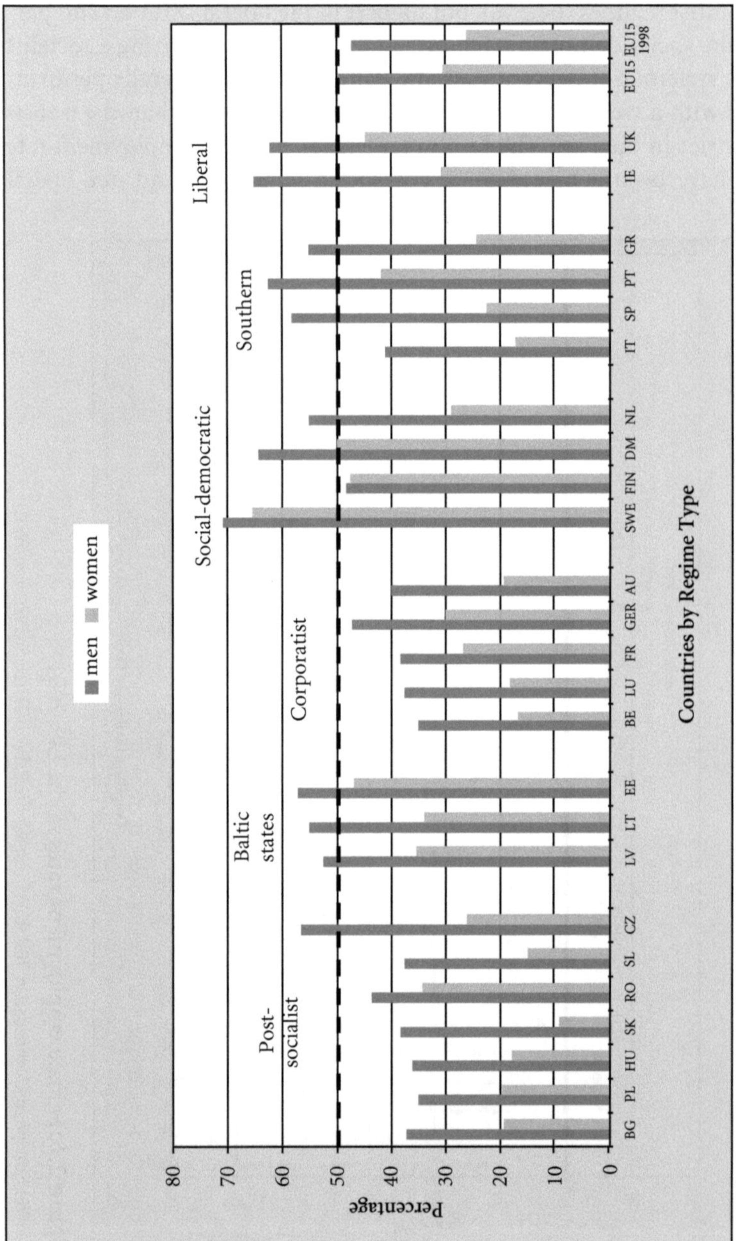

Figure 2.7: Older workers' employment rates in the EU, 2001

Source: European Commission 2003b: 31 (based on Labour Force Survey, Eurostat)

In figure 2.6 and 2.7, we depict the achieved rates for females and for older male and female workers. The picture does not change much. Apart from the social-democratic countries Sweden and Denmark and the corporatist country Austria, also Portugal reached the target of 57 % for 2005 and 60 % for 2010. The Netherlands, as belonging to the social-democratic cluster, still has rather low female participation rates being more in line with the ones for the corporatist countries.

For the older workers the picture emerges to be slightly different. The differences across countries and regimes are even larger. Again, the social-democratic regimes have the highest participation rates for older male workers but also for older female workers although for the latter category their record is less favourable particularly for Finland and the Netherlands. The Southern countries Spain, Greece and Portugal all achieved their target of 50 % for 2010 at least for males; for females they lag behind most other countries. The corporatist countries are performing particularly badly with respect to the integration of older workers into the labour market for male as well as for female workers. They might be set on even par with the transition countries achieving participation levels not much beyond the 30 to 40 % thresholds for male and 20 to 30 % for female workers. Germany reached slightly higher levels for the corporatist countries, as did the Czech Republic for the transition countries.

The differences between the female employment rates for persons and for full-time equivalents show the role of part-time labour in the various countries and regime types. The Netherlands stands out as showing the largest differences between these figures within the social-democratic cluster, because the Netherlands has the highest rate of female part-time work. But also for the corporatist regimes the differences between the two rates are rather small. They are particularly small for the Southern countries indicating that, although not many women work in these countries, when they work they work nearly full-time.

2.2.2.4 The Link between the Social and Economic Agenda: the Role of European Institutions

The debate about employment issues bears a close relationship with the social agenda. This is obvious not only because work forms a major barrier to social exclusion and offers, partly through its income generating effects, important integration opportunities, but also because within the European decision making process the employment issue is discussed along with social issues in the broader Employment, Social Policy, Health and Consumer Affairs Council (ESPHCA). In this forum labour market policy issues are debated, but also the fine-tuning of labour market and social security policies. Issues related to work such as the quality of the job, the work safety, the working conditions and worker's health are on the list of items discussed in this Council. However, due to the predominance of economic and financial matters in the Union, the ESPHCA is fairly unknown and certainly less known among the public than the Economic and Financial Affairs Council (ECOFIN). The ECOFIN is without doubt still the most important decision making body in the European Union, since it deals with economic (the internal market) and financial issues (EMU) which constitute the heart and body of the European Union. From the onset it is clear that on the ECOFIN agenda, as being part of the broad macro-economic policy issues (SGP and the BEPG), employment and labour market issues constitute a major role. In the last ECOFIN meeting in 2004 the need to improve the flexibility of the labour market was one of the major issues. The minutes report on the lack of achievement to meet the Lisbon targets with respect to employment and the need for comprehensive reforms without further delay. Employment issues bear therefore also a close relationship with economic and financial issues. It is probably for this reason that the OMC process started first in the Employment area before it became introduced in the Social domain.

If we indeed believe that part of the EES is to create a better balance between flexibility and work security ('flexicurity'), it would imply that the linkages between economic and social policies have to be strengthened. In terms of institutional reforms it would imply that the ECOFIN would not only involve the economic and financial Ministers but also the Ministers of Social Affairs especially when policies are at stake which have a substantial impact on the balance between flexibility and security. The other side of the coin would of course be that also the Ministers of Economic and Financial Affairs are occasionally invited to the ESPHCA if important and relevant items dealing with the flexibility and security issue are on the agenda. The fine-tuning and coordination of the two different worlds of policy making might then be improved thereby avoiding the enactment of policies, which may be conflicting in terms of the economic and social goals governments are aimed at. The proposals for extending the timing process of the OMC to three years and for a better tuning of it with the Cardiff process on the BEPG and the Internal Market Strategy are valuable in their own right but might not suffice to improve the consistency and coordination needed in the light of the ongoing and rapidly evolving challenges and structural changes on the labour market. First of all, it is beyond doubt that the process of integration will be the more successful if at the same time such an integration and streamlining process will simultaneously take place at the national level and preferable even at lower tiers of government between the national institutions and local/regional bodies responsible for economic and social policies. Secondly, at the European level, we think that the fine-tuning of economic and social policies should preferably be organised within the institutional bodies rather than to leave it to coordination mechanisms between the operating organisations. It means that the international bodies in charge of the coordination process should obtain the authority to manage this coordination and fine-tuning process between economic and social policy in a proper way.

2.2.2.5 Conclusions and Recommendations

In this chapter we addressed the issues of employment and the labour market. The focus has been particularly on the content, the threats and the challenges of the EES in conjunction with the major challenges for the European labour market. The shortcomings of the EES with a view to its achievement to meet the ambitious quantitative targets in 2005 and 2010 have tempted the Commission to modify the system of governance of the OMC process. The proposals to extend the timing and to improve the fine-tuning need in our view be supplemented at the European level with more fundamental changes in the way the OMC is managed and governed. We made a plea for a broadening up of the ECOFIN by inviting the Ministers of Social Affairs in important issues to participate and to be involved in the decision-making process as well as to invite the Ministers of Finance and Economic Affairs to the ESPHCA in matters of concern. This more joined-up approach of economic and social policies might in the end improve the coordination and efficacy of policies in the social domain and might aid to improve the achievements in meeting the ambitious targets for 2010. However, although the contribution of an improved coordination to economic recovery and employment creation is not something to be easily denied, it should not be overstated in its effect either. In the end what counts, is how the main actors on the labour market, employers, employees and governments behave with a view to maintaining a competitive economy without distorting security ('flexicurity' policies). This signals the strength of their economies and the flexibility and efficacy of their policies to deal with economic adversity and underperformance of their labour markets. We suspect that welfare regimes deal with the challenges in a different and not necessarily coordinated way. Regimes that appeared capable of finding a right balance between flexibility and security seem to perform better than regimes that were not. Therefore, we suspect that liberal and social-democratic regimes perform better especially because they give greater room to flexibility without endangering work security.

The experiences with the OMC process show that especially in an economic downturn period there is no reason to believe that the OMC process will automatically lead to improved results. The principle of subsidiarity renders national regimes ample opportunities to downplay the European agenda setting and to stick to minimum targets. In such an event, social policy turns out to be mainly symbolic and pays heed to public resentments about Europe, but has no footing in day-to-day policies and practices. If this scenario turns out to be true, the OMC process has changed into a sort of 'exchange of pleasantries' or even 'beauty contest' in which countries show their successes but hide their failures. The OMC process is then unable to enforce compliance with ambitious target settings and innovative policy endeavours. This is a warning against too optimistic voices on the successes and achievements of the OMC process in the employment and social domain. These shortcomings of the OMC process also paved the road for the establishment of two taskforces on employment who defined proposals for the improvement of the outcomes of the Lisbon process. We believe that the approach to maintain the Lisbon ambition should be preferred to the pragmatic approach of denying the ambitious targets and to transform the mission of the Lisbon process into a much more modest but feasible one.

2.2.3 EU Economic Integration and European Social Policy

2.2.3.1 Introduction and Background

The endeavour to build an economic and monetary union has had and will likely have a strong impact on the economic development in the Union, on the design of the economic policy making in Europe, on the development of social protection systems in Europe and on the way national and European social policies will evolve in the near future. How this process of integration will develop over time remains an open question because the process is strongly affected by economic and social forces and competitive pressures between the dominant economic regions at the global level. It is too simple to assume that the Union's past sets the scene for the future and that the evolution into an integrated Europe will evolve inevitable whatever the contextual changes are. This opinion seems to depart from a rather linear and consensual view on the history of European integration and on its development. The path of development might be paved with many obstacles and beyond that, with unfolding fissures rewriting the course of history. An example might illustrate this. Instead of convergence of national social security systems into mixed public-private systems as the advocates of the European social model sometimes predict, the path of development might also follow a completely different road. Social security systems might also evolve into strictly private systems organised through European wide private social insurance markets or into a balkanised set of nationally oriented social security systems. In either case, convergence is far from being attained, but instead the Union's road will be featured by increased policy competition between the nation states without much Europe or with a Europe of different speeds. The Union's development might also be very dissimilar for the various sub-systems of health, pensions, unemployment and disability. Various chapters in the third part of this book sketching these sub-systems and how they evolve over time confirm the assertion that their development is clearly divergent. The future is inherently mysterious and changes in the various systems tend to develop along unpredicted and unexpected routes.

Therefore, there is no guarantee whatsoever that there is a case for Europe in social space as there is in economic and financial space. There is also little reason to expect that the road the national systems take will be the road of convergence into one European Social Model. The case for Europe especially in the social field (employment, social inclusion) is even rather weak now, since there are solid arguments for social policy to be left to national competence and to be implemented at the national level or even at local tiers of government (Begg and Berghman 2002). It is for the sake of these arguments that the principle of subsidiarity, meaning that the competence is maintained at the national level, has been applied for more than a decade now in the social domain of European policy making. What this means for the evolvement of the various national systems in the future and whether and to what extent convergence will be attained remains unclear. Since Europe will always be internally divided and the divergence is likely to increase due to the enlargement of the Union with 10 new member states, the prediction about which scenario will become true is futile. Our aim therefore is to provide insight into the challenges, pressures and constraints – particularly in the economic domain – the national systems are confronted with and into the viable consequences of national responses to the economic pressures and to European approaches.

In recent years with the advent of the single market and with the EMU, the process of economic integration got a strong impetus and has shown to exert a clear and substantial effect on the economic functioning of the EU as well as on the design of its economic policy. The latter might be illustrated by the acceptance in the framework of EMU and SGP of the BEPG. This has led to an irreversible process of restraining the room for national fiscal policy and of adapting national economic policies particularly to meet the tight budget constraints agreed on in the context of EMU and SGP. The EU regulatory setting might hinder countries to respond swiftly to economic adversity while part of their decision room in economic policy making is given away to European institutions. "The principal concern is that the SGP by capping deficit spending, limits the ability of countries to respond swiftly to adverse economic shocks, thereby exacerbating the consequences of the loss of the exchange rate as a weapon for individual Euro countries" (Mayes and Viren 2002: 29).

Apart from the regulatory context in Europe itself, structural processes of ageing and individualisation – due to dwindling family networks, divorce and separation – are involved, having a profound impact on the social expenditure levels through their effects on the costs of welfare, pension and sickness schemes. Hence, social policies are faced with a double-sided problem which in its policy response is hard to reconcile: that is the problem of an upward pressure on social expenditures due to these long-term trends of ageing and individualisation and on the other hand the inability through the EMU and SGP rules to use the deficit-spending policy instrument to deal with adverse economic shocks. The long-term trends itself might hamper economic growth due to the high labour costs and the downward pressure on innovation due to lack of investments in skill formation and education of an ageing workforce, whereas at the same time economic policies at the EU level hinders national economies to take proper measures. Therefore, governments are forced to reform their labour markets and social systems and to reduce their social security spending which by itself might raise new social problems that need to be resolved.

European predominance with economic and financial coordination policies through the EMU and the SGP raises the question how the national economies deal with asymmetries in the economic shocks they face and what the consequences are for policies in the social domain at the national as well as at the European level. This section will focus on two main issues: first, on the impact of the SGP in conjunction with the EMU, i.e. of the European economic and monetary coordination on European and national social policy, and second, on the more general issue of the relationship between economic and social policies at the national and European level as well as between efficiency and equity goals. The section ends with drawing some conclusions for future European and national social policy.

2.2.3.2 The Effects of Economic Integration

EMU, SGP and the Enlargement

It cannot be a coincidence that the countries facing the strongest problems in adopting the requirements of the EMU and the SGP guidelines, especially with respect to the tight budget rules (a maximum of 3% budget deficit), have recently put to the fore strong retrenchment programmes by trying to reform and to recast the basic principles and values on which their particular social systems were build. In the spring of 2003, the German Bundeskanzler Schröder presented his plan 'Agenda 2010' to save and recalibrate the German welfare state. The basic principles on which these reforms were built can be understood as an appeal to strengthen the role of people's own responsibility in the funding of the social contract and to reduce the part of the responsibility left for the state, which is similar to leaving it more to the market, the family or the civil society. The key words in his proposals were: 'less tasks left for the state and more tasks assigned to the citizen, less easy access to and lower levels of benefits and eventually more and higher levels of contributions for the funding of the social arrangements'. There was a strong public opposition to and debate about these proposals. Eventually, the German government weakened the harshness of its proposals to get it through parliament but the route the government will follow in future was clearly set. Also France facing great difficulties in keeping its budget deficit at the 3% threshold level, presented through its prime-minister Raffarin proposals to cut down the social budget deficit and to reduce the French social expenditure level. In the 'Agenda 2006', proposals were presented for reforming the pension system and for giving more room to the privatisation of the social protection system. On the national holiday, the French President Chirac spoke to the French population with the words: 'Too long, we have lived on the premises that the state is always right. We have to try to escape from such a deadlock situation.

The state can not decide on every thing.' Also in Italy – in the fall of 2003 – prime-minister Berlusconi presented his agenda to reform the pension system, since Italy has one of the most expensive pension systems in the European Union. In the end, in all these countries the majority of the original proposals came through parliament and has been adopted.

We might conclude that within the realm of socio-economic policies a new wind seems to blow in Europe. This started already before the millennium change but the implementation of the proposed changes occurred particularly thereafter. These reform policies consist of rather far-reaching proposals for changing the social contract on which the 'old' welfare state was built. The recalibration entails the principles on which the welfare state in the 20th century was founded and pertains to more than a shift from collective public policies to more market-oriented and market conform policies only. It also involves a shift from classical notions of 'social citizenship' to notions of private and self-responsibility, from compensating people rather generously for the risks they endure to activation of citizens and from redistributing resources between current generations to redistributing resources over the life course and between current and future generations.

Viewing these political developments, one is inclined to believe that they are not so much related to the SGP and EMU rules itself but that they just reflect the ongoing longer-term changes in the economy and the demography. The EMU and SGP rules merely augment the pressure put on national governments already by structural processes of globalisation, individualisation and ageing, to reform and reconfigure their social systems. Of course, in the end governments need to meet the tight budget criteria that they agreed on in the process of monetary and economic integration in the first place, but the cause for it is more structural and fundamental.

That the ECOFIN is generally reluctant to apply the sanction mechanisms that are meant to enforce member states to abide to the rules they themselves have initiated in the first place, reduces the trustworthiness of the EMU and SGP and more generally the economic integration process itself. On the other hand, one might consider it also as the logical consequence of the narrow room that is left to member states – due to lack of discretionary power in the field of monetary, economic and fiscal policies – to respond properly to adverse economic shocks, which hit the national economies in an asymmetric way. When the labour market is not sufficiently flexible to respond to adverse shocks, as is the case in a number of countries such as in Germany, France and some Southern countries, then there is little more room left for a government than to relax its SGP rules and to allow the budget deficit to move beyond the narrow SGP-bounds at least temporarily. In the spring of 2005 agreement has been reached about the relaxation of the strong budgetary rules allowing member states, provided they have objective and undisputable reasons for not meeting the criteria, to temporarily pass the strict limits set for the budget deficit and the share of GDP spent to government debt.

The 'Integration Paradox'

If however, the asymmetries become larger – and they will, due to the enlargement process (Grauwe 2002; Mayes and Viren 2002), there exists a kind of 'integration trap or paradox' meaning that member states need more room for dealing with economic adversity but due to the integration process itself are rendered less room for tackling it properly. There are three solutions to this 'integration paradox'. The first is that the rules themselves are relaxed as happens to be the case recently in the debate about the application of the sanction rules for the countries breaking the SGP budget deficit criteria. In this case countries get more time to adapt their national policies to meet the criteria in the longer run. Secondly, the criteria itself will not be changed but the pace by which these criteria need to be met by particularly the accession countries. One might think of allowing these new member states to meet the access criteria at a different speed. Also two Dutch studies (Dekker et al. 2003; Lejour 2003) conclude that the enlargement issue in parallel with the underlying structural issues and the EU criteria poses the welfare states a major dilemma. Either the Union opts for a Europe of different speeds offering the various welfare states time and room to reform their social systems to make them sustainable for the future, or a process of dismantling of social systems can not be avoided eventually leading to a retreat to minimal American standards. Thirdly, countries in order to adapt to the tight EMU and SGP rules use the only escape routes left to them by reforming their labour market and social security systems as to make them more flexible and responsive to the needs of the economy. Then the question arises as to the interrelations between the economic and the social sector and how reform policies should be designed to make the social sector serve the efficiency goals of the economy without endangering the equity goals they also need to attain. This will be dealt with in the second part of this section.

EU Enlargement and Social Standards

Europe will have to confront the consequences of enlargement, which will lead to the entrance of new member states with much lower social standards in terms of wage levels and social benefit levels. Dependent on the labour flows it will provoke and the response to immigration policy by the individual member states (see section 2.2.1), it might be an autonomous cause for increasing asymmetries resulting from higher unemployment levels and reduced employment growth rates in different parts of the Euro area. The changes might be at the sake of the current and new countries equally but at the same time it might have adverse local and regional impacts leading to new social problems (Mayes et al. 2002).

The new members share a weaker economy, less economic growth, higher unemployment rates, but also higher interest rates than usual in the Euro zone. It is likely to become true as a result of the enlargement that EU-

27 will face larger asymmetric shocks than EU-12. It is also clear that some of the original members will more often then today be outliers in terms of inflation and output compared to the average that the European Central Bank (ECB) will be focusing on (Grauwe 2003). They then will consider the ECB policies to be less responsive to their national needs in terms of the level of interest rates they consider best to serve their national economic interests. The cost-benefit balance of EU integration will be less positive for these members and in general one might expect more tensions to arise both within and outside the Euro zone when more countries consider their interests not well served by the ECB. There is not much the ECB can do to avoid this. As Grauwe (2001) clearly shows, the only way to deal with this is to make sure that individual members have the tools to deal with these asymmetric developments.

Macro-economy and Social Policy

But apart from the effects of economic and social integration of the EU itself there are other macro-economic and social factors involved which have an impact on social policy at the national and EU level. Macro-economic factors are thought to affect social policy in three main ways:[42]

- First the increased pace of structural change stemming from factors such as globalisation is causing existing skills to become redundant more rapidly and is hence resulting in the unemployment of some individuals and the economic decline of some areas in a manner that impedes social integration. There has thus been an upward and sustained increase in the rate at which social disintegration is generated, with no indication that it might be reversed.
- At the same time, it is clear that the process of continuing change is asymmetric in the sense that an adverse shock of a given size in terms of income and wealth tends to result in more unemployment and social disintegration than a favourable shock of the same size reduces them. Thus, to prevent social disintegration from remaining at higher levels a disproportionate response is required. If this does not occur through favourable macro-economic shocks or macro-economic performance then offsetting micro-economic measures will be required.
- Lastly, the EMU in Europe itself affects the scope of macro-economic adjustment through having a single currency and monetary policy, regulating competition in the Common Market,[43] establishing rules for the conduct of fiscal policy through the SGP and coordinating other structural policies (cf. Mayes and Viren 2002).

---

[42] A survey of the whole range of factors influencing social policy issues is to be found in Mayes et al. 2001.

[43] Including the ban on state aids and discrimination in favour of certain groups in the labour market.

How EU integration might affect the performance of Europe in tackling the current social and economic problems is already discussed. The effects of globalisation (first point) and of asymmetries in the way the economy recovers from adverse shocks (second point) are entirely different issues, which in our view are closely related to the question as to how economic and social policies in Europe are combined in offsetting the social disintegration pressures posed by these structural developments. It is through the mix of economic and social policies that social issues such as social exclusion are more or less effectively and efficiently tackled. This is not to say that these problems are easy to solve. The ageing of the population, the rise of more fragile and fragmented work patterns over the life course, the rising numbers of unstable families due to marriage break-ups and the insider-outsider issue related to the advent of the knowledge economy, all need careful and joint treatment of European and national policy makers to address them and solve them humanely and satisfactorily. But, the way these nation states are trying to achieve this is likely to be dependent on the chosen interplay between economic and social policies and the policy mix. The chosen 'policy mix' in the various countries seems to differ widely, not only in the principles on which it is based or in the tools that it applies, but particularly in the distributional outcomes and its performance in terms of the goals it tries to achieve. For this reason we consider the notion of 'welfare regimes' as reflecting this diversity in national social policies in Europe to be important for our treatment of the issues covered in this book: how 'social' Europe is or might be and how 'European' national social policies are or might be.

2.2.3.3 The Relationship between Economic and Social Policy

A particular concern under the EMU umbrella is that much of the policy agenda is predicated upon a narrow definition of remits, with redistribution policy falling almost exclusively under the welfare label, while macro-economic policy is not supposed to pay heed to social policy goals. Employment policy is gradually becoming better integrated with fiscal and monetary policy, the two conventional poles of macro-economics. An excessive compartmentalisation of policy is not very helpful in attaining better fine-tuned and integrated policies (Begg et al. 2002). This is more than a plea for more 'joined-up' government in responding to social issues: the complementary challenge is to develop a conceptual model of policy making that recognises the interplay and reciprocal impact of policy areas in shaping social outcomes. We indeed believe that Sen's capabilities' approach, as explained in chapter 1.3 can serve as a conceptual tool to arrive at some basic understanding of the principles and goals of the so-called European Social Model. Though not very precisely defined, the model seems to hinge upon the idea that integrated economic and social policies are beneficial to the fulfilling of the ambitious goals the European Council agreed on at the Lisbon summit in the year 2000 and which indeed intend to achieve economic and social goals simultaneously (Muffels et al. 2002). However, the notion of a 'European Social Model' seems to disregard the wide variety in social policies and welfare systems in Europe.

Although welfare systems share some common features, they also widely vary in the goals and priorities they pursue as well as in their performance in balancing efficiency and equity goals. In chapter 1.3 and section 2.2.2 we already explained how the notion of welfare regimes relates to the idea of the European Social Model. We concluded that there is no one-size-fits-all European approach to social policy since the challenges and threats in each country need a different treatment due to the differences in the historical paths and the cultural and economic heritage. Therefore, we believe that there are many European Social Models sharing a lot of common features but also differing substantially in the design of their policy mix.

For the same reason the links between economic and social policies are complex and vary from country to country. The challenge for a thorough policy analysis is to find patterns that explain which links have a discernible impact on well-being and social integration (Begg et al. 2001). Structural changes in the economic and social contexts of the matured economies of Europe have affected the way these policy regimes pursued their distinct social welfare objectives through a particular design of the work-welfare policy mix as well as their performance to cope successfully with the strains put on them.

The main question to be addressed now is whether under the influence of European integration these various worlds of welfare will converge or rather diverge. Is Europe shifting into one European world of welfare, into one European Social Model or does it remain a collection of very dissimilar welfare states with common as well as atypical features.

2.2.3.4 To One European Social Model or Many

National or Federal

Leibfried has asked himself in a very early stage of the European debate, whether the separate member states will eventually develop into one federal state (Leibfried 1992). He calls such a federal state the United States of Europe (USE) in analogy with but at the same time with fundamental differentiation from the United States of America (USA). Differences between the USE and USA will reveal themselves particularly in the institutional and cultural dimension. In Europe many countries share a common historical tradition as far as the creation of a comprehensive set of social security arrangements is concerned – a development clearly different from that of the USA. Leibfried poses the question whether Europe will move to a trans-national synthesis of national welfare states of which European citizenship is the backbone, or whether there will be a fragmented whole of divergent forms of federal and national citizenships. One might suspect that in the end Europe will look more similar to the scanty welfare state of the USA than to the generous Scandinavian welfare states. A similar scenario is sketched by others, too (Sykes 1998). In his view, the effect of international competition and globalisation pressures might be that national governments need to shrink and wane the welfare state. What remains is a safety net just like in the USA. Many predict that social protection in the future will be more subdued to market forces and less dependent on the state. The European social model should adapt itself to these market forces instead of counterattacking them. According to this scenario, Europe would be more competitive and more efficient (Palier and Sykes 2001; Prior and Sykes 2001; Sykes 1998). However, even if one believes that attempts to weaken the attainment of a 'Social Europe' are futile, since the social and economic forces at the international level require more integration and convergence of the various national systems, it is still unclear what the integration process would imply. Would it lead to a spiralling downward (a 'social dumping' process) of the level of welfare state benefits to the least generous ones in Europe (Portugal) or an upward levelling process ('California effect') to the most generous ones (Sweden). The process of social dumping implies that countries with a relatively generous social contract would attract people from countries with more unfavourable social conditions. When these additional costs cannot be shifted forward to the wage earners, governments are compelled to lower their social expenditure and benefit levels, particularly in the case that through SGP and EMU there is little room for macro-economic adjustment policies. In such a scenario of a 'race to the bottom' all governments would move into the direction of the least generous systems (Portugal). The spiralling upward or race to the top would arise when there is a large mobility of labour meaning that the best, highest skilled employees start to work in a member state with better labour conditions. Countries with the most generous systems might attract the most productive employees and therewith obtain comparative advantages, which will allow these states to uphold their generous systems or even to improve them. That this is not very unlikely to become true might be seen from the evidence in the 2003 Employment Report of the European Commission on the growth in the proportion of high skilled immigrants (see table 2.5).

In the situation of a race to the top all systems move into the most generous system being the Swedish one. Much of what is likely to happen depends on the economic and social conditions in Europe, the extent by which labour and capital are mobile, and finally also on the role the Commission wants to play and in reality can play given the restraints of the coordination process on social policy.

A Race to the Bottom or a Race to the Top

Krueger distinguishes three reasons why differences across the regimes will continue to exists and a race to the bottom is not very likely to occur (Krueger 2000). First, it is likely to be true that the social contract does not jeopardise the general economic efficiency, as is usually presumed, but instead improves it. The private market is under certain conditions not capable to insure against the consequences of the occurrence of social risks such as unemployment. Government intervention leads in such a case to improvement of the economic efficiency, because of which increased economic integration not necessarily leads to the erosion of social standards.

Secondly, one has to address the limited mobility of labour and capital and of products and services. Even with a common currency this limited mobility will put less pressure on the national governments to lower non-competitive social standards. A good example in this regard is the limited extent of labour migration in Europe that is hardly affected by the release of the restrictions on the free movement of employees. The existing evidence also shows that the flows of capital and final products and services are too low to clear out differences in labour market policies.

Thirdly, he pinpoints to the 'median voter' argument. The rule of law represents the wishes of the median voter. If the median voter requests a high level of social protection, governments will want to take care of that, but they will shift the burden to the employee. The price to be paid for that is a lower wage (through higher taxes and premiums). The result is that national dissimilarities might continue to exist for a rather long period of time. More integration leads necessarily to higher costs for social protection. There is another reason why more integration leads to more social protection. Increased integration leads to rising economic uncertainty for which reason the quest for more social protection rises as well. The earlier mentioned Dutch study supports the conclusion that 'social dumping' is not likely to occur (Dekker et al. 2003).

Also Kleinman argues that the often cited opinion that European welfare states are dismantled through a process of social dumping is not well funded (Kleinman 2002). The facts support a different view according to which there will be stable expenditure levels and lasting divergence in range, form and content of the welfare states in the various countries. There is no evidence that the process of international cooperation will demand a certain type of welfare state to all its members. The evidence shows that welfare states over time show some convergence, especially through processes of monetary and economic integration, but the culturally inherited and historically rooted differences between welfare states are clearly sustained over a longer period of time (Goodin et al. 1999; Muffels et al. 2002).

Others believe that a Europe of the 15, and after the enlargement, a Europe of the 25, provides little reason to expect one European Social Model to develop. Even in the longer run there will not be a single European Welfare State (Hemerijck 2002). The main economic and socio-cultural arguments for such a belief are already mentioned. But there is another more political argument that is that the Union is characterised by a 'democratic deficit'. The Union is not a political union, because of which its political legitimacy is low. The obvious outcome of this lack of political legitimacy is that the citizens are not very committed to further integration. This is very clear in the framework of European social policy where further integration is perceived as a threat to the great feats of the welfare state (Leibfried 2000). This does not mean that there is any movement towards a new European Social Model. European welfare states are increasingly confronted with Europe in terms of the growing importance of European laws and directives. The activities of the EU on social policy have been expanded but remain rather fragmented and concentrated on a few areas, especially those who are related with economic integration and the labour market. One might suspect that a substantial part of social policy will remain national and that Europe will not focus on the nuclear tasks but on the periphery where there is still room to take new initiatives.

Towards a Process of Policy Learning

The conceptual tenets of this study deal with the capability and regime approach. These might, within the framework of the OMC, provide an innovative route for designing policies to tackle the major social issues in the decades to come. The question is how policy regimes try to cope with the major challenges and shortcomings of their labour markets and how successful they seem to or might be in attaining the goals they pursue given the changing economic and social contexts and the role played by the EU in the macro-economic field.

Policy regimes in the member states differ in the way they respond to the advent of the EMU in Europe and its influence on the scope for economic adjustment to adverse shocks through having a single currency and monetary policy, and more or less coordinated economic and fiscal or other structural policies. Eventually, they differ in the way they react to political forces at the European level to achieve more coordination in the social field, as well by establishing a Social Europe along the EMU. To attain a Social Europe might not be taken for granted, since many countries are reluctant to extend the scope for European interference beyond the borders of economic and monetary issues. But it is inevitably true that EMU and the EU enlargement, together with the structural changes in the economic and social domain, will confront the EU with new challenges, which it has to cope with and which ask for clear decisions about how far the EU wants to go in fostering integration beyond the economic and monetary borders. The reluctance of national governments to impair their national authority in the social domain might be held responsible for the advent of the OMC that was adopted in the Luxembourg process, and applied with quite some success, already for three years now in the employment area.

The Open Method of Coordination in the Social Domain

At the Nice summit of the European Council in December 2000 the OMC was also agreed on as the way to proceed for the social agenda of the EU on the road to enlargement. Its success in the policy area might likely be attributed to the process of policy learning that is inherently connected with the OMC process. At the Nice European Council, held in December 2000, it was agreed on to ask member states to develop NAPs for promoting social inclusion. The model for these was, clearly, the EES, which, similarly, requires member states to present and implement NAPs for Employment. The approach underlying both policy areas has come to be known as the Open Method of Coordination since the expression was coined at the Lisbon European Council in March 2000. As we stated earlier, the OMC process represents a new form of policy integration, which by being vested on the subsidiarity principle leaves member states just enough room to implement their own plans and policies.

In the social domain, the Commission's legislative power is indeed governed by the subsidiarity principle implying that the role of the Commission to manage the coordination process is restricted because the member states are considered primary responsible for the design and implementation of welfare state policies. This does not mean that there is no role to play for the European Union in the social domain, because in its positive meaning subsidiarity implies that European interference is asked for when it entails positive effects for the separate member states and for Europe as a whole, which otherwise would not have occurred (Fouarge 2004). Some argue that the coordination process already entered an irreversible stage with the launch of major social initiatives at the Lisbon and Nice European Councils in March and December 2000. During the Belgium presidency in the second half of 2001 the Belgium Minister Vandenbroucke expressed high expectations about the OMC in the social field. It is in his view 'the most promising way to give concrete shape to social Europe' and 'to firmly anchor it into the European coordination process as a common good'.

The challenges Europe faces due to emerging new risks stress the need to adjust the European Social Model and even to formulate a new paradigm for social policy. A recent co-edited book by Esping-Andersen (2002) got the illustrative title "Why We Need a New Welfare State" to express that Europe needs a new social architecture to create an 'active and dynamic welfare state' – as it is coined at the Lisbon summit – to cope with these emerging new risks. Esping-Andersen too sees the OMC as a new promising tool to arrive at this ambitious goal. Some cast doubt what the OMC might contribute to the integration process. Chassard (2001) raises the question of the democratic legitimacy of guidelines drawn-up by unaccountable officials in so politically sensitive a policy area, however well intentioned the proposals and the underlying diagnoses are. A risk in this regard is that the EU would be used as a scapegoat when the process is not accompanied by proper legislation at the European level (Begg et al. 2002). This threat is not unreal in a situation of economic slowdown when national governments are less inclined to improve existing arrangements or to create new arrangements. During the Belgian presidency, a first round of the OMC with respect to social inclusion has been achieved. In the meantime, the method has been applied already for some years in the employment policy domain and since 2001 also in the social domain. The experiences with the application of the method in the employment area are generally positive. The experiences with the OMC in the social domain are too short to draw any serious conclusions. The OMC might serve the goal of motivating the individual member states to reform, modernise and recalibrate their systems of social protection towards more efficiency, flexibility and sustainability in the light of ongoing structural changes. This brings us to the final issue to be dealt with. How would such a European wide reform look alike and what would be the main ingredients of the proposals?

2.2.3.5 Reconfiguration of the Welfare State

The European Commission, as agreed on in Lisbon in 2000, aims at the modernisation of social protection systems so as to make them more responsive to the challenges of the worldwide economic developments and the knowledge economy. In the previous parts of this section we pointed to the various policy scenarios that might be foreseen for the future of a Social Europe. Do we have to fear a retreat of the welfare state to levels similar to the low American standard? Does the welfare state especially the ones in continental and Northern Europe need to reform and to develop in the direction of a subsidiary or residual welfare state because the current levels are not sustainable anymore in the light of the trends to individualisation and the ageing of the population? Does the choice for more Europe in our national social policy mean that the worst-case scenario will be likely to become true, implying that the national welfare states converge to the worst system in Europe (Portugal)? Or do we believe in the best-case scenario according to which the European regulatory process leads to the California effect and the systems converge to the best, most generous one in Europe (Sweden)?

Or do we gamble on the success of the OMC that in the end should lead to the realisation of one European Social Model? The latter option does not imply that everything has to be settled at the European level. Europe needs to formulate the framework but the elaboration and implementation of the guidelines have to be left over to the national member states leaving enough room for national policies and within the member states for policies at subnational or even local tiers of government. Eventually, social policy can be left over entirely to the national member states. The role of the European Commission will be reduced and national systems keep following their national, historically rooted socio-economic development paths. This does not preclude that also in this scenario there will be policy competition between the various systems in Europe. Regardless of which of these routes will be followed in the end, each of them has to define answers to the questions and challenges posed to the European societies of today.

The Transformation into the Welfare Society and the Activating Welfare State

Some observers of the welfare state believe that the welfare state should transform in a 'welfare society' (see chapter 1.3). This perspective paved the way for the notion of the 'activating welfare state', which received support in a number of European countries. The way this debate came up in the 1990s reflects the transformation into modern complex societies of the late 20th century. In chapter 1.3 we asked ourselves whether this means a retreat to 'minimum welfare states' (the 'Night watch' state) or reflects a new configuration of the modern welfare state. In this new stage of policy making in the social field, governments tried to make working more attractive by 'making work pay' and by increasing the flexibility on the labour market. The way the 'active or activating welfare state' is conceived has changed due to the shift into the 'knowledge economy'. This third stadium of the development of the welfare has just been set in motion but it is clear that in some countries this already has inspired politicians to strong reforms.

Implementation and Competence at Sub-national and Local Level

Another issue deals with the level of implementation of the various activities. Social policy asks for implementation at the local and regional level. It implies a flexible approach to issues that need to be tackled at sub-national or decentralised tiers of government. Therefore, discretionary power might be transferred to these lower levels of implementation and local partnerships might be developed to make this endeavour a success. Local partners, just like national partners in the OMC process learn through peer review of each other's 'good local practices'. Such an approach might function as the local pendant of the OMC at national level.

But history has learned us that changes in policy are seldom revolutionary but mostly evolutionary. Institutional inertia is one of the major barriers for fundamental changes (Hemerijck and Visser 1999). Since the preferences and ideas connected to particular welfare state regimes are internalised by the populations of the various member states of the European Union and are also deeply engrained into the institutional structures, the spreading of and adjustment to new forms of social policy will always be gradual and incremental and need to fit within the existing institutional possibilities and policy options. Another barrier is internal disagreement and conflict at the European level and lack of coordination, which will particularly arise when the Union is hit by adverse and asymmetric economic shocks which will hit the various member states in a different way and to which member states will respond differently. What is most likely to happen then is a Europe in which the various clusters of welfare regimes differ in the speed by which they reform or dismantle their social security and caring arrangements or a Europe moving towards the liberal American approach with Spartan levels of benefits of very limited duration.

Quo Vadis?

The future is always shrouded in mystery and changes tend to develop along unpredicted lines. Reforming the welfare state along the lines of activation, investment in people's capabilities and a stronger focus on human and social capital formation might be profitable particularly in the longer run. The reforming of the system along these lines is also required to reinforce the anti-cyclical character of social spending. The basic idea is to build bridges between out-of-work and in-work statuses to smooth the adaptation process to changing economic conditions and to make the system more flexible, responsive and less exclusive (Schmid and Schömann 2003). But in the end, the most important motor of change is undoubtedly the economy itself. If the economic climate is steadily improving and social policy is better tuned with economic policies, the conditions might be set for more optimistic scenarios. Europe has contributed a lot in the past fifty years to make it a better place to live in than it otherwise would have been and Europe will likely play an important role in the future as well. There is no reason to say farewell to the European Social Model as it has been built up in the past even though it is likely to become true that the national systems will remain very different. There will be no single European model but there will be many all with their particular flaws and strengths. Hence, rather than to attain convergence in the near future, Europe will likely remain multifaceted.

2.2.4 European Legal Framework for National Social Policy (in Particular: Fundamental Freedoms and Competition Law)

2.2.4.1 The Legal Framework's Implications for Social Policy

The European Community is – in any event, also – a legal community. It was established by act of law and, with its legal instruments, affects the policies of the Community institutions, the member states and Union citizens, notably through extensive legislation in the form of regulations and directives. Yet the European Community is a legal community also insofar as the actions of its institutions and its member states are reviewed by a court, the ECJ, to ensure their compliance with the – quasi constitutional – foundations of the Community, the European treaties. Community social policy, like any other policy, must observe these provisions, which first of all address the question of competence but also set forth how Community law and the activities of the Community institutions affect the member states. At the same time, the Community can shape this social policy by further developing the framework. This aspect is dealt with in chapter 2.3.

2.2.4.2 On the Way to a European Constitution

The European Communities were established and enlarged by treaties of international law. These treaties have been amended and adapted again and again. Even if the EC Treaty is often referred to as a constitution, it differs from the national constitutions alone in terms of its origin. Above all, however, the European Community lacks statehood. Thus it can only act on the powers conferred upon it by the member states. To date, the European treaties have therefore been restricted to the regulation of individual aspects, conforming to the principle of limited individual authority. In particular, a fundamental rights chapter has been missing so far. The treaties do not set forth citizens' personal liberty rights, nor do they stipulate rights of political and social participation. As for citizens' protection against encroachments on their legal rights through European Union institutions, the ECJ has developed a protective realm by analogously invoking the member states' nationally guaranteed personal liberty rights, thereby also referring to international fundamental rights such as those enshrined in the European Convention on Human Rights.

As for social rights, these still lack legally binding normative substance under European Community law. Notably, the 1989 Community Charter of the Fundamental Social Rights of Workers merely remains a political programmatic declaration by the EU member states from which concrete personal rights cannot be derived.

In the wake of EU enlargement, which, among other things, made it seem imperative to facilitate Community decision-making, a reform of treaty foundations has been embarked upon. This task was prepared by the European Convention in drawing up the draft Constitutional Treaty, which has been approved by the representatives of the governments of the member states and must now be ratified by the states themselves. Following its ratification, the Treaty establishing a Constitution for Europe – a designation that continues to reflect its international law origins – is to enter into force on 1 November 2006 (Article IV-447). The Treaty consists of four parts, with Part I outlining the Union's organisational structures and Part II setting forth the Fundamental Rights Charter, while Part III focuses on the policies and functioning of the Union and essentially reiterates the text of the existing EC Treaty; Part IV summarises the general and final provisions.

Of prime consequence to social policy are the fundamental rights set out in Part II as well as the specific provisions governing Community social policy in Part III. These fundamental rights are already addressed in Part I, Article I-9, which in paragraph 1 declares that the Union "shall recognise the rights, freedoms and principles set out in the Charter of Fundamental Rights which constitutes Part II". In the second paragraph of Article I-9, the Union declares its accession to the European Convention for the Protection of Human Rights and Fundamental Freedoms, and thus also to the procedure whereby these rights can be judicially enforced.

Part II of the Constitutional Treaty begins with fundamental human rights and personal freedoms (rights to privacy from state intervention); under the heading "Solidarity", Title IV proceeds to outline fundamental social rights, for example the legal, economic and social protection of the family (Article II-93), or the Union's recognition and respect of entitlement to social security benefits and social services (Article II-94).

Concerning the field of application of fundamental personal and social rights, Article II-111 states that the provisions of the Charter are addressed to institutions of the Union and to the member states only when they are implementing Union law. It remains to be seen what significance the rights will attain within this scope. A deciding factor will be how the Charter's norms are interpreted by the ECJ. A potential gateway for an extension of Community law may thus have been created, although Article II-111 expressly demands due regard for the principle of subsidiarity.

Part III on Union policies addresses, among others, employment policy (Article III-203 et seq.) and social policy (Article III-209 et seq.). In framing social policy, the fundamental social rights set out in the European Social Charter of 1961 and the 1989 Community Charter of the Fundamental Social Rights of Workers are to be borne in mind by the member states (Article III-209[1]). Here again, it is not quite clear what legal relevance will unfold from this reference in conjunction with the provisions of the Charter of Fundamental Rights of the Union set out in Part II. Moreover, the Constitutional Treaty introduces a new regulation pertaining to social policy in that it expressly lays down the Open Method of Coordination (Article III-213).

2.2.4.3 Lack of General EC Competence in Social Policy

The catalogue of competences conferred upon the EC does not include social policy. The Constitutional Treaty, though regulating employment and social policy issues, basically does not change this. This area of competence, in principle, stays with the member states, meaning that social policy remains a national matter. That is not to say, however, that European Community law is indifferent to social policy issues. On the contrary, a main objective of the Union, set out in Article 2 of the Treaty on European Union (framed in Maastricht), is to promote social progress and a high level of employment and to strengthen cohesion. Also the consolidated version of the Treaty establishing the European Community affirms in its preamble "as the essential objective of [the member states'] efforts the constant improvements of the living and working conditions of their people". Over and above this, Article 2 lays down as Community tasks: to establish a high level of employment and of social protection, the raising of the standard of living and quality of life, and the promotion of economic and social cohesion. Corresponding formulations are found in the Constitutional Treaty.

Furthermore, primary Community law – that is, treaty law – also embodies several social policy regulations governing employment (Title VIII), social policy, notably the European Social Fund (Title XI), and public health (Title XIII). The non-addressed sectors and the non-regulated issues in those sectors which are addressed (these being the large majority) remain within the scope of member state competence, quite apart from the fact that even an existing area of Community regulatory competence can be restricted by the principle of subsidiarity.

The Constitutional Treaty defines the principle of subsidiarity in Article I-11, stating that "in areas which do not fall within its exclusive competence, the Union shall act only if and insofar as the objectives of the proposed action cannot be sufficiently achieved by the member states either at central level or at regional and local level, but can rather, by reason of the scale or effects of the proposed action, be better achieved at Union level." Further procedures on the application of the principles of subsidiarity and proportionality are detailed in a protocol which the national parliaments are called upon to comply with (Article I-11[3] of the Constitutional Treaty). The long-range effects of the subsidiarity principle are scarcely foreseeable at present since it still lacks a precise legal definition.

2.2.4.4 Specific European Legal Regulations Affecting Social Policy

The Community Coordination Rules Governing Migrant Workers

One of the Fundamental Freedoms guaranteed by the EC Treaty is the free movement of workers within the Community. Such freedom of movement from one state to another can only be practised if it does not entail grave disadvantages for individual citizens. With social security systems geared to national territory (territoriality principle), such disadvantages would thus arise for the social protection of migrant workers in the absence of relevant rules. To this end, the European Economic Community (EEC) had quite early on adopted a system of coordination to preclude these disadvantages (EC Regulations 3 and 4, and later Regulations (EEC) No 1408/71 and No 506/72).

Particular regulations apply to frontier workers, thus often forming the basis for the harmonisation of social infrastructures in border regions and serving as a model for the approximation of differing social orders. In the course of time, the coordination of social security for migrants has been extended to include further groups of persons, notably the self-employed. With the accession of new states to the EC, additional systems had to be incorporated into the coordination procedure, making it ever more complex and lengthy. That has made the legal provisions increasingly incomprehensible to the EU citizen and exceedingly difficult for legal authorities to implement. This complexity is reflected in numerous judgments of the ECJ that have repeatedly met with rejection and a lack of understanding in the member states. Indeed, there is the longstanding discussion to simplify coordination legislation; yet the EU simultaneously faces renewed enlargement through the pending accession of the Eastern European states, which will heighten complexity even further. Moreover, the extension of coordination rules to third-country nationals gainfully employed in an EU member state has also been debated for years. If such persons legally reside in the EU, they are now included in the scope of the coordination regulations by way of Regulation (EC) No 859/2003.

The all-embracing aggregate reform of the coordination rules has meanwhile made considerable progress. The hitherto applicable Council Regulation (EEC) No 1408/71 is to be replaced by the new Regulation (EC) No 883/2004 of the European Parliament and the Council of 29 April 2004. It entered into force by law in May 2004, but will not become effective until the new Implementing Regulation has been adopted. One of the most important amendments is the extension of the scope of coordination law to non-employed persons. Also, numerous substantive matters have been newly regulated based on the case law of the Court of Justice and on developments in the member states.

The difficulties of achieving smooth coordination always lead back to the original problem of European social security law. Coordination is thus only required where differing protection schemes exist, meaning that a uniform social security system for the entire Community would make coordination rules within it superfluous. Then only immigration from outside the Community, and vice versa, would have to be regulated. Despite problems posed by the coordination system, the very widespread view in the Community is that harmonisation – a standardisation of schemes – is neither possible nor desirable in the foreseeable future. The European Commission has therefore abandoned its previously envisaged goal of accomplishing such harmonisation. Given independent developments in the social security systems of individual member states and the mental and emotional ties of the respective populations to their own system, general standardisation would meet with extensive resistance in all EU states. Nonetheless, the European Commission has been striving for years to achieve more convergence in social policy developments within the EU. Meant here is that, while basically accepting the different systems, reform measures required in all member states are to be made known through a mutual broad exchange of experience, thus aligning further development – without the need to enforce such development by law. It is to this end that the so-called Open Method of Coordination, discussed below, has recently been introduced. The term is rather ill chosen since the point is not to create rules for reciprocally defining the organisation of social security systems through conflict-of-laws provisions, as in the case of the technical coordination rules adopted for migrant workers to avoid social security disadvantages when moving from one state to another. Rather, the OMC seeks to enrich the social policies of the member states by way of European comparison and benchmarking procedures. Hence, the method constitutes a social policy instrument, not a legal technical one. In the light of the differing usage of terms (e.g. harmonisation, convergence, coordination), it seems recommendable to define them in the respective contexts.

Individual Provisions of Community Law Bearing Direct Social Policy Relevance

As pointed out above, the EC Treaty embodies a host of provisions pertaining specifically to social policy, so that their relevance is easily recognisable. In addition, however, there are a number of provisions whose social policy nature is not clearly discernible at first glance, but which nevertheless unfold direct social policy effects. The most important example here is the principle of equal pay for male and female workers, originally addressed in Article 119 of the EC Treaty and now enshrined in Article 141. This principle is applied also to social benefits, such as occupational pensions and others, and plays a substantial role in the case law of the Court of Justice.

Other Provisions of Community Law Bearing Indirect Social Policy Relevance

Social security is integrated into the overall economic and legal order. This postulate applies without question to national law, even if in reality it is often not fulfilled. At European law level, owing to an only partial assignment of competence to the Community, specific difficulties are apt to arise where Community legislation, such as the Fundamental Freedoms or EC competition law, coincides with legal areas that have remained within national legislative competence, as is often the case with social policy. Community-level legal provisions may thus collide with the structures of nationally regulated areas or, at any rate, fail to interact harmoniously with them. Such conflicts could be avoided by restricting the application of EC legal provisions to areas already subject to Community competence. That, however, would mean giving up the demand for a uniform legal order where it affects the interplay of legal areas regulated by Community law with other legal areas remaining in the national sphere of competence. This difficulty, which at first appears highly abstract, has acquired practical substance and clarity through the case law of the Court of Justice in applying Community law principles to social security issues. Hence, in the following, two main fields of application are to be highlighted: first, the scope of the fundamental freedoms as regards healthcare services rendered by compulsory health insurance and, second, the applicability of EU competition rules to social insurance institutions (notably, statutory sickness insurance funds).

These examples concern social security systems at various levels. One aspect is that social benefits are awarded not only in the form of money but also as benefits in kind and services, especially in the healthcare sector. In such cases, social benefit institutions must acquire the in-kind benefits and services, or – by way of exception – must furnish them themselves. Consequently, they operate as demanders in the healthcare market, which in fact is split up into many sub-markets. As will be shown below, Community law may become effective in these markets. That can also occur at another level, namely when social benefit institutions compete with the suppliers of insurance benefits.

EU Economic Law and Social Policy

- The fundamental freedoms under EU law and their effect on compulsory health insurance:

  The core provisions of European Community law include the Fundamental Freedoms – besides worker mobility, the free movement of goods, persons, and capital, as well as the freedom of establishment and the freedom to provide services. The Fundamental Freedoms were and still are the regulatory focus of the treaty establishing the Common Market. They first of all apply to trade and commerce, but can also be of significance to the social sphere. That becomes especially clear for the healthcare sector. Benefits to health-insured persons are predominantly awarded in the form of in-kind benefits and services. The rules according to which this occurs are laid down by the respective national governments. Thus national law determines who is able to provide health benefits, how these benefits must be structured and whether they may also be claimed in the event of a stay abroad. Proceeding from the territorial limitation of social security systems (so-called territoriality principle), the longstanding accord was that health insurance benefits could only be claimed in the national territory, exceptions being subject to special regulations. Such regulations are found in bilateral social insurance agreements and in the regulations governing social security for migrants (Regulations (EEC) No 1408/71 and No 571/72). According to Article 22 of Regulation 1408/71, medical treatment abroad requires the expressed authorisation of the competent insurance institution. Beyond these specially regulated cases, the territoriality principle was applied strictly in the past. The ECJ has revised this legal situation, having ruled in the Kohll and Decker Cases, for example, that any territorial limitation is admissible only on specific justifiable grounds.

  The Kohll Case (C-158/96) concerned a Luxembourg citizen who without prior approval from his sickness insurance fund had received dental treatment (braces) in Trier, Germany, and then demanded a cost refund from his insurance institution. The Court of Justice held this claim to be justified, referring to the EC rules on the freedom to provide services. In the Decker Case (C-120/95) an insured person, likewise from Luxembourg, applied for a cost refund for spectacles he had purchased from an optician in Belgium. This claim, too, the Court deemed justified under reference to the free movement of goods (regarding both judgments from the extensive literature on the subject see Zerna 2003).

  Restrictions on the claiming of health benefits abroad are acceptable only if they are imperative for the fulfilment of public service obligations to ensure social well-being. Such grounds can be given if, for example, territorial restrictions are required to ensure the financial sustainability of a social security system. The protection of public health in a member state would be a further justification. As to hospital treatment, which in many states is subject to extensive requirements planning, the Court invoked such grounds to justify a restriction of the freedom to provide services in the Case Geraets-Smits and Peerwooms (C-157/99). This restrictive judgment on hospital treatment has recently been confirmed in the Patrizia Inizan Case (C-56/01).

The pertinent case law, though not yet elaborated for all types of benefits, nevertheless clearly indicates that:

- the effects of the Fundamental Freedoms cannot be restricted to the policy areas already entrusted to Community competence, meaning there are no general areas of exemption;
- the differing levels of Community competence can lead to frictions that may restrict the national legislator's freedom of action;
- policy-makers, to a greater extent than previously, must examine whether Community law may come to bear on their decisions or perhaps even thwart them.

New conflicts between the fundamental freedoms and national social law are likely to arise if the proposed Directive on Services in the Common Market enters into force. According to the draft, quality standards for services offered across borders are then to be based on the "country of origin" principle. There are fears that especially in the case of social services, it will not be possible to maintain national quality standards (Europa Social Report 2004: 7).

- Community competition law and social insurance:

Another example of the European Community's indirect influence on national social law is the application of EC competition rules to institutions under social law. Problems arise where initially – in any case, fundamentally – clear-cut distinctions between state social security and the private sector become increasingly blurred, thus prompting questions on the extent to which rules governing market forces (competition law) may also apply to social security. That equally embraces both national and supranational competition legislation. In the main, four trends have led to this obscuring of boundaries:

(1) A privatisation of social security is now advocated with the aim of providing additional private coverage, possibly through tax incentives. Notably the difficulties in securing state pensions in the face of demographic developments generate demands for personal old-age provision through private capital formation. The relevant instruments may take the form of private insurance but also include labour law.

(2) Moreover, statutory social security institutions show a tendency not only to avail themselves of public law instruments but also to operate in the market, for instance by acquiring benefits there to be placed at the disposal of the insured (medical treatment, rehabilitation, medi-

cines etc.), thus acting as demanders of services. In this way they compete with private demanders, which in turn can imply the application of economic law.

(3) The actions of social insurance institutions organised under public law can also be of relevance to competition law if they compete with private insurance companies and must therefore be regarded as undertakings in functional terms. Thus social insurance institutions may offer a form of coverage similar to that furnished by private insurance companies. This can result in a competitive relationship between private and social insurance on the supply side. An example that has occupied the ECJ is job recruitment for executive employees. The monopoly originally enjoyed by the German Federal Employment Office was declared unlawful for Europe in view of the existence of private recruitment consultancy companies (Höfner and Elser Case C-41/90).

A current example is the option available to statutory sickness funds since 1/1/2004 to offer supplementary insurance for benefits no longer covered by the statutory benefit catalogue (cf. v. Maydell and Karl 2003). In future, supplementary insurance for dentures is to be supplied by statutory sickness funds in parallel with private health insurance companies.

Previous distinctions between private and social insurance have been further amended under the Act on the modernisation of compulsory health insurance ("GKV-Modernisierungsgesetz") insofar as statutory sickness funds have been empowered to act as mediators between their insured and private health insurance companies in the field of supplementary insurance.

(4) Finally, the boundary between state social security and private market activities has been made more permeable in that the legislator has given more wide-ranging scope to private forms of protection, such as private insurance, thus minimising differences to state social insurance. The private form of long-term care insurance is an example here.

In principle, all persons insured against sickness are also subject to long-term care insurance. For members of private health insurance schemes, the private sickness fund is also responsible for providing long-term care coverage, with contributions and benefits largely defined by the law-maker. Medical risk assessment, a tool typical of private health insurance, and premium calculation based on concrete risk do not take place.

If state social security is approximated to private forms of provision, the very consistent question is whether the rules governing the market, notably competition law, become applicable. That is relevant to both national and supra-national law, given that German social insurance institutions competing with private insurance companies will also affect foreign competitors' opportunities for accessing the German market. In that case, national rules favouring state institutions, such as the placement monopoly of the German Federal Labour Office, infringe Community competition law.

Also, the financial benefits such institutions receive from the state would have to be qualified as inadmissible aids within the meaning of EC law. In a more recent judgment (Case C-280/00 Altmark Trans), the ECJ developed criteria for assessing when state funding for the provision of services of general economic interest is to be qualified as state aid under Article 87 of the EC Treaty, and when it merely represents "compensation". Yet even after this judgment, drawing the line to state aid remains difficult. Moreover, the aspect of inadmissible cross-subsidisation plays an important role here. This development could obstruct or at least hamper the social aims pursued via social insurance institutions. And that would substantially impair social policy.

This subject is currently under discussion within the European Union, above all under the key term of "services of general interest", which, besides the necessities of life (natural gas, water, communication etc.), include social services extending from voluntary welfare to social insurance. To what extent the European Union, notably the Court of Justice, will in future widen the scope of national rules in this domain is not yet foreseeable. Should they fail to do so, it will become necessary to confine social institutions to their core functions and thus to ensure that social insurance agencies, for example, cannot be qualified as undertakings and, hence, as competitors in the private sector. Consequently, the prerequisite for the application of Community competition law would cease to exist.

In each case, it will be necessary to distinguish the economic frame of reference within which social insurance institutions operate. On the one hand, that may involve their competitive relationship to other insurers. On the other hand, statutory sickness insurance funds, for instance, could be regarded as economic undertakings if they participate in fixing refundable prices for medicinal products. In this context, the ECJ recently denied German statutory sickness funds the status of undertakings in a judgment (Case C-264/01 AOK Bundesverband and Others) concerning the admissibility of setting fixed amounts paid by the funds towards the cost of various types of medicines.

In view of the above-outlined tendencies, differing positions emerge for future social policy activities. It is conceivable that differences between personal insurance and social insurance will be levelled on an increasing scale and ultimately abolished altogether. The development in the Netherlands demonstrates a possible process here. After such a levelling of differences, there would no longer be any cause or option for not applying to public institutions the competition rules governing the private sector. And that would have the advantage of exerting pressure on competitiveness and, hence, on the need to rationalise and operate efficiently. Conversely, it would also be feasible for social insurance to stress its distinctions, in particular its task of social equalisation, and to focus on benefits characterised by this equalising element.

Any mixing with functions customarily exercised on a free-enterprise basis leads to dilution and poses a threat to social policy objectives – with the result that competition law must be applied. The case law of the ECJ vividly illustrates this development, which is also reflected in national law.

2.3 The European Union as a Social Policy Agent

2.3.1 Starting Point

Section 2.2.4 has shown that basic social policy competence is still left to the member states, but that the European Community treaties have created a legal framework which is also relevant to the social legislation and social policies of the member states. This leads to interaction between state and supra-national policies, necessitating an elaboration of rules and principles. Such interaction is decisively influenced by the way in which, and the extent to which, the European Union conducts itself as a social policy actor framing its own social policy. That it is able to do so follows from the Community's primacy over the member states when it acts within the frame of the EC Treaty.

The analysis of the European Union's social policy activities must proceed from several angles of differentiation:

1. The European Union acts through its institutions, whose functions are exercised under differing legal forms but may also differ in terms of content. In so far, distinctions can be drawn between the social policy of the European Parliament, the Council of the European Union, the European Commission and the ECJ, whose rulings may have social policy effects.
2. The EU can shape social policy by adopting legal norms; yet also other activities, located somewhere between soft law and political declarations, may constitute a form of EU social policy action, say, through communications. Further instruments are the use of funding, for instance within the scope of the European Social Fund.
3. On the side of the member states, European Union activities can impact the social policy of the state itself, or of its regions or federal states and municipalities, but also that of relevant institutions organised under private law (private insurance, employers in respect of company social policy etc.).
4. European social policy, if understood as a policy shaped by the Community institutions, is not static but develops in a dynamic process. Nevertheless, this development has not been constant, but sporadic in past decades. This point is elaborated below.

2.3.2 National and/or Supra-national Social Policy – Courses of Development

2.3.2.1 Setting the Course upon Establishing the European Economic Community

As with the European Coal and Steel Community, the prime aim of the Economic Community was to create the Common Market. Whether this was to include the stepwise harmonisation of social policy (in the sense of its standardisation) – as addressed in the final communiqué of the preliminary

Intergovernmental Conference of Messina (1–3 June 1955) – was contested among the founding states. France had voiced this demand on account of its relatively high social security contributions compared with the other states, since its high ancillary wage costs were feared to place the French economy at a competitive disadvantage in the envisaged Common Market. The other founding states, notably Germany, opposed this demand, asserting that social expenditure, besides the tax burden and other location factors, was only one of numerous cost factors to be considered within the Common Market frame.

The dispute was ultimately settled by a compromise elaborated jointly by France and Germany. In essence it was agreed that, apart from freedom of movement, the EEC Treaty would not lay down any competence rules in the social policy sphere, but instead affirm (in the preamble) as the essential objective of the Community the constant improvement of the living and working conditions of its peoples. This objective was to be attained by way of the Common Market, which was hoped to facilitate the coordination of a social order, as well as by the procedures set out in the Treaty (Article 117 EC Treaty) and by approximating legal and administrative provisions. Article 117 of the EC Treaty (in the original version) did not, however, establish any Community competence in the area of social policy legislation.

2.3.2.2 Long-term Trend: Strengthening the Social Policy Activities of the EC

The result of this compromise was that in the ensuing decades until this day, social policy has remained a matter of member state competence; simultaneously, however, manifold activities on the part of the Community institutions can be recorded in this area, their varying intensity depending on the general political parameters. Viewed in retrospect, the social dimension of the EC has nevertheless constantly been enhanced. Decisive to this development was that the non-inclusion of a common social policy, while extending Community competences to ever more policy areas, proved an obstacle to strengthening the Community. Apart from that, the elaboration of EU citizenship was placed at the centre of this development and given a social policy component. Another factor has been that areas already subject to Community competence affect the social sector, thus influencing it indirectly. This influence is especially pronounced in economic and monetary policy. Owing to the Common Market and the single currency, differing costs of production within the Community impact the competitive situation in the Single Market. That applies in particular to social security contributions that increase labour costs and result in higher prices. Consequently, all states with high ancillary wage costs come under pressure to lower these to ensure the competitiveness of their own industries; this can only occur by changing national social security systems. Also the SGP that accompanied the introduction of the single currency brings influence to bear on national social policy by encouraging states to cut social security spending in order to comply with the stability criteria. Finally, the growing complexity of coordination law shows that harmonisation, at least in some instances, could lessen the difficulties inherent in coordination (on the development of European social policy see Schulte 2003). Furthermore, once the Treaty establishing a Constitution for Europe has entered into force, it is expected to strengthen the Community element; the Fundamental Rights Charter could thereby enhance the legal position of Union citizens in terms of social law if it is interpreted accordingly by the ECJ.

The heightened social policy profile of the Community affects all national policy actors by narrowing their field of influence. That applies to the German federal government as well as the regional and municipal authorities. Concerned here, to name just one sector, are the municipal facilities providing services of general economic interest, including electricity, national gas and water utilities, transport facilities and the like, but also social services (e.g. old-age homes and hospitals). The Community Fundamental Freedoms and competition law bear consequences for municipal social benefit provision in that its organisation is being "opened up to competition" (cf. Pitschas 2003).

2.3.2.3 Objectives, Contents, Institutions, Instruments

To further characterise European social policy, one can also ask about its intent and purpose, means and actors.

Objectives

European social policy can serve to accomplish diverse objectives that are often interwoven with each other. An initial concern can be to create the instruments for "Europeanising" social policy. Here, coordination law is the prime tool for achieving the free movement of workers – one of the Community Fundamental Freedoms. Similarly, efforts to attain convergence are not first and foremost geared to substantive social matters, but to reducing disparity between the member states' social security systems, here again constituting an instrumental aspect. The aim can – and, as a rule, ultimately will – be to shape social conditions in their substance. Thus coordination law serves to enhance the social protection of migrant workers. Efforts to set minimum social standards for all citizens of the Community aspire to the same objective.

Contents

In terms of content, the social policy efforts of the Community are concerned with coordination, convergence and the securing of minimum social standards. Coordination establishes rules to permit interaction between the various national social orders so that migrant workers will not forfeit their acquired social status when moving to another member state. Thus a person insured in a member state will receive social health insurance benefits in another member state upon falling ill there. In old-age provision, to name another example, insurance and employment periods are aggregated, say, to determine whether the minimum period of employment has been attained. Coordination requires the adoption of additional legal provisions to safeguard this effect.

European convergence policy seeks to influence national social policies so as, ultimately, to approximate them with the aim of creating uniform living conditions within the Community. Convergence policy may also consist in accepting existing disparities to ensure that at least minimum social standards are complied with – a goal that is likewise pursued by numerous international activities, above all the Conventions framed by the ILO.

Institutions

The EU institutions conform only in part to the political organisation of states. That is because the EU, as a supra-national organisation, is founded on an international act of law and can only exercise state authority if the respective powers have been conferred upon it by the member states. Given that major competences remain with the member states, the Council, consisting of the ministers representing the various governmental policies of the member states, plays a decisive role, above all in the legislative process. By comparison, the European Parliament's importance is less pronounced. Although its competences have been extended in recent years, especially as regards its control function, the position of the Parliament remains rather weak. That applies above all with regard to legislative powers, which largely rest with the Council. What is still lacking is a uniform parliamentary formation of intent within the EU – which is of particular relevance to so central an area as social policy. Without the possibility of debating and decision making in Parliament, a uniform European social policy is scarcely conceivable.

While the member states are represented in the Council, the actual Community element is expressed by the European Commission. Although the Commission depends on its cooperation with the Council and Parliament, it has in the past proven to be the central supra-national institution for conducting, advancing and deciding Community policy. Through its own investigations, the commissioning of experts, and the drafting of programmes and action plans, the Commission has furnished important bases for further developing European social policy.

Finally, European social policy is also determined by the Court of Justice of the European Communities, since the Court is responsible for interpreting the European treaties and other supra-national legislation. The Court of Justice consists of one judge from each EU country. Its rulings are given in chambers (3, 5 or 11 judges) or in the plenary (only in cases of exceptional importance). In 1989, the Court of Justice was reinforced by the Court of

First Instance, which is entrusted with a range of specific competences. Beyond that, there is no further division into panels based on subject matter, meaning that the Court of Justice adjudicates in all legal fields. Its main task is nevertheless to interpret the European treaties, acting on the basis of orders of reference submitted by national courts. More than half of all proceedings deal with such preliminary national rulings. Further types of procedure include: treaty infringement proceedings taken by the Commission against a member state; action against a Community institution to have a decision declared void; and legal action for public liability claims.

On the whole, the Court of Justice resembles a constitutional court – at supra-national level. Similar to constitutional courts, its chief role is not confined to mere legal review; it also exercises a political function. In the past, the Court sometimes perceived itself as a motor of the Community, which frequently gave rise to opposition from member state governments.

Relations between the different Community institutions are, in principle, defined by the EC Treaty and can in so far be regarded as static. In fact, however, the importance of each body, apart from the internal delineation of competences, depends on overall political conditions and is thus certainly subject to changes, these having repeatedly led to Treaty amendments.

Instruments

- Legislation:

  European social policy rests on legislation, which is simultaneously an instrument for fulfilling and implementing social policy. Community law is independent insofar as it is allocable neither to public international law nor national law. Correspondingly, the legal sources are likewise independent, a distinction being made between primary and secondary Community legislation. Primary legislation includes the Community founding treaties. Of notable import to the social sphere is the principle of equal pay for men and women (ex Article 119 EC Treaty) or ex Article 117 EC Treaty.

  The EC Treaty, as well as the other primary Community legislation, applies directly in all member states, without requiring implementation by the national legislator. This distinguishes Community law from international law. The EC Treaty lays down the rights of the Community institutions and also stipulates provisions that establish rights of the citizen. In terms of validity and content, the Treaty is comparable to a national constitution, although the definitional element of its origins was initially of an international law nature.

  Primary Community legislation is flanked by secondary legislation, which embodies the law created by the Community institutions on the basis of the powers conferred upon them in the founding treaties. The treaties set forth certain requirements of form, which must be complied with and do not correspond to the law-making forms that are customary, say, under German national law. The EC Treaty distinguishes the following forms:

(1) Regulations, which can address the member states and the citizens, and are binding in all parts. They are comparable to national laws.
(2) Decisions, which address certain member states or certain citizens and are likewise binding in all parts.
(3) Directives, which address all or certain member states and are binding only in terms of the declared objective. The means by which this objective is accomplished is, as a rule, left to the discretion of the national legislator.
(4) Recommendations, which can be addressed to member states, other Community institutions or individuals, but are not binding.

In addition there are communications and other notifications issued by the Commission, or other bodies such as the Council and Parliament. These, however, do not constitute legislative procedures, although such political activities often form the basis of subsequent legislation.

- Structural promotion:

  The EC Treaty provides the Community with an instrument for structural policies through which social inequalities within the Union, for instance as a result of structural economic measures, can be balanced out. Of prime importance, beside the Regional Development Fund and the Agricultural Guidance and Guarantee Fund, is the European Social Fund, which aims to improve employment opportunities for Community workers. Its resources are allocated from the EU budget, with the Commission responsible for their administration. Together with the other funds, the Social Fund has meanwhile been consolidated to form a comprehensive Structural Fund.

- Political activities:

  At a level below the legislative process, there are action forms – also in the social policy sphere – which, though not having direct binding effect, need not remain without consequence, especially in the long-term perspective. This form of socio-political activity is pursued by the European Parliament through resolutions; or by the Council, for instance under the First Social Policy Action Programme of 21/1/1974 or under the Community Charter of the Fundamental Social Rights of Workers; and above all by the Commission through a host of so-called Green and White Papers and action programmes.

  A special form of action whereby the Commission and the Council cooperate with each other is the OMC, which seeks to achieve convergence in social security system development in selected social policy areas (minimum protection, old age provision and healthcare) (cf. Göbel 2002). Proceeding from the need to align social security systems and given the largely similar challenges facing the member states (e.g. demographic trends), a modernisation of these systems is to be achieved through the joint setting of concerted objectives, through national reports on measures adopted to this end, and through the subsequent evaluation of these reports. The procedures framed under the SGP are exemplary here. In any case, the social policy reform debate is enriched by an institutionalised social law comparison such as that of the OMC, which is designed to expose foreign regulatory patterns and reform ideas. It remains to be seen what effects this new method will unfold in practice. Details on how the OMC works are provided in regard to the EES in section 2.2.2 and in the outline of various policy areas in part 3.

2.3.2.4 Organised Interests at European Level

Social policy impacts people directly, which is why they want to defend their interests against political authorities and see them represented as effectively as possible. To do so, they must join forces to align common interests. This necessity also presents itself at European level, especially as the powers of the European Parliament, which is actually called upon to safeguard the interests of EU citizens, are relatively weakly defined.

In specific cases, the institutional structure of the European Union takes account of interest groups. For instance, there is the European Economic and Social Committee, which has advisory functions and in which the social partners, among others, are represented; but there are also numerous other groups, such as those on behalf of various professions. In addition, the social partners are involved in the legislative process in matters relating to working life. Even if active participation is not expressly provided for, the Commission and Parliament frequently organise hearings in which individual groups can voice their concerns.

Such forms of participation through the EC institutions are flanked by self-organised interest groups at European level. Similar to the national governments, national interest groups maintain numerous representative offices in Brussels (e.g. those of German voluntary welfare associations or German social insurance institutions). To date, the integration of these national interest groups into a single European agency, provided the respective agencies already exist at national level, has progressed to differing extents. Thus, the trade unions are consolidated in a European trade union confederation, as are the employers in their own association, although these alliances are relatively weak compared to their national counterparts. In their relations to the Community institutions the social partners, inter alia, exercise co-determination rights in the drafting of directives within the scope of the European legislative process. In the field of labour law, a representation of the social partners at European level is being debated above all in the light of whether European collective agreements can and should be introduced. Besides the social partners, other organisations have formed similar affiliations at European level. Thus, the social insurance institutions of various member states have joined together under the European Social Insurance Platform. Another example is the association of social service providers under the Platform of European Social NGOs. The activities of these newly formed agencies focus on the representation of their interests vis-à-vis the Brussels authorities. Citizens and their political, economic and social affiliations must likewise seek to align their activities with the European agenda. Hence, the organisation of interests forms part of the endeavour to create a European public, the development of which has recently been advocated by Häberle (2000).

3 National and Supra-national Social Policy: Comparative Case Studies

3.1 Introduction

In the previous part the role of the European Union in the field of social policy and its influences on the member states have been displayed in general terms, i.e. – apart from the emphasis on employment policy – without any in-depth consideration of specific policy areas or countries. Peculiarities have been taken into account only in relation to different welfare regimes, but not to single states. In this way the overall determinants of the European framework for national social policies to a large extend have been traced back to direct and indirect interventions as well as effects of the Union, identified in the form of the economic, legal and growingly political integration, the migration and the Common Market, the supra-national legal foundations, and the direct actions of the institutions at Union level.

With the same broad angle, the analysis was conducted also in part one of this study. The trends of development of the European welfare systems and the present challenges they face – within the European context as well as in matters of the global situation – where analysed mainly with reference to the overall situation, albeit with important distinctions between different classes of welfare regimes. Even more as regarding the ethical foundations the focus was on the commonness of the basic principles rather than on the lower-level oppositional elements of different approaches to the social functions of the state. Finally, the concept of the enabling welfare state, too, was presented in a paradigmatic perspective as a new approach to social policy, towards which the existing welfare arrangements may be re-oriented in the future – requiring more or less radical reforms and changes of different degree and scope depending on the actual situation in the single cases.

The analysis in the following four chapters of part three treats four areas of social policy, in each of which the concrete policies and institutional arrangements of two countries will be compared. The four areas are health care, family policies, old-age security and poverty prevention. The connection to the hitherto applied reflection on welfare regimes is made up by comparing pairs of member states representing different welfare regimes.

The first chapter contrasts health care policies in Germany, as a corporatist welfare regime, and the United Kingdom, representing the liberal model. Subsequently policies of old-age security in Germany and Poland, a transition country, are going to be compared. In the third chapter two approaches to family policy from Finland and Estonia, a social-democratic and another transition state respectively, will be discussed. Finally, the last chapter describes the political and institutional arrangements Belgium and Denmark, a corporatist and a social-democratic welfare regime, have adopted in order to prevent poverty.

Admittedly, the choice of the policy fields and of the states, as well as the limitation to four areas and seven countries is to a certain degree arbitrary and reflects the specific competences and limited resources of the authors of this study. Still the exemplary analysis of the chosen topics may well suffice to cover at least representatively the main issues and the variety of social policies in Europe and allow drawing the conclusions necessary for general evaluations and recommendations.

Starting from the core elements of the concept of the enabling welfare state, the capabilities approach and the life-course perspective, and – as the main adequacy condition of every social system – the requirement of sustainability, the analysis of the four areas of social policy depicted below brings to the fore the main problems and questions. Family policies centred on childhood and parenthood and provisions for old age differentiate, structure and support specific phases of the life course, and enhance the capabilities of the beneficiaries during these periods of life. Health care and poverty prevention systems work as provisions against two basic risks of life, illness and poverty, which are not directly connected to specific life phases but can more or less temporarily and randomly interrupt and prejudice life plans and security expectations. By emphasising preventive more than compensative measures in both fields priorities are again put on the capabilities in the sense of health maintenance, of sustainment of productive and social capabilities, of employabilitiy – especially if poverty prevention is combined with activation policies in employment. The issue of financing health care and pension systems refers to questions of sustainability and re-distribution of means – socially between age cohorts, individually between life phases – with the corresponding problems of expectations about levels of assistance and security, which in turn lead over to matters of distribution of responsibility between the citizen, the state and other social institutions. All these aspects patently demonstrate that just as the elements of the paradigm of the enabling welfare society also the ethical principles underlying it, personal autonomy, social inclusion and distributive justice, are touched in the analysis of every area.

In addition, the confrontation of health care and pensions on the one hand with family and poverty policies on the other exemplifies how social arrangements in different areas are characterised by different grades of institutionalisation and more or less narrowly defined scopes of purposes and action. While health care and pensions are rather clearly defined, legally determined and institutionalised with a specific and narrow range of purposes, rights and entitlements, policies against poverty and in support of families are less structured, more multidimensional and transversal, connected in many ways to other areas, especially employment, and influential on various fields of social security through spill over effects. The aim of facilitating the reconciling of work and care, for instance, can be obtained through both measures of employment and family policy, which may exert reciprocally sustaining effects on each other and have at the same time positive effects on poverty prevention and on social inclusion, too. The close connection with employment policy indicates also the broader interlacement of social and economic policies. Obviously, changes in the health and pension sectors are intertwined with other areas too, but their institutional structure predefines narrower priorities in the purposes they have to serve.

If the four areas of social policy chosen for the analysis are not meant to cover completely the various fields of welfare policy, but rather to represent exemplarily the most important dimensions thereof, the same holds for the investigated and compared countries. The intention of the study is neither to give a full picture and comparison of the policies of all member states of the EU, nor to give a complete overview of the various policy arrangements, which can be found in Europe in each of the four areas. Instead the analysis is also in this regard mainly exemplary, contrasting for each area two countries only, which however represent more or less distinctly different welfare regimes and allow demonstrating differences and similarities from a static comparative as well as a dynamic developmental, reform oriented perspective.

Based upon common base lines of analysis the four chapters share the same structure. In the starting points a review on the main issues of the respective area of social policy is given. Thereupon a description of the policy approaches of the two countries in the given field follows, illustrating the institutions, methods, developments, challenges and policy goals in each of them. Subsequently a comparison between the two countries is given, stressing similarities and peculiarities, strengths and weaknesses. Criteria for the comparison are the two basic elements of the enabling welfare state, the capability approach and the life-course perspective, and additionally the principle of sustainability. By this way comparison and evaluation of the two countries are made not only in respect to each other, but also in regard to the degree they either already implement measures supporting the paradigm or are open to reforms and changes necessary for realising the underlying elements. In another section, the European dimensions and elements of the policy area are highlighted. On the one hand, the focus is on the interaction between the supra-national and the national level, i.e. the role of the Union in form of its various institutions (the Council, the Common Market, the Union law, the ECJ) and policies in other areas, which exert significant influences on the specific field of social policy and the national policies of the two considered countries therein. On the other hand, attention is given to the relations between member states, i.e. the effects of national policies on other member states by direct interaction, comparison and competition as well as via the institutions and mechanisms at Union level, above all the OMC. Finally, each chapter finishes with drawing conclusions and formulating recommendations. The final remarks regard the evaluation of the systems in respect to their capacity to respond to the challenges they face, their needs and their potentials of innovation and reform towards the enabling welfare society, the extend to which the ethical principles of personal autonomy, social inclusion and distributive justice are respected, the possibilities of reciprocal policy learning and adaptation of best practices, the options available for action at the supra-national level and the expected effects of the most promising interventions.

The overall picture resulting from the detailed studies of the four areas and the seven countries is characterised above all by heterogeneity. Although the main purpose of the comparisons is not the confrontation of good and bad examples, relatively clear judgements will be drawn in regard to the better and worse practices and to the most recommendable models for policy learning and innovation – obviously without neglecting the difficulties of transferring measures and methods from one context to another and the specificity of some urgent challenges the single states face.

With reference to health care Germany and the United Kingdom are confronted. As the two systems differ in many also fundamental regards, to say that the former for instance outperforms the latter in respect to quality standards, responsiveness and equal access, does not at all automatically mean that the British system should be changed according to the would-be German model. From the perspective of the enabling paradigm it will therefore be more important to elucidate, in which regards the two systems face different challenges with various degrees of urgency, i.e. how the priorities should be set amongst cost containment, efficiency and long-term sustainability, enhancement of equal access, enlargement of patients' freedom to choose and self-responsibility, in order to implement an enabling approach to health care. In addition, it will be of particular interest to analyse how the two systems are going to be affected by different means through the European Union, in particular the jurisdiction of the ECJ, the competition law and the common market, and how the interrelations between them are going to change due to these influences.

In the field of old-age security Poland and Germany both aim at adequate and sustainable pensions and both face similar challenges deriving from the tendencies in demography and on the labour market. One of the main issues is how public pension systems can be adjusted towards sustainability and complemented with private provisions for the sake of adequacy. The question is, to which degree public systems should eventually suffice only for some minimum income in old age while the maintenance of the living standard before retirement has to be provided by additional private provisions, for which the individuals not only are responsible themselves but also have a wider range of options available. Whether the Polish three-pillar system could serve as a model, depends not only on its success but also on the specific conditions for the implementation of more or less radical reforms. What can be learned from the extraordinary transition situation in Poland, where the financial sustainability of the formerly existing pension schemes couldn't be guaranteed in the long term, for the specific requirements of the present situation in Germany, where by now rather slow and gradual policy shifts have been put into action?

In the area of family policy at first glance the conclusions to draw from the comparison seem clearer. The family policy in Finland is definitely more advanced towards the enabling paradigm and may hence serve as a model for policy learning. But the comparison between Finland and Estonia will demonstrate above all how fundamentally divergent aims can be connected to family policies. In view of the main goals of family policy in the Scandinavian country, which are the reconciling of family and work, especially for fathers too, and the protection and enhancement of the childhood over a comparatively long time span, the question arises whether the Estonian policies given the dramatic demographic situation in the smaller country should strive for the same goals too. Or will a rather different priority setting with a clear pro-natality orientation, which supports especially birth giving and the first year of children's life, be acceptable and reasonable as well – eventually at least for the time the specific circumstances persist?

More promising in regard to policy learning seems to be the situation between Denmark and Belgium. Albeit both countries are quite successful in keeping poverty at low levels, Denmark performs better on the labour market due to its pronounced active labour policy centred on investment in human capital, flexibilisation of the labour market and social inclusion, especially of woman, young and elder people. Reducing early retirement options and unemployment benefit duration and subsidising low skilled workers are some of the concrete measures implemented by Denmark with great success. However, if this is so plainly evident, why is Belgium not simply following the example of its neighbour? What are the institutional and political barriers and path dependencies that are hampering political learning and hindering the adoption of elsewhere successfully implemented measures?

With respect to the supra-national level the analysis will bring up a pretty unbalanced and complex picture. The influence of the European elements on national policies is growing, in direct as well as in indirect ways. The importance of the supra-national level therefore will have to be taken into account in a comprehensive way on both levels of policy making. However as long as the principle of subsidiarity is regulating the competences in the social area, only the soft instrument of the OMC seems to be feasible. Yet up to now it is still discussed which purposes it should best serve, and the conclusions will be different in the four areas analysed here.

The questions opened up in the following investigation and the intermediary results obtained in regard to the exemplary areas and member states will be brought together and summarised in part four. In connection with the central issues treated in part one and two they are going to serve as the basis for the recommendations of strategies of actions for an enabling social policy in the European Union.

3.2 Health Care: Germany/United Kingdom

3.2.1 Starting Point

3.2.1.1 Functional and Institutional Perspectives of Financing and Purchasing Structures in Health Services

The allocation of economic resources in health care on four levels can be seen in figure 3.1. It illustrates at the same time functional features of a health care system. By means of political decisions, market forces, self-governmental processes etc., societies have to decide how much they want to spend on health care on different levels.

Within the health care sector (level 2 in figure 3.1), decisions are likewise necessary in order to decide how scarce resources should be allocated to:

1. disease prevention and health promotion,
2. curative treatment for individuals with acute and chronic diseases and for healthy individuals, including:
   - pharmaceutical therapy,
   - physical therapy,
   - Medical devices;
3. rehabilitation,
4. nursing care,
5. psychosocial care,
6. dental care,
7. sick pay.

What is called for in this context is a diversified and flexible health care system. Such a system should respond to given needs with cost-effective services, which are coordinated across sectors and focus on patients' preferences. Apart from the overall options of financing, which are quite differently used throughout Europe and which describe different ways of collecting money ("external financing"), the purchase of health services from hospitals, rehabilitation institutions, and office-based physicians as well as the purchase of drugs, remedies, medical appliances and so forth is the second important aspect. This aspect can be called "internal financing", i.e. the reimbursement or payment of each health care service. Ideally, all services should be delivered according to medical guidelines, best practices and in regard to outcome measures. While the reimbursement systems should be less revenue-oriented and more outcome-driven, they should also refrain from reimbursements on a fee-for-service-basis.

The overall goal is to overcome segmentation in health care and to work on an integrated and quality assured medical care network. In order to achieve this target, a functional approach to health care is indispensable for upcoming reforms. New forms of selective contracting between providers of health services and sickness funds are needed for an integrated care delivery system. The provisions of medical treatment and nursing care, including rehabilitation, systematically belong together and should be covered through joint remuneration.

This could be achieved through network-budgeting and new kinds of fee-per-case-payments. Comprehensive "all-round-care" is the new subject of financing. So far, no golden rule for purchasing all these services has been found, so that probably more competition is the answer to this problem. However, proposing a network is much easier than accomplishing it. Pricing, purchasing

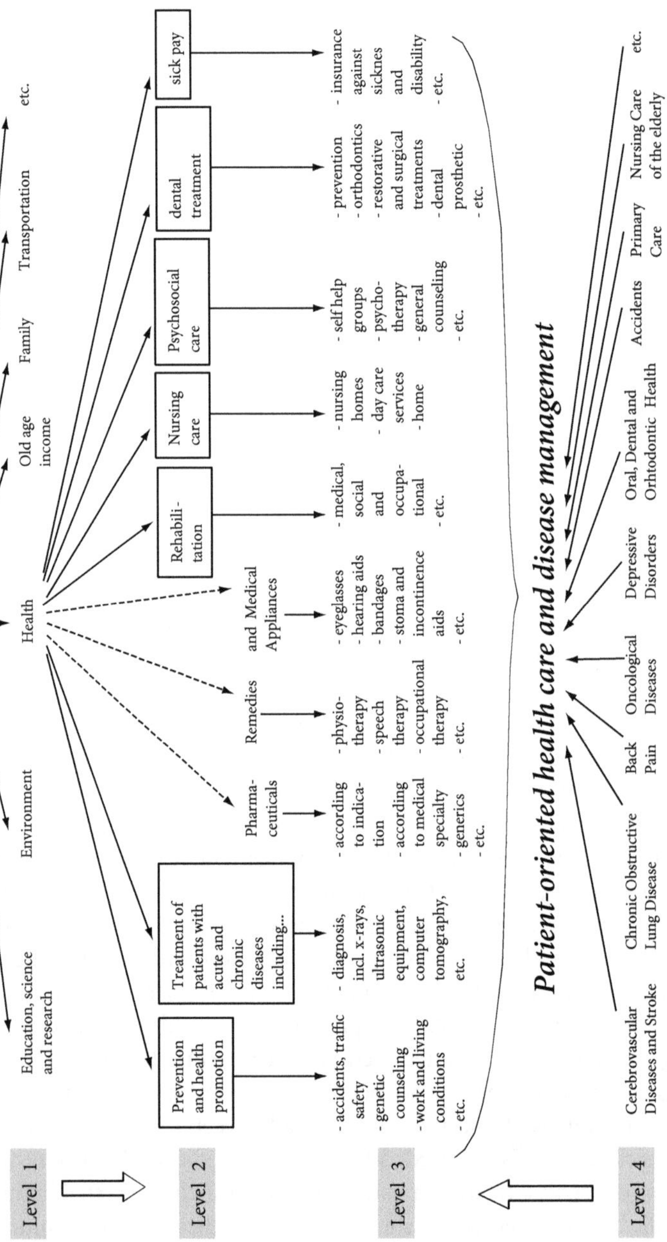

Figure 3.1: Four levels of a health care system
Source: Henke 2004: 13

(e.g. through Diagnosis Related Groups (DRGs), reference prices or on the basis of a fee schedule), spending, and financing (taxes, contributions, premiums etc.) of health services represent an extremely complex picture for all participants. It raises more questions than it gives answers and, perhaps, a socially-bounded competition may help to further develop the institutional details in providing, funding and purchasing the required health care for the entire population (Henke et al. 2004a; Henke et al. 2004c; Strang and Schulze 2004).

From an institutional perspective, it is important to differentiate between the sectors and sources of funding health care systems. In Germany, a system relying on Social Insurances shows eight different expenditure carriers with their specific characteristics. Four branches of the social security system are responsible for health care, pension, care and the statutory accident insurance. Besides private insurances, the state government itself, out-of pocket and employers' expenditures exist.

In Germany, the Social Health Insurance (SHI) system is responsible for over 50 % of health care spending. However, it must be recognised so that all the other payers also contribute to caring for the sick and providing health care to the public. Health insurance, long-term care insurance, worker's compensation, and the rehabilitation benefits of the social pension scheme are closely intertwined and cannot be easily or meaningfully separated. In Germany, the continued payment of wages, which is initially the responsibility of employers, is usually not included in this sum, which also omits the out-of-pocket payments of private households and the public funding of capital costs in the hospital sector.

The National Health Service (NHS) in the United Kingdom can be characterised as a centralised system for providing medical care over defined regions with mainly a tax-financed system. The proportion of private health insurances in the UK is small but increasing. Hence, universal access as one main goal turns out to be questionable in the light of long waiting times.

This chapter explores the question of state interventions in health care, which takes place all over Europe. It becomes clear that there are several aspects, which distinguish health care services and the health care markets from other services.

3.2.1.2 Allocational and Distributional Reasons for Interventions by the State

From an economic point of view, three major stakeholder categories constituting a health care system can be described: the health care providers, the patient/insuree/taxpayer and the health insurance funds resp. health authorities. Consequently, different sub-markets exist, namely, a market of health care services and goods between the health care providers and the patient, a market of services between the health care providers and the health insurances or health authorities, and a market of health insurances between the patient/insuree/taxpayer and the health insurance funds.

Due to the fact that one side is always better informed than the other, several settings of asymmetric information exist between all stakeholders included in figure 3.2. This results in corresponding sets of incentives and disincentives. Malpractice and misconduct are the consequence of asymmetric distribution of information and appear in different constellations, which are to be described in the following sections.

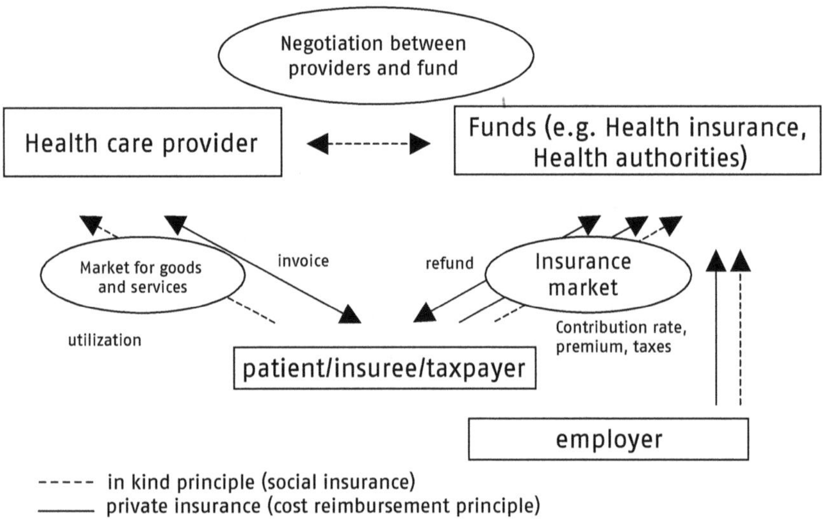

Figure 3.2: Health care from an economist's point of view

On the market for health care services and goods, a principle-agent-relation between the physician and the patient means that neither all actions are visible nor are the results always related to the actions performed by the agent. The physician (agent) as a specialist has a prominence according to the choice of treating the patient (principal). In this context, one refers to the so called "supply orientated demand" because the physician (agent) can prescribe a certain therapy for the patient. However, the physician-patient relation has changed during the last decades. As a result of the developing information society, the patient is able to overcome certain information gaps; examples include the usage of the internet and organisation in self-help groups with regard to chronical diseases or scarce illnesses. From this point of view, the formerly missing sovereignty has been improved.[44] That is why,

[44] In some cases, e.g. coma, dementia or Alzheimer or severe illnesses the consumer sovereignty is missing since the affected individuals in these circumstances are not capable anymore of rational decisions, which have to be take by others, e.g. the relatives.

in some cases, the patient can be seen as an expert and the physicians functions more as competent consultant and less as the only decision maker. Therefore, the scope of the physician's action is partially bounded. Additionally, patient's rights are nowadays discussed in all European member states (Gethmann et al. 2004: 75 et seqq.). On the other side, the physician has only partial information about the patient's behaviour and compliance, as the patient will only inform the physician to a certain degree. In this context, one has to bear in mind that the results of a therapy are not solely dependent on the agent's (physician's) action. More so, the health status is influenced by a set of complex multi-causal factors, which are of individual, structural, demographic, socio-economic and medical nature.

On the provider market between the health care providers and the health insurances funds, asymmetric information can be found insofar as the health insurance funds reimburse the health care providers, but have no possibility to fully judge the necessity and extent of the performed services. Nowadays, with the introduction of the DRGs system, guidelines and best practice, the insurance has a better chance to estimate the effective costs and hence reimbursements of the health services.

Asymmetric distribution of information has an influence on the functioning of the market for health insurance, too. Two phenomena that lead to market failures have to be mentioned:

- moral hazard by patients and health care providers,
- adverse selection by health insurances.

Moral hazard describes an individual's opportunistic behaviour before or after contracting with health insurance funds. Likewise, the existence of insurance and its offered benefits will impact on the individual's behaviour as well. Ex-ante moral hazard, which is based on hidden information and hidden action, describes the behaviour of individuals who provoke the occurrence of the insured event, e.g. due to unhealthy life-style. After the event, causes of a disease are sometimes no more detectible for insurance companies. The severeness of a disease is another factor the insurance cannot judge.

Ex-post moral hazard in health care comes into play when the insured event has happened and the patient is ill. The patient and the physician care little about cost containment and its negative effects on the insurance. Again, the physician (specialist) is better informed about the medical and technical spectrum to cure illnesses and might have an interest in widening his services. Similarly, the patient's demand for health services and goods grows as long as the insurance contract provides full coverage. In order to reduce the latter problem of ex-post moral hazard, insurances could introduce deductibles and co-payments and special tariffs for the insurees. The physician's behaviour might be influenced through guidelines and the application of evidence based medicine. Another possibility would be to ration services with reference to quantity or type of service.

Another form of misconduct is called adverse selection. On an unregulated insurance market without risk selection instruments, individuals with good risks are left uninsured, while the bad risks receive a contract with fairly high premiums. Such a result stems from the process of risk selection, as the insurance cannot differentiate between individuals with good and bad risks. Hence, insurances would offer contracts based on the average risk calculation. The corresponding premium, however, would be too high for good risks and too low for bad risks. Thus, only the bad risks would demand such an offer. Consequently, the risk composition of the insurance would require higher premiums to cover all costs. Again, individuals would drop out, knowing that in comparison to their risks the requested payment is too high. However, the technical aspects of gaining information about individual risks touches upon some ethical issues, as some risks can only be discovered by genetic screening.

In order to combat adverse selection, the state could introduce compulsory insurances for all individuals. Additionally, the obligation to contract all risks without discrimination could impede risk selection. Lastly, a risk compensation scheme could be introduced to ensure the competition between health insurance funds.

3.2.1.3 Capability Approach and Life-course Perspective as a Basis for Comparing Health Care Systems

The underlying principles of the paradigm of the enabling welfare state are the capability approach and the life-course perspective. The main features of these concepts are their interdisciplinary character and their multidimensional approach to factors influencing individual well-being. The application of Sen's capabilities approach as a framework for evaluating health care systems is rather unusual and new. The "approach highlights the difference between means and ends, and between substantive freedoms (capabilities) and outcomes (achieved functionings)" (Robeyns, 2004: 1). The capabilities describe what a person is able to do and to be. In this context health is an integral part of individual well-being and serves to build capacities to work and to function. Thereby there exists a close link between capabilities and abilities to lead and manage one's own life. In order to achieve such a fulfilled or satisfied life an activating welfare policy would aim at offering pluralistic possibilities to act and freedom to choose. The capability approach "asks whether people are being healthy, and whether the means or resources necessary for this capability, such as clean water, access to medical doctors, protection from infections and diseases, and basic knowledge on health issues, are present" (Robeyns, 2004: 5). In health care this clearly means more than just the absence of pain: for instance, to guarantee adequate rescue time in cases of accidents, offering acute and preventive medicine, adequate access to health care services and responsiveness. Individual health is not only influenced by medical facilities and treatment, but also by a lot of other factors, e.g. economic prosperity, employment, social cohesion but also economic and social inequalities.[45] Age-related changes in capabilities and differences between men and women with regard to life expectancy and illness pattern are to be taken into consideration as well.

Offering opportunities to participate and act and therefore strengthening the personal autonomy is part of the concept of individual empowerment. The notion of offering opportunities or possibilities to act can be seen in health care as providing choices. Choices in health care could comprise the choice of insurances, of providers and of treatments. Additionally and related to questions of justice, the problem of inequalities of opportunities arises. Choices are always influenced or limited by age, educational level, geographical distribution of facilities, societal norms and the willingness or ability to pay. The level of choices in the European health care systems varies to a large degree and can be illustrated by comparing Germany and the UK as two examples. In general an equal access for all to health care services is struggled for in both countries, yet knowing that the wealthy always profit from a wider degree of options.

The life-course perspective traditionally distinguishes three periods: education, work life and retirement, but refers to the intergenerational side of social policy, too. The demographic challenges, particularly the changing age structures require reforms of pension and health insurance schemes everywhere. Pay-as-you-go systems and fully-funded systems point to specific strengths and weaknesses. The important point is that they cover different risks and this is an argument to establish a two-track system of social insurance. Any measures which take more capital-forming pension schemes into consideration are desirable but careful deliberation must be given to their institutional structure. Furthermore, the reform of health care financing has to be combined with necessary adjustments on the supply side of the system.

Regarding the problem of rationing, the best way would be the allocation of resources via market competition because that would serve individuals' and consumers' sovereignty. In contrast, direct and indirect forms of rationing in health care are carried out through state interventions. To treat similar cases in similar ways the emergence of evidence-based medicine can be interpreted as another form of addressing certain inequalities. As all European health care systems need to tackle the mentioned problems. In the following part of this chapter, we will firstly describe major differences and similarities of the European health care systems. Thereby, special emphasis will be put on the comparison between Germany and the United Kingdom as they represent two opposing models of financing and purchasing structures in health care services. The next sections will mainly describe and

[45] In addition, the possibility of enabling individuals to achieve functioning is bounded by the individual's abilities and differences to benefit from treatments and interventions.

analyse the German and British health care system by referring to the mentioned capability approach and the life-course perspective.[46]

3.2.1.4 Introduction to Health Care Systems in Europe

Health care systems in the European Union differ e.g. in terms of financing, the scope of benefits and the purchasing structures in health services, the provision of health care, the quality of care, the degree of freedom to choose and the distribution of responsibilities, the skill-mix of health professionals, the education and training of health care professionals as well as the set of incentives for the production of health services (Henke 1992).

To some degree, waiting lists or partial bottle necks in the supply of medical treatments exist in several countries, especially in national and tax-financed health services. Countries operating on waiting lists are not only the UK, but also Norway, Denmark, Sweden, and Finland (Preusker 1999). Inequalities in access to health care services result as a consequence of fiscal constraints in the form of restricted or excluded health care services.

Due to rising additional private (co-)payments or private co-insurances for the patients in payroll taxed related or tax based health care systems, the claim of equal access for all becomes questionable. In terms of future challenges for the European health care, it can be observed that all systems face structural changes (integrated health care instead of segmentation, fiscal sustainability) and demographic challenges (aging, shrinking birth rates and decreasing total population). This is combined with the wish for medical-technical progress, which ultimately brings along cost effects and expenditure growth in general. This situation coincides with restrictive financing margins, increasing national budgetary deficits, as well as increasing claims from the patients in medical care and the consumers in the area of wellness and fitness.

Due to continuous reforms and different historical conditions in all member states of the European Union, any classification of health care systems becomes increasingly difficult and remains inaccurate. Each system has its own history, culture and set of institutional, organisational and political elements, which determine path dependencies for future developments.

The traditional differentiation between the Bismarck model and the Beveridge model becomes challenging, as both systems show more and more a mix of different modes of financing. Nevertheless, it can be stated that the Beveridge model aims at complete equality of supply. However, the majority of national health care systems operate with waiting lists, restrictions in the

---

46 For an extensive and more traditional and system-oriented description of the detailed structures of the German health care system see Busse and Riesberg 2004; Henke and Schreyögg 2004; Gethmann et al. 2004; and for the UK see Healthcare Commission 2004; European Observatory on Health Care Systems 1999.

freedom to choose doctors, as well as pronounced additional insurances to be financed privately (Mossialos and McKee 2002), all of which stand in sharp contrast with the goal of equality of supply. Comparing the Bismarckian type countries, a heterogenic picture can be observed: some countries follow a more competitive and others a more corporatist model. The new Middle and Eastern European member states were formerly influenced by the so-called Siemiaszko model (Figueras et al. 2004) which was based on a centralised socialist model with equality, central planning, and national property of all health care facilities. After the political and economic breakdown of the former socialist countries, the Middle and Eastern European countries followed different pathways and can no longer be described by one model. Some countries tried to establish a Bismarckian type of model, while others implemented a Beveridge model or adopted elements of both.

To sum up, the European member states share a common set of fundamental principles and values. These include providing necessary health care services to their population, practicing the solidarity principle (though on different levels), and understanding health care as an investment in human capital. Nonetheless, systems with elements of competition claim to pursuit regulated socially bounded competition. In comparison to the United States of America, which spend 15 % of the GDP on health care but at the same time have 42 million of uninsured, one major difference lies in Europe's claim to allow equal access to health care and to offer a certain minimum standard to all. The American health care model of self-help, on the other side, is associated with more individual responsibility. This includes looking after one's own health, paying on an individual level, and having a restricted extent of public sector payments for certain groups, e.g. through Medicare and Medicaid. This stands in sharp contrast to the European approach with its almost universal coverage based on solidarity and society's responsibility for individual health (Henke 1999; Brown and Amelung 1999; Shalala and Reinhardt 1999; Mossialos and McKee 2002; Leidl 1999).

The method of system comparison in the EU was not only used to compare health care systems during the last years. For health care systems in Europe, the application of the OMC in health care means to work out a feasible set of indicators as a basis for a good benchmark system. Unfortunately, there is no agreement on a fixed set of indicators and that is why a wide range of different analyses and choice in indicators exists and rankings lead to different results. Table 3.1 tries to give a general overview about frequently used indicators for comparisons of health care systems.

To a large extent, the choice of indicators for a comparison depends on the experts setting the agenda. The selection of indicators thereby can be described as a key for a benchmark system and explains why the results of ranking health care systems differ so much.

3.2.2 The Case of Germany

3.2.2.1 Coverage

Approximately 90% of the population are covered by the German SHI system financed through employer and employee contributions based on wages and salaries. Private health insurances and other carriers cover the remaining 10%. Through the fact that payroll taxes to the SHI are independent of individual, medical, and social risks of the insured and co-insured (spouse, children as their family dependents) this high degree in coverage represents a basis for equal access to health care for the overall population.

Co-insured dependents earning less than 400 Euro per month or children involved in education, training or studies can be co-insured until their 25th birthday (§ 10 Social Security Code Book V). Students are insured on a mandatory basis until they reach the age limit and pay only half of the contribution rate, while the other part is financed through the federal government. Since 2004, equal treatment of all pensioners is introduced: now pensioners have to pay full contribution rates based on all pension payments including company pensions and non-statutory pensions.

The special arrangements for co-insured and students respectively individuals with low income (less than 400 EUR per month) embody one dimension of the social character of the health insurance system. Additionally, the coverage and contributions to the SHI insurance are financed according to the ability to pay on the basis of income and salaries with a fixed rate. There exists an income threshold for contributions to the SHI. Up to the threshold contributions are calculated as a percentage rate. In 2005, the average contribution rate is close to 15%.

During the working life, people have to contribute according to their income to the SHI or can alternatively choose a private health insurance with risk-equivalent premiums and capital funding, if their income is above a certain income threshold (monthly income of 3.825 Euro in the year 2004). This provision of choice can be offered because individuals with higher income are considered capable to manage their own provision of health care. Those who do not want to change to a private health insurance can opt to stay in the SHI at a maximum fixed contribution rate – which obviously has regressive effects. Thus there is a kind of freedom to choose above a certain income level between private insurances and the statutory funds. For civil servants too there are opportunities to select additional coverage by private insurances.

3.2.2.2 Benefits: Scope and Structure

The benefits (services) of the SHI and the private insurances are more or less the same for all insured and co-insured. The SHI Insurance Law requires medical care to be "sufficient and effective according to the standards of medical practice". Due to only one criterion – medical care must be "neces-

Table 3.1: Parameters for comparisons of health care systems

1) Economic, demographic and epidemiological parameters
Health Status
– Life expectancy, standardised ratio of mortality, loss of potential years of life by the main causes of death (ischemic heart conditions, illnesses of the cerebro-vascular system and cancer illnesses), indicators for dental diseases, alcohol consumption, tobacco consumption – Inability to work, work accidents, street road casualties – Morbidity, recovered years of life – Age distribution, sex structure, birth-rate, deaths
2) Structural parameters
Basic Structure of the Health System
– Structure of the health care system, planning, regulation and management, decentralisation of the health care system, structure of providers
Financing and Purchasing in Health Services
– Financing structures of the system – Analysis of the health insurance system (insurance form, calculation of premiums and/or contributions, participation, subsidies) – State share, social insurance contributions, co-payments, private insurance contributions, etc.
Supply and Demand of Health Services
– Supply forms, hospital beds, length of stay, cases in- and outpatient care, health care personnel – Number of consultations with doctors and dentists, drug prescriptions, hospital cases, acute hospital cases, hospital days – Resources: number of doctors, dentists, pharmacies, pharmacists, hospital staff, nurses and midwives – Public health service, in- and outpatient supply, supply with specialists, long-term care supply, personnel resources and education, drug, technology assessment in the health system – Structure and organisation of the supplier side, admittance, reimbursement, etc. – Measures of the prevention
Purchasing in Health Care Services
– Expenditures for in- and outpatient care, pharmaceutical expenditures, etc. – Costs and their development, development of causes, etc.
Quality Assurance of Health Care Services
– Doctors' education, training and education system in medicine, specialist education, etc. – Further education, training and quality assessment, etc.
3) New parameters
Capabilities Approach, Life-course Perspective, Enabling Welfare State
– Freedom to choose between insurances, providers and treatment – Intergenerational equity – Satisfaction of the Population

sary" – the built-in tendency for expansion of benefits is further strengthened. In general, approximately 95 % of the SHI benefits are the same in all statutory funds. The benefits in the private funds are more or less the same, except for additional qualities in hospital care and in some cases quicker access to medical treatment. Generally, the range of services of the statutory health insurances is determined by the regulation of the Social Code Book V (especially third chapter, section 11, kinds of benefits). The entitlement includes treatment in case of illness, health screening, contraception, as well as rehabilitation of illnesses. The German SHI offers:

- almost free ambulatory and hospital care,
- freedom to choose a general practitioners and specialists (including dentists),
- certain kinds of preventive care,
- family planning services,
- medical services in case of rehabilitation.

The medical specialists may work in outpatient as well as inpatient settings. Since the beginning of 2004, visits to office-based physicians cost 10 Euro per quarter. Without a referral of the family doctor, additional 10 Euro have to be paid by the patient when consulting a medical specialist. The uniform assessment scale defines reimbursements between providers and sickness funds. The individual contribution payments are paid to the different types of funds. They collectively transfer global sums to the doctors associations which pay the individual doctors according to certain rules. In comparison to other EU states, waiting lists and/or open rationing is still a rare exception in Germany, e.g. with organ transplantations.

The sick benefit, which is paid when a hospital or rehabilitation treatment hinders a patient from working, is being reformed. This benefit is recently excluded from the SHI, and replaced by an obligatory insurance on an individual level, i.e. the employer is not longer paying half of it.

The described scope and structure of benefits is an integral part of the German health care system. The insuree may choose between different types of funds, providers and treatments. As a result an equal and fast access to health care services is given, personal autonomy and freedom to choose is achieved as well as a certain degree of competition between the providers. With the newly debated integration of more preventive matters into the SHI not only the treatment of illnesses is considered necessary, but the investment in the citizens' health potential as well. This can be seen as a sign that capabilities and abilities of the individuals are respected in German health care.

3.2.2.3 Financing and Purchasing in Health Services

In Germany, the SHI is only responsible for approximately 57 % of total health care spending. Figure 3.3 shows the different sources of financing the German health care system.

During the last health care reforms a trend towards more private expenditures is visible. It is not only the result of the cost-containment policies. At the same time it can be seen as a sign of more self-responsibility on the side of the patient. The purchase of health services from office based physicians, hospitals and rehabilitation institutions, as well as the purchase of drugs, remedies, medical appliances can be considered as "internal financing" with many different markets. Additionally, Germany also strives for overcoming segmentation in health care, and to work on an integrated and quality

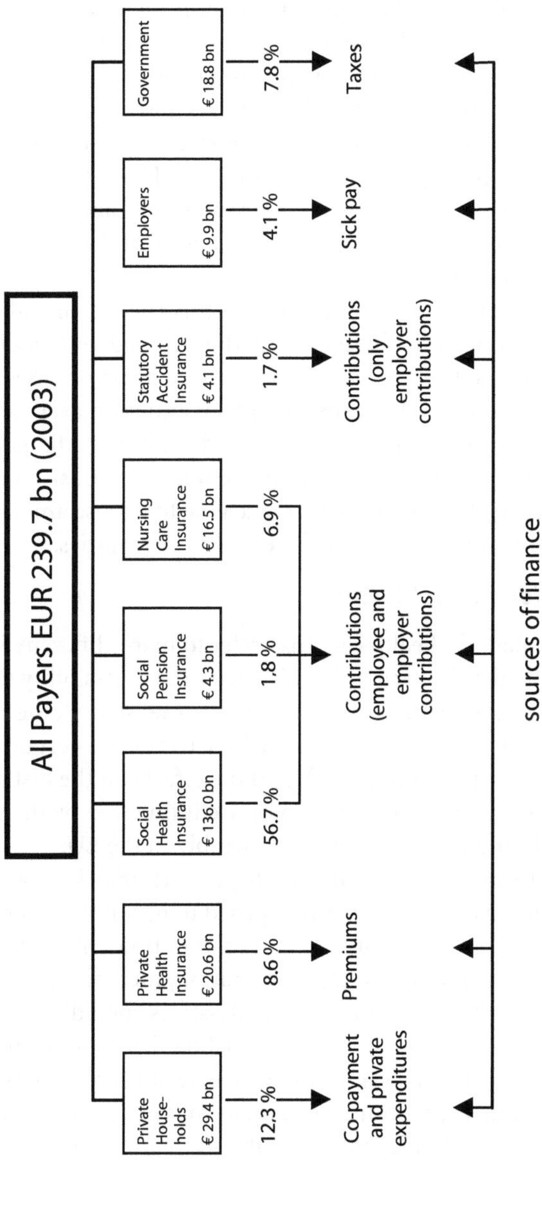

Figure 3.3: Payers and forms of finance in the German health care system, 2003
Source: Federal Statistical Office 2005, www.destatis.de

assured medical care network. In order to achieve this target, a functional approach to health care is indispensable for necessary reforms. Such an integrated care delivery system demands new approaches (i.e. new forms of selective contracting by the sickness funds), which have been introduced during health care reforms in Germany since 1997. The provision of all kind of out- and inpatient medical treatment, nursing home care and rehabilitation belong together. Comprehensive "all-round-care" is the new subject of financing. However, proposing a network is much easier than accomplishing it.

The focus on an integrated health care system can at the same be interpreted as an investment in individual health and capacities, since such a system provides a continuum of treatments and avoids unnecessary or doubling of services. In achieving this goal, monetary resources could be saved. However, such integration could as well mean restricted choices, depending on the kind of managed care society is choosing. The orientation on evidence-based medicine can be seen in the context of capabilities and as a means to balance inequalities of opportunities and treat the same symptoms in a standardised manner.

Germany holds a high density of medical infrastructure including a dual infrastructure of specialised physicians in the in- and outpatient sector. Additionally, the rescue time in cases of severe accidents or cases of emergency is quite short compared to other countries and is on average less than 20 minutes. Thus, with regard to capacity and treatment the good infrastructure and density of providers help to provide a responsive health care system, which enables the patient to a large degree to choose among alternatives and at the same time gives room for equal opportunities.

3.2.2.4 Distribution in Health Care

Distributional elements in health care may refer to many different issues, e.g. the socio-economic distribution of health status, the accessibility to medical care (independent from income, residence, and social status), the pattern of utilisation of certain types of care (e.g. early detection of certain illnesses, regular check-ups and preventive care), and the effects on the distribution if income. Some of these effects depend on or are influenced by the provision and financing of health care and vary considerably in the EU.

However, health as well as wealth is disproportionately distributed in every society. Individuals are differently exposed to harmful strains, i.e. from nutrition, smoking or chemicals. At the same time individual health resources differ due to inherited factors, education, income, self-esteem and participation. Additionally, the individual life-styles including risk behaviour or symptom tolerance are unequally distributed among the population (Rosenbrock and Gerlinger 2004). A responsive health care system therefore should address health promotion, preventive matters, cure, rehabilitation and long-term care and should not rely only on treating illnesses. By doing so it contributes to the individuals' capabilities and enlarges the possibilities of individuals to lead a satisfied life on their own. Another distributional aspect refers to the already mentioned effects on the financing side of the health insurance. In this respect, intergenerational aspects and sustainable financing of the system are key factors.

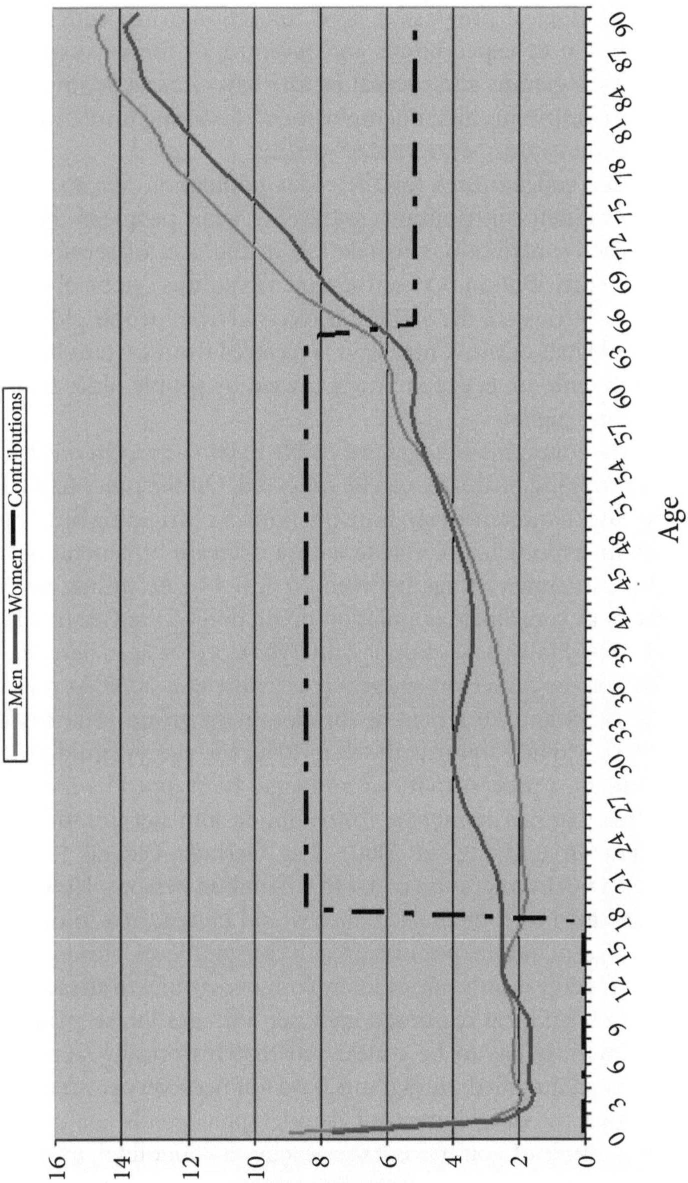

Figure 3.4: Standardised health expenditures of German Sickness funds in 2002

Source: Henke 2005: 98

Based on a pay-as-you-go system, the payroll tax contributions in Germany contain elements of redistribution. This represents to a large portion a withdrawal from the insurance principle, since the benefit entitlement is carried out independently of the contribution. The processes of redistribution of income through the SHI system are constantly researched in terms of quality and quantity, particularly when the life-cycle approach is considered. Along with the increasing demographic ageing of the German population and medical technological progress, a need for action arises with respect towards reorientation of expenditures and revenues of the pay-as-you-go system. Similarly, exogenous and cyclical factors have a negative impact on revenues, such as continuous high unemployment, declining future working force, and the trend towards new forms of work.

In addition, the expenditures for the older population are, to a large extend, higher than their contributions to the SHI, while people in the middle ages pay higher contributions in relation to the cost of services they demand. Hence, redistribution between several age groups can be observed. In Germany, contributions to the SHI funds received from people older than 60 amounts to only half of the contributions received from citizens between 20 and 60. At the same time, expenditures caused by people older than 60 years are three times higher.

In Germany, demographic changes are visible in two ways. On one hand, a continuously dropping birth rate can be observed. On the other hand, life expectancy increases simultaneously. Both phenomena (dropping birth rates and increasing life expectancy) coincide with a decrease of generations of potential mothers (women of age between 20 and 44). According to estimates from the 10th coordinate population prediction of the German Federal Statistics Office (Statistisches Bundesamt 2003), people aged between 50 and 60 years will be the dominant age group as by the year 2020. At present, the aged between 35 and 40 represent the dominant group (Statistisches Bundesamt 2003). Moving towards the year 2050, the age pyramid will be turned upside-down, a process that can no longer be stopped by immigration (OECD 2003; Zimmermann et al. 2001; Bundesministerium für Familie, Senioren, Frauen und Jugend 2000). The German Federal Statistics Office calculated a net immigration of 4.9 to 9.3 million persons. However, a net immigration three or four times higher would be required in order to keep the present German total population at a constant level. However, due to the fact that Germany is only one of many countries trying to attract qualified people and for political reasons, a high net immigration seems rather unlikely. In addition, it has to be considered that historically Germany's approaches to attract qualified immigrants have not been very successful.

Considering that the development of demographic trends is a creeping process, one can think of society as experiencing a "doubling" in ageing. Effectively, this means that the current group of citizens between the age of 20 and 60 is not large enough to sustain the social insurance system of welfare states (Henke and Schreyögg 2004). Keeping the mentioned development of the German population in mind, pay-as-you-go systems are dramatically challenged. The burden of financing the SHI system is laid upon the working population, which, however, will most likely decline in the future.

Regarding the life-course, intergenerational justice is thought about comprising two dimensions: it can be a comparison of age groups at a given moment (cross-sectional analysis) or alternatively concentrate upon cohorts (longitudinal analysis) during their life span. Distributional justice is considered to be a normative notion and represents a highly debated issue. For achieving sustainable financing an equal treatment of generations could be targeted with a balance between received and offered benefits over the generations; however there exists no database for a cost-benefit-access (Schmähl 2005; Henke 2005; on distributive justice see section 1.2.3).

3.2.2.5 Current Problems and Long-term Perspectives

On an international level, the German healthcare system stands out by virtue of its high standard of coverage and its comprehensive safeguard against the vicissitudes of life. It continuously ensures access to a high standard of medical and health care services for every person, irrespective of income, social status, and place of residence. Despite this overall assessment, there is a growing pressure towards reforming the system. Everyone concerned would like to preclude the threat of rationing, master the demographic challenges and safeguard this personnel-intensive, future-oriented growth sector. Therefore, a transparent and sustainable high quality oriented health care system of protection against health risks is required. A new framework of order, comprising pertinent incentives, must be seen as a prerequisite to this end.

If there is a consensus to improve the existing health care system and thus to remain within the given framework, the following requirements would be generally recognised as valid (further see e.g. Henke and Schreyögg 2004):

- more competition among health insurers and among health care providers,
- increased efficiency in hospitals and privatisation of ownership at local level,
- more prevention, health promotion and self-responsibility,
- more projects for the promotion of approaches to health care and modes of finance that cross sectoral boundaries and focus on outcomes,
- more health education, information, transparency and thus empowerment,
- more quality assurance based on the certification of health care facilities and evidence-based medicine.

In addition, there is also general agreement about the permanent mobilisation of the so-called 'efficiency reserves'; and the slogan 'rationalisation before rationing' implies more efficiency before the utilisation of new sources of finance.

Finally, if the self-governmental system remains intact – which, from an economic perspective, can best be justified on the basis of the fact that the system is funded equally by employers and workers, but which leads to a situation in which both parties strive to influence the health care system – there is only a little chance that this reform approach will be much more than 'muddling through on a relatively high level'.

Due to increases of payroll tax rates or contributions, the question of cost containment in the SHI is constantly on the political agenda. In the past, measures included an increase of contributions and co-payments, budget cuts and price controls as well as moderate exclusions of ineffective medicine from refunding by the statutory sickness fund. Hardly ever the system structures were changed so that the underlying power and incentives and disincentives of the different stakeholders remained.

During the year 2004, numerous concepts for reforming the financing of the German health care system were under discussion aiming to achieve more fiscal sustainability. To downsize the distributive elements of social health insurances the current balance between benefits and premiums could be adjusted on the basis of the insurance principle and thus create more transparency. Additionally, broadening the range of persons and the assessment base of the SHI – as proposed in a citizen's insurance scheme – is not considered a solution to reach more sustainability in financing. If a system based on more competition between the health insurances (private and statutory) is desired, structural changes become necessary, i.e. to allow the SHI more differentiation of their services packages. Therefore, changing to a flat-rate payment system (capitation or lump sum fee) could be desirable (Advisory Board of the Ministry of Finance 2004).

In respect to a fundamental financing reform, public debate particularly focuses on two concepts, namely, citizen insurance versus community rating (per capita premiums). Other reform efforts mostly concentrate upon improving conditions for an integrated medical care. In order to create new incentive structures, the possibility for selective contracting and integrated care of sickness funds with providers was introduced and is currently in working progress.

3.2.3 The Case of the United Kingdom

3.2.3.1 Coverage

The main objective is to provide the same healthcare to all citizens in the United Kingdom, without reference to their ability to pay or social status (Healthcare Commission 2004; European Observatory on Health Care Systems 1999). Approximately 11% of the population have however an additional and comparatively expensive voluntary health insurance, of which 59

% are employment-based, 31 % individual and 10 % employee-paid-based. The share of voluntary health insurances was quite stable over the last few years. The authors of the European Observatory Report explain the stable rate with the fact that these contracts are offered at very high prices. Taking a closer look at the voluntary private health insurance reveals that mostly higher socio-economic groups hold these insurances. Avoidance of waiting lists (queue jumping), the choice for hospitals, and a better accommodation are regarded as leading motives for more private health insurances in the UK (European Observatory on Health Care Systems 1999; Busse and Schlette 2003).

It can be stated that the consideration of the capabilities in this context reveals that the equal access to health care for all citizens is endangered by the fact that a growing proportion of the population is demanding private insurances to encompass waiting lists and receive full benefits. It is a double standard system although the rich pay into the NHS according to their tax share. In case the English tax system is progressive, the rich pay relatively more into the NHS than the poor.

3.2.3.2 Benefits: Scope and Structure

In general, the NHS does not have an explicit list of services and operates on a vague definition in order to provide an acceptable level of services. It refers to all necessary and approved personal medical services of the type usually offered by general practitioners. The authors of the European Observatory Report further state that a large degree of discretion is exercised with regard to the range of provided services. The only defined service covers only the pupil's medical examination (European Observatory on Health Care Systems 1999).

In general, three pillars of medical care services can be described: public health care provided by the communities, ambulatory care through general practitioners (as a core element), and hospital care mainly under the control of Health Authorities. As one exception, university teaching hospitals are controlled by the Ministry of Health. With regard to medical services, a large part of services functions with co-payment arrangements, i.e. for dental services 80 % of private co-payments are normal, ophthalmic services are

Table 3.2: Number of people waiting for hospital admission, England, 1997–1999, in thousands

Waiting time	Mar 1997	Jun 1997	Mar 1998	Dec 1998	Feb 1999
Total	1158	1190	1298	1174	1120
< 12 months	1127	1143	1230	1118	1068
12–18 months	31	46	68	56	51

Source: Europeam Observatory on health Care Systems 1999: 38

almost private and prescription charges are used regarding pharmaceuticals (European Observatory on Health Care Systems 1999).

At the beginning of the 1990s, a charter of patient rights was introduced and later reviewed in order to include local rights. Anyhow, these rights are so far not enforceable legal rights (European Observatory on Health Care Systems 1999). Examining the scope and structure of benefits, the waiting time problem of the NHS needs to be addressed as one major field of current actions (see also 3.2.3.5 below). Normally, patients wait more than 18 weeks for a hospital admission or 13 weeks for their first appointment for outpatient care. Table 3.2 shows that since the beginning of 1999 the number of people waiting decreased slightly. However, the extent of the problem is still quite large. The latest reforms address exactly these problems; however, waiting lists are an instrument of rationing in times of tight tax-financed budgets.

For pharmaceuticals, a list of drugs, which cannot be prescribed by general practitioners, and another list with predefined conditions for the prescription of special drugs exist. In addition, the Department of Health distributes charts among the physicians, which highlight cost comparisons of alternative products to promote moderate prescription handling. A task force carried out by the National Institute for Clinical Excellence should deal with the evaluation of costs and benefits, especially in regard to pharmaceuticals. In the United Kingdom, prices for medicine are agreed between the association of the pharmaceutical industry and the Ministry of Health (European Observatory on Health Care Systems 1999).

Furthermore setting priorities should be mentioned as an instrument to allocate scarce resources in health care (European Observatory on Health Care Systems 1999). The National Health framework sets out priorities in 2004 for cancer, paediatric intensive care, mental health, coronary heart diseases, and older people, wile diabetes, renal services, children's services and long-term conditions remain in a preoperational stage (Healthcare Commission 2004: 127). Success in reaching target goals has to be considered with caution, as there are some strategies to accomplish a target without really solving the issue. According to the Healthcare Commission, patients reported short notice arrangements. That means if a patient refused an offered date for an appointment, waiting time was automatically rescheduled down to zero (Healthcare Commission 2004: 41).

In the area of quality assurance, the National Institute for Clinical Excellence was founded. It is supposed to deal with guidelines as well as assessments of average cost effectiveness. The evaluation of costs and benefits, especially of pharmaceuticals, has been introduced in several European countries. The Commission for Health Improvement in the UK shall regularly examine health care facilities in reference to quality assurance measures. Lead should be taken by health authorities in cooperation with the local agencies. Besides this, the instrument of the health technology assessment is used (for more details see European Observatory on Health Care Systems 1999).

The loose definition of benefits and the displayed waiting time problem illustrate that the responsiveness of the English health care system is questionable. Due to the waiting lists, people without a private health insurance may not reach their treatment in time and are therefore limited in timely access. Considering health, benefits as an investment in the potential of individuals to lead an independent live and to address their capabilities the British system is chronically underfinanced and is endangering the live of individuals. The Common Market with its freedom of services may change the situation slightly because more and more citizens go to Continental Europe for treatment.

3.2.3.3 Financing and Purchasing in Health Services

The NHS in the UK is regarded as a relatively inexpensive health care system (share of health care expenditures in GDP is approx. 7 %). The health care system is mainly financed through general revenue, i.e. mainly direct and indirect taxes. The funding base comprises all forms of income and is as progressive as the tax system is. According to the OECD health data, the share of private expenditures was 17.7 % in 2003. The private share consists of voluntary private medical insurances, out-of-pocket expenditures for private medical care, long-term care, pharmaceuticals, dental care and ophthalmic services. In 2003, the British government spent £ 65 billion on health care, which amounts to 15 % of the overall central government spending. In comparison to other budgets like education and defence, health care topped them all and even was twice the amount of the defence budget. Per capita and year it equals £ 1.100 (Healthcare Commission 2004: 127).

For the future, the government planned to further increase spending to a per capita of £ 1.800, in order to reduce the length of waiting times. Most of the budget is allocated to local hospitals (42 %), community and general practioneer services, and pharmaceuticals.

The global budget of a certain region or district is administered by the corresponding ministry official. The Medical Practice Committee plans a nationwide distribution of family doctors according to the number of patients in the corresponding districts. Service agreement contracts are signed with family doctors, freelance active general physicians and dentists, ophthalmic opticians, pharmacists, as well as hospitals (NHS hospital trusts) and Primary Care Groups. The Primary Care Groups are subordinate to the superior health authorities (District Health Authorities and the NHS Executives (cf. European Observatory on Health Care Systems 1999).

Control of the health care system is achieved through regional budgets and management by the Department of Health. The latter is in charge of general health policy in addition to the identification of major issues. The NHS structure has to implement and carry out healthcare to the population. The benefit entitlement of the patients is restricted, and, moreover, the freedom to choose physicians is very restricted. In the primary care sector, the gatekeeper principle, enrolment with a family doctor and general practitioners, plays the major role in accessing health care services. The primary physician refers patients to in-patient care. In order to limit consultations with a physician, a patient telephone was set up. This can be used in cases of minor illness, as first help can then be exercised directly by the patient.

Looking at the described pattern it becomes visible that the British health care system takes a different approach in comparison to social health insurance systems. The health care budget is distributed to the regions on the basis of an allocation formula and from there to the local providers. Most actions taken centrally reflect more the idea of cost reduction than the consideration of a responsive health care system. However, as most British do only know their own system, they accept it. Only as the waiting lists were targeted by ECJ the government acted to invest more in the health care facilities to reduce waiting times.

The general practitioners are predominantly working as employees in health care centres and since 1999 in networks called Primary Care Groups (later Primary Care Trusts). The remuneration of general doctors is made by capitation fees in principle on the basis of the number of their registered patients and can be increased by surcharges or individual remuneration, i.e. in form of additional pay for long-term treatment of chronically ill. Bonus payments to the doctors are used to remunerate them for active participation in advanced and specialised training. Medical specialists working in the in-patient care sector are remunerated from the global budgets as well as per case. Physicians can increase their salary through private consultations, as physicians who work in the public sector are allowed to work a limited amount of their time independently for the private sector (European Observatory on Health Care Systems 1999). According to the Healthcare Commission Report 2004, private and public sectors work closely together in England. In 2002 and 2003, for example, a number of 55.000 operations were carried out by private providers and funded through the NHS. To mention a few, one third of psychiatric beds are in independent facilities, 70 % of the public staff carry out private work and 37 out of 41 new hospitals where built up by private financing initiatives (for further details see European Observatory on Health Care Systems 1999).

3.2.3.4 Distribution in Health Care

Due to inequalities of people's health, wealth, and age, the allocation formula is adjusted to meet regional differences. This is also known as target allocation. However, differences occur between the real allocation and targets, measured in distance from target. Nevertheless, until 2010, targets should be obeyed (Healthcare Commission 2004: 129). According to the report of the European Observatory on Health Care 1999, a major field of action comprises the priority to reduce inequality and deprivation by collaborative cooperation between all relevant authorities and organisations.

With the help of the Resource Allocation Working Group, a resource allocation formula adjusted to different regions was worked out and constantly modified. In principle, a weighted capitation payment based on needs includes a region's population size, age, composition of sex, and level of morbidity (European Observatory on Health Care Systems 1999).

As the NHS is mainly tax-financed, all tax payers finance their health care system according to the absolute amount of their tax payments. Whether the English tax system is proportional, regressive or progressive is an open question as it is in other countries as well. With the increasing number of individuals insured in voluntary medical insurances, the equality in access and benefits is undermined.

3.2.3.5 Current Problems and Long-term Perspectives

The key priority of the NHS deals with tackling waiting times and waiting lists. Since summer 2004, a new initiative offers patients, who have been waiting more than six months, the choice of being treated at another hospital or from a different provider. As a target, by December 2005 the NHS wants to reach the goal of offering four to five choices of hospitals or providers for planned surgeries (http://www.nhs.uk).

According to the Healthcare Commission 2004 in England, the Department of Health gave out a target plan to be achieved by March 2004. A major goal was to lower waiting time in case of elective surgery to less than nine months for admission to hospitals. In comparison to the previous five years a sharp decrease was achieved. Another goal was to decrease waiting time for a first outpatient appointment to less than four months. Even though most patients were still waiting longer by March 2004, the total number decreased over the past 18 months. In at least 90% of the cases, access to accident and emergency treatment takes no longer than four hours. (In Germany, an average of 15 to 20 minutes is normal. Keeping in mind that in case of a heart attack the first two hours are predetermining survival chances, four hours seems quite long.) General practioneer appointments are offered within two working days and cancer specialists can be seen in a time period of no longer than two weeks. However, the picture does not look equally as satisfying all across the United Kingdom. In Wales, for example, 11% of the patients were still waiting longer than 12 months for hospital admissions. For outpatient treatment, 10% of the patients referred to specialists reported waiting times of more than a year. Two major arguments can be found to explain this situation: firstly, differences in population and their health status, and, secondly, differences in the way services are organised. In general, a shortage of resources (infrastructure and manpower), inefficient organisation and planning, as well as inefficient use can be observed. Since summer of 2004, all patients waiting longer than six months are offered the possibility of treatment at another hospital (Healthcare Commission 2004).

Another and major challenge is the shortage in workforce. As the NHS is heavily recruiting labour workforce from abroad, a shortage of manpower in health care is evident. This situation is enhanced even further as the European Working Time Directives creates new demand of nurses and physicians. Additionally, migrants are not necessarily staying in the NHS or United Kingdom because they might return home or go to other countries. Moreover, British workforce turns to more part-time arrangements and early retirement (nurses are aged 45 to 55 on average now) (Healthcare Commission 2004: 37).

The reform debate in Great Britain concentrates on increasing expenditures for health care systems, for the overall health sector within the next five years, as well as the development and modernisation of the corresponding infrastructure. The budget of the NHS will most likely increase from 93.5 bn Euros (2003/2004) to 151 bn Euros (2007/2008) (Busse and Schlette 2003). Stevens (2004) mentions that expenditures of the NHS in proportion to the complete GDP will most likely reach 9,4 % by 2007/2008.

3.2.4 Comparison of the Two Countries

3.2.4.1 General Remarks on System Comparisons

Not only in health care comparing and benchmarking systems became very popular. Regarding health care systems, the public became particularly aware since the ranking of the World Health Organisation in the year 2001. The comparison of health care systems wins a far-reaching meaning through the application of the OMC at a European level in the fields of employment, pension, and health care. OMC provides a frame and opportunity for a target oriented European Health care policy, reflecting the social political goals of the Lisbon-Strategy (to become the most competitive market worldwide). As far as health care is concerned, the OMC process is still at the beginning and momentarily aims at agreeing upon a set of European targets and indicators. Principles like equal access for all, high quality standards, and sustainable financing could be chosen. With regard to the application of the OMC in health care, EU member states are rather hesitant, as the method should not undermine the principle of subsidiary in health care.

The comparison of health care systems not only offers an overview of features, characteristics, and structures of a system. Benchmarks also place attention (and sometimes pressure) on highlighted health care service arrangements. Furthermore, comparisons can serve as basis to initiate a mutual learning process for the improvement of national systemic arrangements, and support its main goal of improving the population's health situation.

Due to the complexity, historical path dependencies, different measuring and assessment methods, political, historical, cultural, and socio-economic constellations of the different health care system, a systematic comparison is inappropriate. Moreover, so far hardly any international comparison was carried out with common approaches, indicators or aims (Riesberg et al. 2003; Henke and Schreyögg 2004). To a large degree, this is due to key actors setting the agenda. Most benchmarks use indicators, which do not accurately reflect distinctiveness and individuality of the national systems in comparison (Schneider 2004).

3.2.4.2 Comparing Germany and the United Kingdom Based on Selected Indicators

One classical approach to health care system comparison would begin with analysing the share of health care expenditures measured as ratio of the GDP (ratio of health care expenditure) to show the size of the health care sector. The standardised OECD data show a ratio of health care expenditures for European member states between approx. 11 % (Germany) and approx. 6 % (Ireland). Analysing the ratio of health care expenditure of the Middle and Eastern European Countries, 4.5 % to almost 7 % can be recorded. However, many Middle and Eastern European Countries still exhibit some degree of informal payments, which is difficult to assess.

Comparing health care expenditures as a share of GDP from 1992–2001 the trend in Germany and Great Britain points in the same direction. However, the starting levels remain different. Health care expenditures measured in USD purchasing-power parities show that UK expenditures per capita are lower by USD 816, which corresponds to 71 % of German expenditures. Considering the renewal of infrastructure and the dismantlement of waiting lists the future development of health care expenditures in the UK will rise.

The mere comparison of ratios in health care expenditures is not meaningful, as the level of ratios is influenced by the growth rate of the GDP (Mossialos et al. 1996). Moreover, a trend of slowly rising health care expenditures does not signal a mandatory deterioration of the state of health of the population. It could also be explained by changing consumer habits. The structural composition of the population therefore gains meaning, especially in regard to ageing, multi-morbidity, and increasing chronic illnesses. However, benchmarking so far neglects a population's (possible) willingness to pay, e.g. for shorter waiting list (Schneider 2004). To sum up, the comparison of ratios is limited and further analysis is necessary. Confronted with decreasing financial resources and budgets as well as long waiting lists, countries are eager to allocate resources more efficiently.

A better indicator for evaluating health care systems could be the avoidable mortality, as these deaths are influenced by prevention and cure. Other indicators would be infant mortality or the most frequent causes of death (Riesberg et al. 2003). Comparing these indicators, the German position is better than the UK in most cases. Likewise, Germany exhibits a smaller number of age adjusted deaths. From this point of view, the rationing of the NHS seems to lead to comparatively lower output.

Apart from the discussion of figures, quality assurance is a huge topic in most European member states. In Germany, the discussion about quality of care takes place continuously and initial improvements are targeted by the introduction of disease management programmes for certain illnesses, i.e. diabetes and breast cancer. The UK has a unique body of technology appraisal: the National Institute for Clinical Excellence. The external review of the National Institute for Clinical Excellence (carried out by international experts at the World Health Organisation, Regional Office for Europe in Copenhagen) highlighted the commitment of the Institute and its work as "an important model of technology appraisals internationally" (Hill et al. 2003: 3). So far, Germany started with the introduction of a similar institute ("Institut für Qualität und Wirtschaftlichkeit im Gesundheitswesen"), in order to target at improving quality of care.

It is believed that a good performing health care system is a system in which the patient seeks treatment and the appropriate treatment is delivered in a suitable time period. This situation would at the same time correspond with a health care system targeting capabilities and the empowerment of individuals. In this light, fair financing and equal access could be considered as indicators. This reveals the fact that, in order to compensate for the unsatisfactory tax financed NHS service provision, private health insurances arose as supplementary insurances in the UK. The entitlement of a complete equality of medical care access is thereby undermined, as private insurances only offer very rigid and expensive rates. Also, persons with low income and/or bad risk profiles are excluded from the utilisation of private health services. On the other hand, the tax financing approach comprises the overall population. In Germany, the mode of financing through pay role taxes covers only 90 % of the population, however, the other 10 % of the population are considered being able to privately provide for their health risks. Additionally, the SHI system with its social elements, i.e. non-contributory co-insured and contribution ceiling, realises a certain degree of redistribution of income within the SHI. Considering the idea that the social insurance should reflect the life-course the German SHI pays attention to it. The working population supports the non-working, family dependants and children. As illustrated, the retired contribute to the SHI to a smaller part compared to their benefits. However, supplementary private health care insurances are not yet necessary to cover basic health care services. Considering rescue times and the density and quality of health care facilities in Germany (including the number of beds per room in a hospital), full-coverage with equal access for all is better achieved there than in the UK.

In addition, the German health care system allows for a range of choices with regard to insurances, providers and treatments. Every SHI insured may choose insurances and change to another SHI once in a year or by changing employer. In cases of illness the patient is free to contact any physician, general practioneer or even specialist. That means that there is no obligation to contact the general practioneer or family physician first. However, the latest health care reforms try to redirect patient flows by the introduction of a 10 Euro charge, if the patient did not collect a referral before seeing a specialist. To refer again to the capability approach in Germany the empowerment and responsiveness of the health care system seems to be higher.

Both countries, Germany and the UK, had to react upon the rulings of the ECJ. The NHS was put under pressure in regard to the waiting list problem, whereas the German health care system was affected differently. The institutional framework of the German health care system seems to become vulnerable to the European Competition Law and the "Aquis communitaires" with its strong emphasis on competition. There is a huge debate on how to assess the influence of the European Competition Law and ECJ rulings on National social law (for further details see section 2.2.4).

Regarding future challenges, the two systems differ from each other. The main issue for the British system is to solve the waiting time problem and to implement the programme for reforming the NHS. Especially, patients' choice of specialist care and hospitals, which is limited in the UK, should be targeted in order to achieve a more patient oriented health care system. The European dimension of the issue of waiting lists is still unresolved, as the ECJ was asked to take a decision in that matter. In a European context, this could be the basis for introducing a European wide minimum standard with regard to waiting times. The European Law could possibly serve as a tool of privatising the NHS (Henke 2002). Therefore, the actions of the Labour government need to be monitored, as it tries to draw back the market orientation introduced by the Thatcher government (European Observatory 1999). In comparison to the UK, Germany is affected more by the European Law, which interferes with the German Social Law (Busse 2004; Eichenhofer 1994; Henke 2004; 2002; Henke et al. 2004; Karl and v. Maydell 2003; Marhold 2001; Mossialos et al. 2001; Schneider 2003; Schulte 2002; Pitschas 2002; see also section 2.2.4).

Since the health care sector is a growing market, health services should not be reduced to mere cost factors. Moreover, health care markets provide contributions to the value of society's human capital and productivity. Thus, chances for economic development and the population's health status can be influenced positively. Another aspect deals with the importance of health care markets and the production of health care products and services. Additionally, various suppliers, i.e. pharmaceutical industry, medical device manufacturers and biotechnology industry, guarantee employment opportunities and contribute to a growing economy through research and development activities. The denotation of "health care services" covers a broad set of actors, e.g. the providers of inpatient care (acute-care hospitals, nursing homes, hospices and prevention and rehabilitation facilities), providers of outpatient care (doctors and dentists in private practice, therapists and practitioners of natural medicine, ambulatory nursing services), providers of rescue and transportation services, as well as providers in the areas of business-to-consumer and business-to-business. "Retail sales and trades" encompasses pharmacies, distributors, retail medical supply stores, dental technicians, opticians, orthopaedic technicians, and hearing aid acousticians. In this context, the development of the health related fitness and wellness industry is worth mentioning. The employment opportunities of health care service providers and in health related sectors can be illustrated by envisioning that approx. 10 % of all employees work in this sector. Overall, the creation of many new jobs can be ascribed to the health care sector in a broader sense (Henke et al. 2004b). In this regard the NHS risks to loose large employment opportunities because the under financing of the system has led to a shortage of health care professionals. At the same time, shortages in a labour intensive market affect the quality of care and service provisions.

Summing up, the cases of Germany and the UK show that each system has its own way of financing, organisational structure, and instruments. However, some similar trends can be extracted as well. These include the targeting of issues like quality assurance, priority setting through guidelines, certifications, and evidence-based medicine. Another aspect is the approach of a mixed system in financing: both systems share co-payments. While Germany has elements of tax financing, the British system shows a growing part of voluntary insurances. Both health care systems, in Germany and the UK need to pay careful attention to future (reform) actions. The more the two systems introduce market elements and in particular in the German case downsize redistributive elements, the more they will end up applying European Competition Law as well as the Four Freedoms of the Common Market with all their consequences for organising the systems.

Considering again the book's principles the degree of freedom is larger in the German health care system in various ways. Empowerment of the insuree and the patient together with the introduction of more preventive medicine may boost the health potential of the population. The way in which the German SHI is financed displays that the abilities of the citizens are paid attention to and that it corresponds at least from a theoretical point of view to the requirements of the life-course approach. Reform efforts in the direction of a more sustainable financing of the health care system show at least that the aspect of intergenerational distribution is seen. Furthermore and more important in respect to capabilities within the population is the fact, that the statutory fund system is offering special bonus packages to encourage patients to participate in preventive health care programmes or sport activities. Even if this measure is categorised as a marketing instrument to attract health insurees at the end it can help to invest in people's health abilities, too. Of course, the empowerment has it limits; however, every possible (even monetary) incentive should be applied to preserve a healthy life and well-being.

In contrary, the NHS in the UK offers only limited choices and often makes a supplementary private health insurance necessary to provide for basic health services and encompass waiting lists. The rather long rescue times prejudice health potentials because a long waiting time can impair the individual's abilities to recover. Empowerment and enabling do not seem to be in the centre of attention. Additionally, due to open rationing a person in higher ages is risking to face long waiting times.

Last but not least, the education of health professionals could be a new topic on the list of indicators to compare the health care systems. And here the different curricula at the universities are to be compared. The quality of the education of the health professionals is another topic with growing importance in the common market. And finally the quality of medical treatment itself might depend on the scarceness of resources.

3.2.5 European Dimension and Elements Affecting Health Care

3.2.5.1 Integrating European Health Care by Different Concepts: Coordination, Convergence, Subsidiarity, Harmonisation, Competition

Strictly speaking the European Commission's competences in the field of health care within the overall framework of subsidiarity are mainly reduced to public health issues (Art. 152 EC) as communicable diseases or health promotion. The competences in organisation and in financing of health care systems are in the hands of the EU-member states. However, Wismar and Busse (1999) found out that from the European level a number of 250 community interventions affected the organisation of the national health care systems. In general, all actions taken in any policy fields at the European level need to secure a high level of human health protection [Art. 3 (1), Art. 95 (3)]. The fact that approximately, 80% of the interventions are provisions/directives or guidelines, illustrates the binding character of these actions at supra-national level.

In contrary to the public perception, the first interventions with impacts on the organisation of national health care systems can be traced back to the late 1950s as the Common Market was agreed upon and conditions for migrant workers were established (Wismar and Busse 1999). The latter is nowadays better known as the Directive on Coordination 1408/71.

The main influence on the national health care system can be ascribed to the implementation of the European Single Market and derives additionally from the European Competition Law (further see chapter 2.2.4). The Fundamental Freedoms of the Common Market regarding free movement for goods (Art. 30), services, persons (including workers) (Art. 39 (3) and Art. 46 (1)) and capital comprise the major fields of actions of the European Community, however other policy fields, i.e. environmental policy (Art. 174 (1)), agricultural policy, consumer protection (Art. 153) and food safety or safety at work (Art. 137) impose pressure on the EU member states and their health care systems.

Of special interest for the organisation and financing of the national health care systems are Community actions dealing with regulations of health care professionals, of access of patients and providers to health services and of pharmaceuticals. The movement of patients and access to health care abroad was highly debated since the ECJ rulings. Ever since, the access of patients to ambulatory care is given. Only with regard to hospital care a restricted access was agreed upon because of the special national planning procedures and capacities of the inpatient care sector. Additionally, the possibility of cross-border negotiation with health care providers was implemented. All these decisions display that the assurance of access has to be reflected by the organisation of the national health care systems. For instance, for the NHS in the UK prices of goods and services had to be calculated. Furthermore, mainly states with a chronically under financed health care system were asked to reduce waiting time lists.

Equally, the regulations regarding health care professionals in Europe and the mutual recognition/approval of degrees and diplomas affected the organisation of the national health care systems and provided a ground for a free movement of individuals working in regulated occupations. In this context the training and education frameworks of the EU member states were analysed and certain minimum standards could be agreed upon. Nowadays several directives provide the European framework and even the implementation of the Bologna Process reflect this coordinated policy approaches.

Looking at the mentioned examples and the character of the European elements in health care it becomes clear that a wide range of different actions takes place representing several instruments and methods. However, in general, a harmonisation of the national health care systems is not the target goal and would be heavily opposed by various stakeholders. Nevertheless, the decisions of the ECJ gave rise to the tension between the principles of the Common Market and the principle of subsidiarity because more and more the organisation and/or financing aspects of the national health care systems are affected. Another more soft way in influencing the national health care system is followed by the Commission's proposals, green and white papers as well as communications and memorandums. To sum up, the European activities can be characterised as an interlinked patchwork.

In parallel to the mentioned developments and apart from the European Community action characterised as a top-down approach, a convergence of the national health care systems can be observed. How far the concept of convergence goes remains still open. Convergence could mean a slow phasing in along similar trends yet coming from different levels, or it could also imply convergence towards one common target. The latter would probably be the case if there European quality standards and guidelines for medical procedures were to be introduced. Nowadays, convergence is visible as the EU-member states are facing similar challenges, i.e. demographic aging and structural changes, increasing costs pressures of emerging new medical-technical innovations or tight public spending budgets. Of course the member states are nationally approaching their reform needs, but in some cases similar methods are applied. The latter can be seen regarding the quality assurance efforts, the introduction of health technology assessment or the implementation of cost-benefit calculations as a condition for an allowance for reimbursement with the social health insurances or the national health services. Another form of convergence can be recognised with regard to financing health care systems. On one hand, the tax based systems strive for more private financed arrangement, as the public systems get more under pressure of cost containment. On the other hand, the payroll-tax based systems are incorporating more tax financed solutions or co-payments.

The OMC in some ways is inextricably associated with the concept of convergence because the foreseen procedures could impact the health care systems in a kind of process oriented convergence. At the moment the EU member states are struggling to reach a feasible set of indicators for an effective comparison of health care systems. Conceivable and sufficiently vague indicators could be the following: equal access for all, fair financing, and a high quality level of health care. However, in the general the various national stakeholders are closely watching the process and prefer a rather loose instrument application, since the overall principle of subsidiarity should be maintained in health care and thus the European Community actions remain limited. The OMC serves additionally as a bottom-up approach to stimulate policy learning processes and system competition. As the application of OMC in health care is quite recent the outcome or utility cannot be fully judged at this point. Certainly, the OMC has a political potential of influencing the organisation of the national health care systems and bringing forward more convergence in this policy field at the European level. However, the extend of its application and impacts is carefully watched by the national stakeholders.

Apart from the considerations above the Common Market gave rise to European-wide markets in health care. Even if the development of an integrated European market in health care is only restricted the European framework together with the ECJ rulings significantly enhanced its formation. A more accurate picture appears below, where several submarkets will be analysed, e.g. the market for health care goods and services, the market for providers of health care services and the market for health care insurances. The following sub-section will illustrate additional European elements in health care with regards to specific markets and bring to mind the complexity of the European dimension.

3.2.5.2 Emergence of Different Health Care Submarkets in Europe

Market for Health Care Goods and Services

The market for pharmaceuticals and medical products, also considered to be a market of goods, is most likely an entire, yet segmented, market using different strategies in pricing and distributing. On the supplier's side, the industry, wholesalers, pharmacies and physicians follow there own set of incentives. To counterbalance the different sets of incentives, state regulation interferes in all EU member states, i.e. by applying different VAT rates or by operating with different co-payment regulations. Only the market of prescription-free drugs works according to a 'real' market of goods. Especially internet pharmacies take advantage of this difference.

On the side of the demand, European citizens have free access to European wide markets for pharmaceuticals and medical products. With regard to pharmaceuticals, a uniform admittance by the EU drug authority in London could be established through a consent of the European Council. This would stand in accordance to the adopted reorganisation of the drug right by the European Parliament for newly developed drugs against cancer, AIDS or diabetes mellitus (Nink and Schröder 2004).

The markets for health care services are part of the category of general services. However, they display certain features which distinguish them from ordinary goods and services (for more details see 2.2.4). As a rule, a patient treatment as close as possible to the patient's residence is important in acute case. In terms of planned resp. elective surgeries, a Europe-wide supply plays an even more important role. From an economic perspective, a cost-benefit calculation is carried out, considering transaction costs as time, control costs, safety, and average contract costs. Moreover, waiting costs, co-payments, and language barriers have an additional influence on an individual's decision. The market opening for in- and outpatient care was reached through ECJ rulings and through the influence of the European Competition Law in combination with the progress of European integration and increasing mobility (see 2.2.4). As a result, the ECJ insisted in all recent jurisdictions upon a free access to outpatient care without having to ask for permissions in advance. In the area of outpatient health care services, cost analysis considers transaction costs as co-payment rates, which are differently regulated from country to country, e.g. differences of co-payments for the consultation of the doctor. Another reason for the demand of health care services outside of one's own country could rise, as health care services in other systems, like new diagnoses and therapies or special types of treatment, are offered at all or more cost-effectively.

Nevertheless, a restricted access for inpatient care exists, due to necessary permissions to the corresponding health insurances. That the inpatient health care sector enjoys special protection can be justified because of the requirements of capacity planning. Since the introduction of DRGs in Germany, as well as standardised procedures, services of inpatient care can be compared more effectively on a European scale. Even the NHS is now moving towards price building efforts, as they need to reimburse medical treatments obtained Europe wide in the context of reducing waiting time lists. Some people say it is only a matter of time until the hospital sector will be completely liberalised Europe wide. However, for most patients from countries with extensive waiting times, access to Europe-wide inpatient care means an improvement in the quality of life. In some extreme case, Europe wide inpatient care might even prolong life. From a sickness fund's point of view, cheaper providers at the same quality could be very interesting, too.

Market of Health Care Providers

Another submarket is the market of health care providers. The dimensions to be examined here are the freedom to practice, the cross-border contracting, and the question of quality assurance. In general, providers of health care services have the full freedom to practice according to the Common Market in the EU.

Along with rising cross-border demand for health services, regulations regarding cross-border contracts (particularly with regard to quality assurance), accreditation, and transparency play an essential role. In this context, the new plan for a General Directive on Services in Europe based on the principle of origin can be criticised, since in regard to quality assurance it would imply a second best solution. The principle of origin would mean that the home country of service providers is responsible for controlling.

Beside the possibilities of Europe wide service, new forms of cooperation between providers can already be observed among dentists in Mallorca, e.g. patients receive addresses from the corresponding cooperation partners in the home country to contact in cases of after-care (Schaub 2001). Cross-border service agreement contracts are becoming increasingly interesting for Germany, if neighbouring Middle and Eastern European Countries offer well priced and high quality care. Caused by growing demand, new forms of Europe-wide cooperation between stakeholders could emerge. This could lead to uniform treatment guidelines or uniform accreditation proceedings of providers. Nevertheless, a system of Europe-wide planning of capacities in inpatient care seems unrealistic.

Market of Health Care Insurance

In recent years, the health care insurance market was opened for Europe-wide activities in the area of private health insurance. Therefore, private insurance packages can be offered unhindered everywhere. Nonetheless, no greater changes have arisen till now. This could change, however, when national social health insurance systems exclude more health care services from their reimbursable service packages (Greß et al. 2003). The question of "fair financing" and distribution of equal access could become a highly debated topic. For the overall German Statutory health insurances, the Europe-wide health insurance market is still relatively insignificant.

At this point in time, it remains rather unattractive for individuals to search for health insurances outside their own country. One reason could be that, due to the ECJ ruling, it is possible to reimburse health care services obtained abroad. No real necessity to have a Europe wide voluntary health insurance seems to exist. Another possible reason is the fact that the acceptance to contract with foreign insurances is low from the side of the consumer.

The last section covers both remarks and recommendations with regard to the two compared systems. It also gives a brief review on how much the capabilities approach and the life-course perspective, the bases of an enabling welfare state, can be observed with regard to health care.

3.2.6 Concluding Remarks and Recommendations

Boosting the self-responsibility and empowerment of individuals through co-payments serves at the same time to combat the growing budget constraints in the national health care systems as well as in social health insurance systems. However, in Germany self-responsibility is additionally accomplished through offering patients the possibility to ask for an invoice in order to recognise the real costs behind their treatment. Another method to be mentioned are the efforts of health insurances to offer benefits in kind or cash, if a patient is ascribed in a fitness club or can prove the participation in prevention programmes. Lastly, the possibilities for patients to have a voluntary private health insurance for special desired benefits can be counted as well. Strong patient rights give the patients an additional tool to pursue their own interests.

For the UK, only a little is known about targeting self-responsibility and empowerment. The existing patient rights are quite weak. Perhaps, the efforts performed in order to offer a limited choice to the patients, if waiting too long on the list for elective surgery, can be seen as a positive reaction towards the establishment of a more patient oriented health care system. The aspect of empowerment is also pursued on a European level, as an information base is to provide the European citizens with information about the other health care systems. In addition, the consumer protection is a European task. It was noticeable for all European citizens during the "mad cow" crisis, and also with regard to food and drug safety.

The life-course perspective embraces different aspects, as it refers to the individual level on the one hand, and to the generational dimension on the other. The individual dimension can be observed by boosting preventive and active life-styles, in order to prevent or delay the outbreak of diseases and to stabilise the health status conditions. On the generational dimension, targets include the sustainability of the health care system and also its financing. In this context, the demographic change of the German society is an important topic. Generally, the existing system could broaden the financing base in the direction from salaries and wages to taxable income and include the entire population in the SHI. However, this would only be a short minded and not sustainable solution. Other models (see section 1.1.1) plea for a system with partially capital funded elements. The alternative would be to reduce benefits or to exclude them from being reimbursable with the SHI.

The British system is based on strict global budgeting and operates with open rationing. Therefore, another model is followed. Every citizen is covered by the NHS, and if these benefits are reduced, the population is demanding a larger degree of voluntary private health insurances. With the growing share of the private market in health care, equal access is questionable and from a generational aspect their distribution is based on underlying tax system.

At the European level, the intergenerational aspect could be emphasised by demanding a system with equal access, fair financing, and a high quality level of health care. These three aspects might serve as the underlying principle for the OMC in health care. At present, the process has only started in the field of health care. Working groups are now assessing existing indicators for health care system comparisons. So far, an overall applicable approach has not yet been agreed upon. The existing comparisons, i.e. carried out from the World Health Organisation, are highly criticised because of their complexity and the influence of agenda setting.

Looking at other political fields of application of the OMC it can be debated if in the case of health care this instrument will lead to more than an overall declaration of indicators. The European member states are in favour of a rather slim OMC process, without a tight system of reviews and reports. They favour the OMC being more a platform for exchanging information. Thus no new "Maastricht Criteria" will be developed for health care.

Regarding the EU Commission's ambitious strategic programme with installing the Directive on Services, an approach for more market orientation can be observed. Without improvements (especially regarding sectors of social security), these directives could jeopardise the principle of subsidiarity and national competence in the field of health care. That is why the European actors try to reach an exception of the health care sector or the consideration of especially social policy principles and issues. At present, the European influence on the national health care system derives to a large degree from the ECJ rulings (see chapter 2.2.4). However, harmonisation in the organisation of national health care systems is not strived for and politically not wanted (further see Bundesministerium der Finanzen 2000). However, it seems clear at this point, that if the national health care systems will drive away from their social elements of redistribution, they will get vulnerable by the European Competition law, and liberalisation of the health care markets will continue.

From benchmarking and country comparison it should become clear that elements working in one country do not necessarily evoke the same effects in another national setting. More so, a need for adaptation to the respective country system exists. Nevertheless, comparing activates a process of learning from each other. It becomes clear that no matter what organisational structure a system has chosen, the challenges (e.g. demographic changes and technological progress) are mainly the same all over Europe.

However, all systems seem to have in common the shift to more private financing of health care. This is especially due to tightening national budgets caused by unemployment and lower economic growth.

At this point it is still too early to judge if the standardisation and introduction of similar means (medical guidelines, evidence-based medicine, certifications of providers) will lead to a European minimum standard in health care (not necessarily as a race to the bottom). For sure, comparing systems consequently leads to more competition between them. Low performing countries could come under pressure, and the media could play a great role in using rankings. Priority setting, competition and quality assurance will gain more importance for all European health care systems.

3.3 Old-age Security: Germany/Poland

3.3.1 Starting Point

3.3.1.1 Objectives and Methods of a System of Old-age Security

All institutionalised sources of income security for old age can be understood as a comprehensive old-age security system. The system can consist of many elements, which are often combined together and share the primary common objective to secure income. The old-age security system may be understood in an even wider sense, as the aggregation of state institutions and benefits which are related to the social situation of "old age" (Schmähl 1986). In reality, however, the term only includes systems of cash benefits or even only the general old-age pension system (Igl 1988; Williamson and Pampel 1993). Concentration on pension systems does not of course mean that other components of the life situation in old age are underestimated, both material (housing, health protection, infrastructure) and immaterial (social integration).

If only formal income sources are counted as old-age security system, other "informal" sources are excluded, for example through family or neighbours, which used to be the most important form of old-age security before the development of pension systems. Today family is still a crucial source of security in old age in less developed countries in the world. Throughout their development in Europe in the 20th century old-age security systems have replaced family support and thus increased personal autonomy in old age. Usually only such sources are included into old-age security, which explicitly serve the function 'old-age'. This means that social assistance is usually not included, and there are problems with proper positioning especially the third tier of old-age security (see further below). To what extent, for instance, do different forms of saving like bank deposits or housing serve as an income security for old age? The main function of an old-age security system is often subdivided into two objectives. On the one hand, old-age security should guarantee a minimum income (prevent poverty), on the other it should also make it possible to keep the living standard in old age which was reached during working life. Various old-age security systems weigh both objectives differently.

The two general methods of financing an old-age security scheme are pay-as-you go and funding. Pay-as-you go, the method used today in most public systems, means that current pensions are financed by current revenues from pension contributions or taxes. Funding, mostly used in private schemes, is accumulation of assets in advance to meet future pension liabilities. As moving to funding has been treated as one of the ways to increase sustainability, many recent reforms of old-age security systems have included the development or strengthening of funded schemes. However, both methods have benefits and problems and there are many myths concerning these issues (Barr 2002; 2004).

"Providers" of old-age security include, apart from the state, insurance companies, banks, employers and the individual him/herself. The traditional economic argument for state involvement in organising pensions has been negative external effects of non-insurance. People who choose not to insure themselves for their old age will have to rely on others in financing their consumption in old age, thus imposing external costs on the rest of society. In this sense, an obligatory pension system can be seen as a protection of society against short-sightedness or deliberate 'free-rider behaviour' of a minority. In fact, this is an argument for an obligatory system, not necessarily organised by the state. Secondly, it can be a justification for a pension system, which guarantees only a minimum protection.

Economic arguments which traditionally have been raised in favour of state involvement, including state production, in some other social areas like health, not always concern pensions. The strong development of the state's role in pension systems can thus be rather attributed to equity than to efficiency arguments. (For a classical presentation of efficiency and equity arguments for social policy see Barr 2004.) All aspects of justice are relevant in the context of pension systems (see section 1.2.3). Old-age security systems redistribute income over the individual's life course, between individuals, between age groups and between generations.

Recently, problems with information have been raised as an economic efficiency argument for the state's role in old-age security. They can be a justification for a larger state pension system as well as for an important role of the state in organising and supervising private provision for old age. The welfare state in this sense not only fulfils its 'Robin Hood' function, through poverty relief, redistribution of income and wealth and the reduction of social exclusion, but it also has the 'piggy-bank' function of ensuring mechanisms for insurance and for redistribution over the life course (Barr 2001). In fact, from their invention on, old-age security systems have decisively contributed to the standardisation and institutionalisation of the life course (see 3.3.1.3).

3.3.1.2 Diversity of Solutions in Old-age Security Systems in EU Member States

The concept of three pillars – or tiers – is used to describe the old-age security systems in the EU. The three tiers are distinguished as follows:

- state (social) old-age security systems: compulsory systems organised by the state on national scope and embedded in the general social security system;
- occupational systems: deriving from the initiative of employers organising old-age security for their workers, ranging in scope from one single company to a whole industry or sector;
- individual systems: derived from individual initiatives in organising old-age security through, for example, buying a life insurance or saving in a bank.

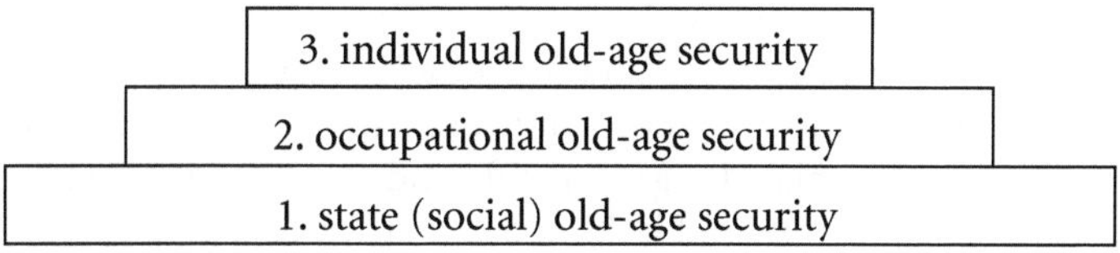

Figure 3.5: Tiers of old-age security

For the EU, the dominant role of state systems is typical. Member states differ very much as far as the scope of occupational and individual systems is concerned. In most member states (see table 3.3) the general state old-age security system provides employees (or, in some cases, all economically active) with earnings-related old-age pensions. Pay-as-you-go defined benefit schemes prevail – they exist in 16 member states. In recent years a new solution has been introduced in Italy, Latvia, Sweden and Poland: pay-as-you-go 'notional defined contribution' schemes. In countries with earnings-

Table 3.3: Types of old-age security systems in the EU-25

General state pension system (first tier)		Supplementary pension schemes (second and third tiers)	
Pension level	Basis of entitlement/ pension assessment	Voluntary	Obligatory *s: state; o: occupational; i: individual*
Individually assessed	Insurance/earnings (DB)	1/ Austria, Belgium, Cyprus, Czech Republic, Finland (2), Germany, Greece, Lithuania, Luxembourg, Malta, Slovakia, Slovenia, Portugal, Spain	2/ France *(o)* Hungary *(i)* Estonia *(i)*
	Insurance/ contributions (NDC)	3/ Italy Sweden (a)	4/ Latvia *(i)* Poland *(i)*
Flat-rate	Insurance (paying contributions)	5/ Ireland	6/ Denmark (2) *(o)* UK (*s*, *o* or *i*)
	Residence	7/ Finland (1)	8/ Denmark (1) *(o)* Netherlands *(o)*

General state pension systems in Denmark and Finland consist of two parts (see also table A.1 in Annex).

(1) National pension based on residence;

(2) Social insurance pension, in Finland earnings-related, in Denmark flat-rate;

(a) In Sweden, both parts of the new general system are classified as first tier: the pay-as-you-go part (NDC = notional defined contribution) and the fully funded part.

related (or contribution-related) pension systems, supplementary, mostly funded schemes are usually voluntary (France had been an exception). A new solution, an obligatory general funded scheme, has been recently introduced in Sweden, Estonia, Hungary, Latvia and Poland. In Slovakia, such a scheme has started in 2005, and in Lithuania, a partial switching to a funded scheme has been introduced as an option, not an obligation (Holzmann and Hinz 2005). As can be seen from this list, almost exclusively the new EU-member states have so far decided to reform their pension systems in that way. Objectives and methods of such reforms will be analysed below in regard to the case of Poland (see 3.3.3).

Five countries from Northern Europe (Denmark, Finland, Ireland, Netherlands and the UK) have old-age security systems, which provide flat-rate benefits to all residents or insured. These again may be subdivided into three groups. Finland has a mixed system: a flat-rate national pension for residents and a defined-benefit social insurance scheme. Ireland and the UK have social insurance systems financed by contributions, which however only determine entitlement but not the level of benefits. In Denmark and the Netherlands there are national pensions, which are based on residence, however they are financed by contributions in the Netherlands. (Table A.1 in the Annex presents the main elements of construction of general (regular) old-age security systems within social security in EU-25, according to the situation on 1 May 2004.)

In the majority of countries, there is a pension insurance contribution, which in most cases finances old age, invalidity and survivors. Almost in all cases both employees and employers pay contributions, the employer's part is usually higher. There is a ceiling on contributions in most countries. The state also takes part in financing of pensions. In all countries, both reaching the legal retirement age and a prescribed "waiting period" are conditions for drawing pensions. The waiting period consists of contributory periods or (in Denmark, Finland and the Netherlands) of years of residence. For drawing full pension usually some 40 years of insurance (or residence) are required. Legal retirement age is equal for women and men in most cases, usually 65 years. In almost all the countries, which still have a lower retirement age for women, equalising laws have already been enacted.[47] In almost all countries, pensions are adjusted automatically, usually on the basis of price development.

As compared to other parts of the world, also to other highly developed industrialised countries, the EU is characterised by a high importance of state old-age security systems within social security. In all member states state (social) systems are the main source of income in old age (Żukowski

[47] The legal retirement age for women will reach that for men in 2009 in Belgium, 2015 in the Czech Republic, 2016 in Estonia, 2008 in Latvia, 2020 in the UK and 2033 in Austria. Correspondingly, also waiting periods will be equalised. In Lithuania, the retirement age for women will be raised to 60 by 2006.

1997). Spending on all pensions (old age, invalidity and survivors) accounted for 12.5 % of GDP in 2001 or 47.5 % of all social benefits in the EU-15. Old-age (retirement) pensions alone amounted to 10.1 % of GDP or 38.5 % of all social benefits (EUROSTAT 2004b). Of course, as there is no general Social Model for all European states, the situation in various countries differs. As data from the first Joint Report on Pensions of 2003 demonstrate, for example in Austria, only some 10 % of insured in social insurance have earned a right to an occupational pension and only 2 % of current pensioners receive an occupational pension. On the other hand, in the Netherlands 91 % of all employees belong to a pension system of the second tier, and 83 % of pensioners' households receive an additional pension (Council of the European Union 2003: 31–32).

Other institutions, outside of pension policy, also contribute substantially to social security of elderly people in the EU member states – again with different solutions in every country: housing, health care, social services, social assistance, long-term care. In this way, the welfare state guarantees social integration in old age well beyond the pension system only.[48]

3.3.1.3 New Challenges to Old-age Security: Flexible Life-course and Activation

The fact that different old-age security systems in the EU member states face similar problems which could be solved easier by learning from each other, has been at the root of the application of the OMC (see chapter 3.3.5). From their invention in the end of the 19th and throughout the 20th century, old-age security systems have decisively helped to construct and consolidate the 'tri-partition' of the life course, that is its division into: childhood, adulthood and old age (Kohli 1987). Childhood was the time for education and dependence on the family, adulthood was the time of work, and old age a period of rest after a life of work. The welfare state has standardised life events and institutionalised the life course. However, for the last decades, the environment in which old-age security systems function has changed decisively. The main changes include:

- demographic developments, short labelled as 'ageing population': 'from below' (little births) and 'from above' (growing life expectancy);
- social processes: growing individualism, changes in family life, growing mobility;
- economic developments: slower economic growth, higher unemployment, changes in working life, globalisation;
- European integration: EMU, SGP, EES.

---

[48] The confinement on pension systems in this chapter is a result of an institutional approach chosen in the whole book where social problems are analysed by dealing with separate areas of social policy. This choice does not mean underestimation of the importance of other policy areas for the living conditions in old age.

The life course has become more flexible and people are expected to assume more self-responsibility and exercise personal autonomy. One of the main challenges for the pension reforms has been to adjust pension systems to these expectations and situations: individual, diversified and flexible life-course trajectories. The methods used to adjust to new and more flexible life course patterns fall to a large extent under the broader tendency of activation. Stress has been put on employment, as the crucial method both to secure adequate income security in old age as well as to maintain financial sustainability of old-age security in the long run. Particularly important has been the prolongation of the working life, to increase (real) retirement age, especially through restricting early retirement. Flexible retirement age or partial pensions support both flexible life courses as well as the financial sustainability of pension systems.

A stricter earnings-relation of pensions, a clear tendency in many pension reforms, may also be counted as activation. As an extreme example of that tendency, the new systems of notional defined contributions clearly motivate people to work longer and make pensions strictly related to earnings. Notional defined contributions or a similar earnings-related pension system with individual accounts is partly an answer to individualised life courses. However, it creates additional risks for the employed, too: because of fragmented careers, many people will end with an insufficient income in old age from such quasi-actuarial old-age security systems. Activation measures often go in hand with more individual choice and responsibility: flexible retirement age, partial pensions, choice of pension fund (Hinrichs 2004). Both case studies presented below illustrate the tendencies to adjust old-age security systems to new life and work patterns, to raise activity rates, and to increase individual choice and responsibility.

3.3.2 The Case of Germany

In view of the book's framework, a description of the German old-age security system is not necessary here. Very briefly, only the following elements will be presented: the present system, the direction of the newest reform, the character and objectives of the system. The first data collected within the OMC process (see chapter 3.3.5.2) will be used.

In Germany, the first tier, i.e. the general pay-as-you-go, earnings-related pension system is clearly the main institution of old-age security. It covers around 82 % of the employed population and accounts for 78 % of all income of people aged over 65. The voluntary second tier are occupational pension schemes which contribute 7 % of the total income of senior citizens. The individual third tier (mainly life insurance) accounts for around 10 % of total income in old age (Council of the European Union 2003: 118). Minimum security is guaranteed by social assistance.

An analysis of the German old-age security system as compared to other EU member states in the framework of the OMC highlights the following characteristics (see also table 3.4):

1. The income situation of elderly persons is better in Germany than on average in the EU. So is adequacy of the old-age security system, which accounts the importance of the German system for the vast majority of incomes in old age.
2. Elderly people in Germany are threatened by poverty exactly as seldom as those under 65. The risk of poverty in Germany is also noticeably lower than in the EU (at 60 % of median it was 11 % in Germany and 17 % in the EU-15).
3. The income distribution in Germany is more equal than in the EU and almost identical for those under and above 65. This is a result of a much stricter earnings-relation in the German system than in many other European pension systems, which are much more redistributive.
4. Unlike adequacy, financial sustainability of the German system is to be ranked as below the EU-average. Public pensions expenditure is already higher in Germany than in the EU-15 and the difference will grow in future, even after the newest pension reform (by 2050 it should increase to 34.7 % of GDP in Germany compared to 27.9 % of GDP in EU-15).
5. The demographic situation and projection until 2050 does not differ much between Germany and the EU: in both cases old-age dependency ratio will double.
6. The employment situation in Germany is at the EU-average.
7. Public finance is a bigger challenge for the German old-age security system than for the EU-15.

Thus, the old-age security system in Germany fulfils its social function well. It is clearly oriented towards securing the living standard, which has been reached throughout working life, and not minimum security. However, the German system fits comparatively very well on poverty protection, too. Older people are relatively seldom threatened by social exclusion (European Commission, Directorate General for Employment and Social Affairs 2002a). The pension system contributes considerably to this aim. Only a small fraction of elderly has to rely on social assistance.[49] Other elements around old-age security, too, contribute to social integration of older people (housing, health system, long-term-care, social services).

The success, however, has its price: financial problems of the system (NSB 2002). Many factors contribute to them, especially: ageing of the population, continuing unemployment and the German reunification. In the last decade of the 20th century, the problems were increasingly analysed in the context of labour costs and their impact on competitiveness in the era of globalisation, as well as in the context of the Maastricht-criteria for the introduction of the Common Currency.[50] Some, albeit limited, influence on the discussions

---

[49] Only less than 2 % of all elderly received social assistance in the end of 1998 (Schmähl 2003a: 7).

[50] For a review of problems see Schmähl 2001a.

Table 3.4: Background statistics on the old age security system in Germany in the framework of the OMC

	Germany						EU-15					
Recent income situation (ECHP 1999)												
	Total		Men		Women		Total		Men		Women	
	0-64	65+	0-64	65+	0-64	65+	0-64	65+	0-64	65+	0-64	65+
At-risk-of-poverty rate (at 60% of median)	11	11	10	9	11	13	15	17	15	15	16	19
Inequality of income distribution	3.5	3.6					4.6	4.1				
Long-term projections of public pensions spending (EPC 2001)												
	Level			% increase			Level			% increase		
	2000	2030	2050	2000–2050			2000	2030	2050	2000–2050		
Old-age dependency ratio (65+ / 15-64) 100	23.8	33.5	49.0	101.0			24.2	32.2	49.0	2102.5		
Public pensions expenditure % of GDP	11.8	12.6	16.9	43.2			10.4	11.5	13.3	27.9		
Up-dated projection	10.8	12.1	14.9	34.7								
Scope for policies to ensure sustainable pensions												
Employment (2001)	Total		Men		Women		Total		Men		Women	
Employment rate (15–64)	65.8		72.6		58.8		64.1		73.0		55.0	
Employment rate (55–64)	37.7		46.1		29.3		38.8		48.3		28.7	
Effective labour market exit age	60.4		60.9		60.7		59.1		60.5		59.9	
Public finances (2001)												
Public debt, % of GDP	59.5						63					
Budget balance, % of GDP	- 2.8						- 0.8					

Source: Council of the European Union 2003: 122

came also from developments and debates in other countries and from the World Bank strategy of pension reforms (World Bank 1994b).

On this background it is understandable that reforms of the German pension system which were implemented in the last years mainly aim at improving its financial sustainability in the long run. The adjustments include reductions of future benefits, moving from security of accustomed living standard towards security of a minimum income.

After having broken with a long-standing tradition of political consensus in pension issues, the ruling coalition passed a pension reform in 2001 ("Riester reform"). The objective of the reform has been not to let the contribution rate for the general pension insurance rise above 20 % by 2020 and 22 % by 2030. This has been assessed as a paradigmatic change in the German old-age security systems: from the spending-oriented revenue policy to the revenue-oriented spending policy. Now, benefits are being adjusted in order to stabilise contribution rates (Schmähl 2001b).

The level of benefits in the general pension insurance should be reduced. After 45 insurance years, the 'standard insured' should receive 64 %, instead of 70, of his/her net income before retirement. This reduction can be partly compensated by supplementary voluntary fully funded old-age security arrangements, which are subsidised through tax subsidies ("Riester-Rente"). The additional contributions are paid by the insured. Also the pension adjustment formula has been changed towards gross wages and salaries and a notional contribution rate for private pensions has been introduced into the adjustment formula in the general pension insurance.

In 2001 an important change in social assistance for older citizens has been introduced as well. Income and property of family members will not be considered any more while assessing means for granting social assistance. The aim of this change has been to reduce the phenomenon of 'shameful poverty'. In order to improve old-age security for women, in the 2001 reform the recognition of years spent for bringing up children has been increased and an option of splitting pension rights acquired during married life has been introduced.

Shortly after the reform of 2001 further changes appeared already necessary, especially due to the difficult situation on the labour market and in public finance. The reform measures which were included in the act on "pension insurance sustainability" ('Rentenversicherungs-Nachhaltigkeitsgesetz') accepted by the Bundestag on March 11, 2004 were meant again to make pension finance secure in the middle and long perspective. The objective of pension contribution stabilisation was additionally pursued through:[51]

- the introduction on July 1, 2005 of a new "sustainability factor" into the pension adjustment formula, reflecting the ratio of contributors to beneficiaries;

[51] www.die-rente.info/print/356.php.

- the increase of retirement age for the earliest possible retirement due to unemployment from 60 to 63 between 2006 and 2008;
- the cancellation – after 4 years of transitory regulations – of the recognition of periods of general education as well as of higher education.

In 2005 the "Rürup-Rente", a new form of life insurance with tax incentives, was started, especially attractive for self-employed who are not covered by the general pension insurance.

To sum up, the policy on old-age security, which has been realised in Germany in recent years, is an important attempt to guarantee the long-run financing of pensions. The main objective has been the stability of pension finances, both of the contribution rate as well as of public subsidies. The policy also includes the strengthening of funding as compared to pay-as-you-go financing. Proponents of this redirection see it as a necessary modernisation of the German old-age security system (Neumann 1998; Siebert 1998).

The policy will lead to reductions of benefits. Guarantee of a minimum income and thus poverty prevention as objectives of old-age security are not threatened by the reforms. However, the policy will lead to a paradigmatic change in the pension system. More and more, the explicit or real objective of state pension system will be to secure a minimum income. The level of pensions from social insurance will come closer to social assistance. For securing the accustomed living standards, more and more additional fully funded old-age security arrangements will be necessary. It is an open question whether these will become obligatory in future. In this way, the German pension system may come back to its roots in the era of Bismarck (Schmähl 2003a).

3.3.3 The Case of Poland

As Poland is a new EU member state whose social policy is much less known, its old-age security system will be presented in some more detail (cf. Żukowski 2003). Unfortunately, it is not possible yet to use the same framework of the OMC for the analysis. Poland has a long tradition of social insurance, which continued also under socialism, although with some important elements of a state redistribution system. The Polish pension system was in a sense between the traditions of "Bismarck" and "Beveridge" (Żukowski 1994). The transformation process influenced the pension system clearly: the number of contributors fell and contemporarily the number of pensioners rose – partly as a result of special early retirement schemes connected with unemployment.[52] This, together with an increase in pension levels, led to a financial crisis (see table 3.5). However, these were the costs of a policy that successfully safeguarded incomes of retirees in the difficult time of an economic and social transformation.

[52] Unemployment rate was 6.5% in 1990, 14.9% in 1995, 15.1% in 2000 and 19.1% in January 2005.

After transition, several reform plans met with political resistance and the few changes that have been introduced concerned only some parameters of the system, without any structural reforms (for a review see Żukowski 1996). Unlike many other areas, the pension system was reformed only in the 'second wave' of the reforms.[53] There are several explanations of the fact that the pension reform was made only some ten years after the beginning of transformation. Firstly, Poland had inherited from the socialist time an old-age security, which was able to function under the changed circumstances too, unlike many other areas, which had to be built from the beginning, like taxes, banks, capital market or – in the social policy area – labour market policy. Secondly and because of these deficiencies, at the beginning of the transformation some important preconditions for functioning pension funds, which were an element of almost every reform concept, were absent (capital market, banks, insurance). Thirdly, a political consensus necessary for such a deep reform was absent in Poland for a longer period. However, with time the understanding of the problem, especially of the systematic burden of the system, has grown.[54]

Only in 1996, the work which started on the reform concept "Security through Diversity" (Office of the Government Plenipotentiary for Social Security Reform, 1997) led to a success. In 1997, the parliament with the left majority enacted two first acts on the reform, concerning the second and the third pillars of the new system. The centre-right government coming out of the elections in September 1997 completed the reform by enacting two major acts on the system of social insurance and on pensions from the Social Insurance Fund at the end of 1998. The new system came in force, as originally planned, on January 1, 1999. The start of the pension funds (second pillar) was postponed, however, until April 1, 1999.

Several factors can be mentioned which enabled such a structural change in the old age security system. The first was the critique of the old system, the second the reform concept and, finally, the appropriate organisation of the work on the reform, including political consensus. The main objectives of the reform were both micro-economic and macro-economic. The first micro-economic concern was to create a far tighter link between contributions and pensions, thus strengthening the incentive to work and the disincentive to evade. The other micro-economic objective was to lower – in the

53 Alongside the pension reform, in 1999 also three other structural reforms were implemented: the health insurance was introduced, the educational system was restructured and the structure and constitution of the state was changed, through the introduction of the second (powiat) and third (voivodship) levels of governance.

54 The demographic situation at present in Poland is – from the point of view of old-age security – better than in most EU member states: old-age dependency ratio (65+/15-64) was in Poland in 2000 with 17.8 % much lower than, for instance, in Germany (23.8 %).

Table 3.5: Data on the old-age security system in Poland 1989–2003

	1989	1992	1995	1998	1999	2000	2001	2002	2003
Insured (ZUS) in 1000	14696	13199	13206	12737	13271	13060	12851	12761	12739
Insured (KRUS) in 1000		1,663	1,390	1,371	1,383	1,404	1,448	1,541	1,589
Pensioners (ZUS), including	5,471	6,282	6,779	7,184	7,231	7,217	7,156	7,122	7,129
Retirement pensioners (ZUS)	2,264	2,826	3,046	3,303	3,333	3,365	3,401	3,479	3,590
Invalidity pensioners (ZUS)	2,152	2,402	2,602	2,702	2,704	2,640	2,526	2,400	2,284
Pensioners (KRUS)		1,990	2,049	1,969	1,929	1,887	1,842	1,798	1,755
Average retirement pension as % of average national wage (ZUS)	58.1	68.6	69.2	65.0	62.3	59.9	61.8	63.7	65.0
Spending on pensions (ZUS and KRUS) as % of GDP	6.5	15.0	14.6	14.0	14.1	13.5	13.6	14.0	14.1

Source: www.zus.pl/statyst; www.krus.gov.pl; ZUS 1992: 8, 19, 29, 31, 40; Ministerstwo Polityki Społennej 2005: 11; author's estimates

longer term – social insurance contributions paid by the employer, in order to reduce labour costs and to increase employment. The key macro-economic aim was to lower the level of public expenditures on pensions, as a proportion of the GDP, to relieve public finance for other aims towards growth. The other aim was to induce people to voluntarily save more.

The new old-age pension system covered younger insured (under 30) in full. Those aged between 30 and 50 years were given the choice until the end of 1999 either to participate in both new pillars (pay-as-you-go and funded, and split pension contribution accordingly) or to stay in the new pay-as-you-go scheme with the entire contribution. The insured older than 50 were not covered by the reform and will retire according to the old rules.

The pension reform in Poland has replaced a one-pillar scheme by a multi-pillar one. This change has, however, not concerned farmers who are still covered by the separate scheme of the Agricultural Social Insurance Fund (KRUS).[55] The new system, which replaces different regulations for various groups and covers all employed persons outside agriculture, consists

[55] Kasa Rolniczego Ubezpieczenia Społecznego (KRUS). Problems of old-age security for farmers are closely related to the transformation of the Polish agriculture (cf. Chlon 2000).

of two obligatory parts, called in the reform "pillars". The first pillar is a pay-as-you-go scheme administered by the Social Insurance Institution (ZUS)[56], while the second one is fully funded and privately managed. Additional sources of income security, among them the employees pension programmes (occupational pension schemes) constitute the third, voluntary pillar.

Pensions from the first pillar will be based on the principle of notional defined contributions whereas the old pensions have been with defined benefit. The new pension formula includes only two components: the sum of indexed contributions paid, divided by the average life expectancy at retirement age in the calendar year of retirement. For persons born after December 31, 1948 who had been insured in social insurance before January 1, 1999, a 'starting capital' according to the old pension rules will be assessed and recorded on the individual account in ZUS.

The same defined-contribution formula (with real capital) will also be used in the second pillar. The newly created open-ended pension funds are administered by private pension fund societies, organised as joint stock companies. The insured may choose a fund and change the choice. The funds are supervised by a state agency and there are strict regulations concerning the functioning of the funds. A multi-step procedure is foreseen in case of fund insolvency up to the taking over of the fund's management by another pension fund society. Every fund has to achieve a minimum rate of return, relative to the results of all funds. After reaching the retirement age the insured will exchange the capital they have accumulated for an annuity from a separate old-age insurance company that is not constituted yet (the first pensions from the second pillar will be paid starting in 2009).

In the new system the risk of old age has been separated from the risks of invalidity and death of breadwinner. There are two separate social insurance branches with respective contributions: old-age insurance (and contribution) and "pension" insurance, covering invalidity and survivors. The rate of old-age insurance contribution is 19.52 % of the income up to a ceiling on the level of 2.5 times the average national wage and salary. For employees it is paid in equal shares by employees and employers. For members of pension funds, a part of old-age insurance contribution equal to 7.3 % of income goes via ZUS into the chosen pension fund. As in the old, there is a minimum pension also in the new system. It is now, however, financed from state household and not from contributions. It will be paid under the condition of fulfilling a minimum insurance period of 20 (women) or 25 (men) years as an increase on the sum accumulated on both accounts – in the first and in the second pillar. The third pillar is made of various forms of a voluntary supplementary old- age security. A special form are the – within the pension

[56] Zakład Ubezpieczeń Społecznych (ZUS).

Table 3.6: Main old and new pension system characteristics

	Old system	New system
Structure of the system	Only social insurance (ZUS and KRUS)	3 pillars of which 2 obligatory for all insured outside of agriculture
Financing	Pay-as-you-go	1. pillar: Pay-as-you-go 2. pillar: Funding
Contributions	Total contribution for the whole social insurance, 45 % of wage paid (for employees) only by employers	Separate old age insurance contribution, 19.52 %, equally paid by employer and employee
Administration Contribution collection	Social Insurance Institution (ZUS)	ZUS collects contributions for all social insurance branches, including 2nd pillar
PAYG pillar	Social Insurance Institution	Social Insurance Institution
Second pillar (accumulation)	n.a.	Open-ended pension funds and pension fund societies, supervised by the State Supervision Agency
Minimum period of insurance	20 years women 25 years men	No For a minimum pension: 20 years women 25 years men
Legal retirement age	60 years for women 65 years for men. Many possibilities of early retirement	60 years for women 65 years for men. No early retirement possible
Pension formula	P = 0.24 W + W*I*0.013*L + W*I*0.007*S W – national average wage for previous quarter I – individual wage index L – total length of service S – additional years accepted for insurance	P = K / G K – pension capital of insured, composed of imputed, registered old-age contributions G – life expectancy coefficient at pension allotment

reform specially designed – 'employee pension programmes', which are often considered as a synonym for the third pillar thereby underestimating other forms of additional provision for old age, like life insurance, saving, capital investments or others.

The new system differs clearly from the old one (see table 3.6). On the one hand, it consists of two pillars, which are based on two different methods of financing and will be obligatory for all people in the end. On the other hand, the new pay-as-you-go system is based on quite new principles compared to the old one. Although both parts are differently financed, the pension formula and the whole logic of the system are the same. The new system has been functioning for six years now. An assessment can therefore only concern the reform concept and the start of the system, and should be very provisional, taking into account the long-term character of an old-age security system, especially of a funded one.

Provided that the old pension system in Poland was unsustainable, its removal through the reform is a success, although of course only if the new system is more efficient than the old one. The new system seems to have some clear advantages compared to the old one. First, the system has introduced a risk diversification because it is partly based on pay-as-you-go and partly on funding, and through this it partly relies on the labour market and partly on the capital market. Although an obligatory funded scheme is not part of the European tradition of social security, funded pension systems, on a voluntary basis, have been developed in many European countries for decades. The situation in Poland used to be different, and the reform can be considered as a method to accelerate the process of risk diversification in old-age security.

The new system acknowledges the self-responsibility and self-provision for one's own old age. This is clearly to be seen in the voluntary third pillar, which has been presented in the reform programme as an important element of the old-age security. A certain degree of choice, even if very restricted, is also included in the second pillar: the free choice of a fund. Additionally, at least in the first year of the reform, the 30 to 50 year-old could choose between an option with and one without pension funds.

Rather unquestionable are the micro-economic advantages of the new system in the first pillar, which are also directly related to the problems of the old system. Through a strict link between contributions and pensions and the abolition of early retirement, the new system should create, unlike the old one, positive incentives to contribute and to stay in employment longer. So far, an earlier retirement has been even supported by the old pension system, which has been one of the largest threats both to the adequacy and sustainability of pensions.[57] Thus, the new system aims clearly at activation.

A key role in the reform was played by the assumption that the new pension funds would contribute to economic growth, through savings and investment. These advantages, theoretically debatable, can of course not yet be analysed empirically. From the point of view of security, the first experiences with the new pension funds are rather positive, most likely because of legal regulations and state supervision.

There are, however, also disadvantages and risks of the reform programme. They concern mainly the second obligatory pillar. The doubts, well

[57] Employment rate of persons between 55 and 64 was in Poland in 2002 with 26.6% much below the EU average of 39.8% and the EU objective of 50% in 2010.

known from the international pension debate, mainly concern the level of future pensions, the profitability and security of investments of funds, the coverage of contributors and, last but not least, the transition costs. Pensions from the new system will be lower than those from the old one. The reform programme stated that the replacement rate from both obligatory pillars should be on average 50–60 %. There is, in addition, a much higher risk as far as the level of pension is concerned, and the risk is carried by the insured, which is an element of the defined-contribution scheme. The above-mentioned stronger link between contributions and pensions will lead to bigger differences in pension level, which will lead in many cases to under-provision. Unemployment will also lead to low pensions.

A questionable element of the new system is the obligatory character of the second pillar. Thereby the new Polish pension system combines a pension based on contributions (and through these on earnings) to a pay-as-you-go system with a defined-contribution annuity from the pension fund. This solution differs from the prevailing pattern according to which an obligatory second pillar is usually combined with a universal minimum pension from the first pillar (see table 3.3). The reform was based on the intent to strengthen the citizens' self-responsibility and free choice in the old-age security. However, the total scope of the obligatory old-age security has not been reduced by the reform. Furthermore, the state obliges the insured to a high level of insurance or to save for their old age, although the state itself partially escapes from direct responsibility. The accordingly large scale of the obligatory system is probably also the main reason for the poor development of the voluntary third pillar. For example, till the end of 2003 only 259 employee pension programmes have been registered.

In Poland, almost 10 million people became members of new pension funds in 1999. The fact that more people have decided in 1999 to join pension funds can, on the one hand, be seen as a success of the reform. On the other hand, the question arises whether the reform was not "too successful". The transition costs and their influence on the public finance are probably the biggest weakness of the reform (see table 3.7). High transition costs will make it difficult for a longer period to decrease the contributions to the old-age insurance, which was one of the aims of the reform.

The implementation of the reform, especially of the new information system, was very weak. This has to a large extent contributed to a negative perception of the pension reform among the public. Partly this was caused by the extremely short time between the enacting of the last legal acts in October and December 1998 and the start of the reform on January 1, 1999. Secondly, the leadership of ZUS at that time is responsible for the delay of works at the information system. But thirdly, also the authors of the reform are responsible for underestimating the problems of implementation, as if the work on the reform had been completed with the enacting of the legal acts.

Table 3.7: Sources of revenues of the Social Insurance Fund 1998–2003 (billion of Zloty)

	1998	1999	2000	2001	2002	2003
Total revenues (billion of Zloty) *= 100 %*	72.0 *(100.0)*	73.7 *(100.0)*	81.3 *(100.0)*	91.7 *(100.0)*	95.3 *(100.0)*	98.6 *(100.0)*
Social insurance contributions *(as % of total revenues)*	62.8 *(87.3)*	64.1 *(86.9)*	65.6 *(80.7)*	69.9 *(76.3)*	68.2 *(71.6)*	70.3 *(71.3)*
Systematic subvention for non-insurance benefits *(as % of total revenues)*	2.7 *(3.8)*	3.2 *(4.3)*	3.3 *(4.0)*	3.7 *(4.0)*	3.4 *(3.6)*	3.5 *(3.6)*
Additional subvention covering the deficit of contributions *(as % of total revenues)* of which	5.6 *(7.8)*	6.2 *(8.4)*	12.1 *(14.9)*	17.5 *(19.1)*	23.5 *(24.7)*	24.7 *(25.1)*
subvention to cover the deficit resulting from directing contributions to pension funds *(as % of total revenues)*	–	2.3 *(3.1)*	7.5 *(9.2)*	8.7 *(9.4)*	9.5 *(10.0)*	9.9 *(10.0)*
Other revenues *as % of total revenues*	0.4 *(0.5)*	0.2 *(0.2)*	0.3 *(0.4)*	0.6 *(0.7)*	0.1 *(0.1)*	0.1 *(0.1)*

Source: www.zus.pl/statyst; ZUS 1997: 29; author's estimates

Certainly, the new system is better than the old one as far as its inner rationality is concerned. Open are, however, its future consequences for the level of pensions and therefore for fulfilling the main function of the system. The adequacy of old-age pensions may be a problem in the long-run. The new system is based on a strict equivalence between contributions and benefits, which however may lead to weaker solidarity and inadequate minimum security for those who will not fulfil the strict conditions for a minimum pension (see table 3.6). As a result, the new system may cause social exclusion of some people who will not be socially insured long enough.

The fear stems also from the difficult environment of old-age security systems. The following problems cannot be analysed here, since they are not part of old-age security as a cash transfer system, but they are nevertheless crucial for the social situation of older people (cf. GVG 2003):

- limited access to health care due to raising financial participation of patients (mainly for medicines);
- housing, which in many cases and especially more often for older people than on average lies under a socially acceptable standard;
- insufficient and underfinanced care facilities and services, which to most elderly leave no alternative but to seek care by family members.

In the case of Poland, social integration in old age is often more threatened by these circumstances than by the old-age security system alone, which delivers relatively adequate pensions. In the medium term, high transition costs and their influence on public finance are a big deficit of the reform. Although the reform programme can awake interest from abroad,[58] transition costs and implementation problems should rather be a warning.

The pension reform in Poland, as the other reforms of social policy in most countries of Central and Eastern Europe, is an attempt to find a new answer to the questions about the role of the state in social policy, its relationship to people's self responsibility for their own life and the cooperation with the market (public-private mix). Even if this is a long-run reform and should therefore be assessed only in such perspective, evidence is indicating that this question has not been answered finally by the reform of 1999.

3.3.4 Comparison

The comparison of old-age security systems in Germany and Poland allows formulating the following statements:

- Old and new EU member states often have common traditions of social security.
- State old-age security systems are well developed in both countries and spending is similar in proportion of GDP.
- In both countries, public old-age security systems are the most important source of income in old age.
- Old-age pensioners in Germany enjoy a much higher welfare level than pensioners in Poland, which is mainly explained by the much higher general income level in Germany, but partly also by a better general social situation (health system, housing, care etc.).
- General objectives and directions of pension reforms, which have been implemented in both countries, are similar: financial sustainability of the systems in the long-run, adjustment to the demographic developments, shifting from pay-as-you-go financing to funding.
- Both reforms aim at activating people through discouraging earlier retirement, and stricter earnings-relation of pensions (with other methods

[58] It is even sometimes seen as a model for Europe (cf. Bräuninger 2002).

(notional defined contributions), Poland has gone in this respect the German way).
- In both cases, the division of financing sources has been strengthened: contributions should finance intra-personal distribution over time (insurance) and taxes interpersonal redistribution.
- The Polish structural reform goes further than the German reform what may be explained by the special circumstances of a reform in the framework of a deep transformation of the entire political and economic system.
- The 'Swedish' pension formula, introduced in Poland into the first pillar, seems a clear and reasonable consideration of the demographic problems as well as a clear strengthening of incentives to prolong the working lifetime.
- Poland has opted for a radical abolition of early retirement options, which – together with the new pension formula – should increase employment of older workers significantly.

Both countries have opted for different methods, to combine the pay-as-you-go old-age security system with funded schemes. Poland has introduced an obligatory second pillar, which secured a wider application but is associated with higher transition costs. Germany has opted for a voluntary funded old-age security, however highly subsidised.

Although the distinction between "parametric" and "paradigmatic" pension reforms is not clear-cut (Hinrichs and Kangas 2004), the Polish reform of 1999 certainly was much deeper or structural than the German reform of 2001. This can be mostly related to specific circumstances of transition and can stand for the whole range of structural pension reforms implemented in Central and Eastern European countries (see section 3.3.6).

3.3.5 European Dimensions and Elements

3.3.5.1 National Competence and Role of the EU

Old-age security systems are an area in which member states have a clear competence. They decide freely about objectives, structures and design of their systems. A harmonisation in sense of a unification of solutions in this field, although raised sometimes in the past, has never become a political objective of the Community. Such a harmonisation would be very difficult due to different traditions deeply rooted and protected by societies. For the rest, its necessity for a unified Europe is even questionable (Schmähl 2002).

A Europe wide solution concerning old age and more generally social security has been needed however for persons moving within the Community. The rules of coordination, which have been introduced to solve the problem, have supported the realisation of a treaty freedom: the free movement of workers. This law system is an example of the realisation of the principle of subsidiarity: every member state decides about its own old-age security system, while Community rules are needed only for solving problems going beyond borders of member states. Thus the EU influences old-age security systems of the member states mainly when a person moves within the Community. This a very restricted "Europeanisation" of old-age security (Eichenhofer 2002).

Yet this is a well known picture of division of powers in social security. However, it may be somewhat different if we understand old-age security broader, and not only as obligatory public social security systems, as is often the case in the EU. In areas important for the functioning of private and funded pension systems, competences of the EU are stronger than in social security. 'Competition' has been classified as an area in which decisions are taken at both national and EU level, and 'capital flows' as the one in which most policy decisions are taken at EU level (Longo 2003). Thus, the economic EU law has started to affect old-age security in the EU through the Competition law and the freedom of services (see section 2.2.4).

3.3.5.2 Open Method of Coordination for Pensions in the EU

A new element in the EU policy on pensions is the Open Method of Coordination of pensions systems. This method, which had been earlier adopted in the areas of EMU, employment policy and social exclusion, started in the area of pensions in Lisbon in March 2000. The European Council asked the European Commission to collect information on the development of social security in a longer perspective, with special focus on the sustainability of old-age pension systems.[59]

In October 2000, the European Commission prepared a framework for analysing pension systems published in a communication with the heading: "The future evolution of social protection from a long-term point of view: safe and sustainable pensions". It was stressed that for the sustainability of pension systems isolated reforms will not be enough, since also permanent economic and employment growth is needed. Every country decides which pension system it wants to have. However, given that all countries face the same problems, a coordination of these efforts as well as an exchange of information about current or planned reforms seem to be reasonable. Such a cooperation would be easier if objectives and methods would be explicitly formulated.

At the European Council in Göteborg in June 2001, the application of OMC in the area of pensions has been officially decided. The corresponding communication of the Commission was entitled: "Supporting national strategies for safe and sustainable pensions through an integrated approach",

---

[59] All the documents and reports which are prepared within the OMC of pension systems can be found on the website: http://europa.eu.int/comm/employment_social/soc-prot/pensions/index_en.htm.

wherein three broad principles (objectives) for pension systems in the long-term perspective have been formulated in the following terms:

- to safeguard the capacity of pension systems to meet their social aims;
- to ensure the financial sustainability of pension systems;
- to enhance the ability of pension systems to respond to the changing needs of society and individuals.

In November 2001 the "Joint report of the Social Protection Committee and the Economic Policy Committee on objectives and working methods in the area of pensions: applying the Open Method of Coordination" was published and afterwards accepted by the European Council in Laeken in December 2001. The report formulated eleven broad common objectives within the three broad principles. The next crucial step was the preparation of National Strategy Reports by member states until October 2002 on the future of their pension systems, under the heading "adequate and sustainable pension systems". The reports include a diagnosis of important challenges, information on realised and planned reforms and data for analysis of consequences of present policy and reforms.

After the analysis of the national reports and the identification of innovative approaches and best practices, the European Commission together with the Council prepared the Joint Report on Adequate and Sustainable Pensions for the European Council in March 2003 (Council of the European Union 2003). In a communication published in May 2003 the European Commission proposed a "streamlining of open coordination in the field of social protection" through integration of open coordination in the areas of social exclusion, pensions and (a new field of OMC) health. The new objective is the preparation – in a three-year cycle – of a report on social protection, integrating the three areas (European Commission 2003a).[60] The presented country study on Germany has been partly based on the data presented in the Joint Report of 2003, which also serves as an illustration of the OMC.

3.3.6 Concluding Remarks and Recommendations

Older people in EU member states are well protected from poverty and also have a broad access to systems which enable them to maintain an accustomed living standard. Old-age security systems in Europe thus support social integration in old age. A crucial position is taken by state pension systems in the framework of social security. Europe is characterised by a high degree of 'nationalisation' of income security in old age. Other sources like occupational pension schemes or different forms of individual provision for

[60] In the field of pension systems the future plans include: by the middle of 2005 preparation by the new member states of national strategy reports and by the old member states implementation reports and an update for the period 2002–2005; in 2006 a Joint Report on Social Protection should focus on pensions. European Commission 2003a.

old age play a supplementary role. Very important for social integration in old age are also other areas of social policy, especially health care, social services, social care or social assistance.

In many European countries, the scale of obligatory pension systems has become that large throughout the development of social security systems that it has replaced personal responsibility for old age. Recent reforms in many countries have aimed at increasing individual choice and responsibility. Although the state pay-as-you-go earnings-related pension systems do clearly prevail so far, there is a clear tendency in the pension reforms to increase the role of private, funded systems. This may partly mean that in future the role of public systems will be much more limited to minimum security. Even in Germany, the country with the oldest social insurance system in the world, whose function is clearly security of the living standard, the newest reforms align with that tendency.

Pension reforms in most of the new EU member states in Central and Eastern Europe have gone further than those in the old EU by partially replacing previous pay-as-you-go systems with fully funded, privately managed systems of individual accounts,. In most countries, these reforms have been combined with a deep restructuring of the main pay-as-you-go pillars. Altogether, the new EU countries have opted for more paradigmatic reforms, whereas the old EU member states have mostly introduced parametric changes. As was illustrated for the case of Poland, this can be mainly related to specific circumstances of the transition: extraordinary conditions of a transformation of almost all economic, social and political institutions. Political resistance to change was in this situation weaker and the political will to deep reforms bigger. By the way, it is also a clear case of an international policy learning process, mainly from Latin America (Müller 2003).

Both in the old and new member states, recent reforms of old-age security systems have aimed at a better adaptation to more flexible life courses. This has been realised, inter alia, through individualisation of public systems (individual accounts, notional defined contributions), moving from defined benefit to defined contribution or through development of fully funded private systems. Also activation, mainly through stricter earnings- or contribution-related pensions and restricting earlier retirement, has been a crucial objective of recent reforms, since employment is seen as a key factor for achieving both adequate and sustainable old-age security.

Old-age security remains in the competence of EU member states. Different solutions result from different traditions and conditions. The role of the EU has been mainly concentrated so far – according to the logic of subsidiarity – on coordination of different systems for migrant workers. Cross-border portability of pension rights under the statutory public schemes is thus not an issue for people migrating within the EU. In contrast, cross-border portability of occupational pension rights should be further improved also at the European level.

As long as old-age security remains publicly organised and with little elements of markets, it will stay in the competence of the member states. However, both internal problems of sustainability of pension systems as well as external influences of the globalisation may rather lead to an increasing role of market and competition in that sphere, and as a consequence, to an increasing role of the EU as well.

The EU Enlargement has further increased diversity of old-age security within the Community. For the purposes of this study it is important to note that the new member states also have well-developed old-age security systems which generally contribute to the social integration of older people. The case of Poland may be an example here. The main differences between the new and the old member states consist in the welfare level, whereas the pension systems themselves often guarantee similar replacement rates. The only way to improve the living standard of the elderly is welfare improvement through economic growth, which is exactly one of the main objectives of EU integration of these countries. Finally also migration, which will increase due to the enlargement, may be an additional argument in favour of more structural pension reforms on a EU scale (Holzmann 2004).

Holzmann (2004) favours a development towards a pan-European pension structure which could both help to solve internal problems of pension systems as well as facilitate mobility between countries with different structures of old-age security systems. The suggested structure is a multi-pillar system, with a notional defined contributions system at its core, and coordinated supplementary funded pensions and social pensions at its wings. The author points to the fact that some of the new EU member states already have such a desired structure of the old-age security. Holzmann finds the approach initiated and led by the EU Commission not promising and suggests instead a cross-country government-led approach.

The main problem of old age security systems, both in old and new member states is to maintain pensions financially sustainable in the long-run without loosing their main social function. The pressure comes partly from economic developments within the EU, whereby also the EU influences old-age security systems. The OMC may be a useful tool to reconcile both sides of old-age security: adequate and sustainable pensions, through a process of common learning while maintaining at the same time national competences. The main strength of an OMC of pension systems lies in facilitating the policy learning. The structural reforms introduced in many new EU member states can be an example of successful policy learning even without a formal structure. The OMC can deliver such a structure and thus create an even better ground for the national reforms. The method should be developed as it began in the first "round" 2000–2003. Common objectives should be preferably expressed in terms of indicators, even if the work so far has shown how difficult this is. The objectives should even further take into account interrelations between old-age security and other areas, especially where the OMC has already been applied. In addition to the areas already included: employment and public finance, also some other areas crucial for actual security in old age could be added, like health care, social services or housing. To facilitate the policy learning process, a better structure for analysing the stakeholders in the old-age security in every country could be suggested. A more unified structure of analysing and presenting national strategies and policies also with a set of commonly agreed indicators, could be proposed. The OMC should however still be only a tool in policy learning. Policy making, in contrast, should remain the exclusive competence of the member states. It is to be seen whether this method will lead to move competences in the social security area towards the European level (Schmähl 2003b: 24).

3.4 Family policy: Estonia/Finland

3.4.1 Starting Points

3.4.1.1 Family Policies

Family structures belong to the societal phenomena, which have drastically changed during the 20th century, and especially after the World War II. Urbanisation, industrialisation and modern (non-agrarian) social mentality, new moral codes included, are the factors, which have had remarkable influences on the family formation and family policies. In the same process, the two-generational nuclear family – parent(s) with children under 18 years – has become the dominating family type and household formation in the Western world. As a parallel process, the divorce rates have increased, and non-marital unions are usual today also in the catholic regions of Europe, and affiliated to that, there appears a rise in births outside marriage. As an implication of the increasing divorce rates, the new formation called 'step-family' has entered the stage. At the same time, the family as the locus of the most intimate living sphere has remained persistent despite the changes of the family itself and the social environment (e.g. Finch 1989; Therborn 2004). The nuclear family is interpreted as the cradle for developing individual capabilities, however, the social institutions have increased their influence on the first phases of a personal life-course.

The most influential factor of the family policies today is the waged employment of men and women. The traditional roles of men as bread-winners and women as house-wives have been replaced by a social practice in which both men and women are expected to educate themselves and participate at the labour market, and, similarly, to form a nuclear family. The mentioned trend has been strengthened through the gender equality policies launched in several European countries since the 1960s, and in the socialist countries it was a norm for both genders to participate at the labour market. As a consequence of the European labour market and gender policies, the reconciling of work and family is the foremost issue of the contemporary family policy in the member states of the European Union. Family issues are mostly debated in the context of the current employment policies: how to ensure the equal opportunities for education and waged employment for men and women, and how to facilitate the everyday care of children.

The first modern family-policy innovations and measures were closely connected to population policy understood as improving the living conditions by environmental hygiene and by better housing, by organising health care for pregnant women, by giving intensive and professional help in childbirth, and by offering special maternal social benefits in cash and kind (Myrdal 1941). After the baby boom at the end of the 1940s, the fertility stabilised in Northern Europe in the 1960s and in Southern Europe in the 1970s on the level on which it still remains, although a gradual decreasing of the number of children has been in progress ever since. In the year 2001, the Total Fertility Rate[61] was 1.5 in the EU-15 countries.[62] The European societies have adjusted to the low number of children during the last decades and to the nuclear family as one of the most prominent social characteristics in the post-modern society.

It is reasonable to conclude that the very first ideas of the family policy connected to monitoring the fertility rates have been rejected and replaced by approaches, which emphasise the supporting of parenthood in order to ensure the equal opportunities for the growing children. If a pure avant-garde example of the enabling policies is asked for, the modern family policy with its one hundred years' history of measures focused on all children, not only on the poor ones would suit very well. The public support for the children in the form of care services (kindergarten; primary school) and most of the child allowances were launched on a universalistic basis from their very beginning (Rauhala 1996).

In the EU-15 countries, 46 % of all families in the year 2001 were first-marriage based nuclear families with two parents and their children. In Europe, roughly 75 % of all under 18-years old children live in a family with two adults, of which the one is the biological mother or father of the child(ren); 10 % of children of EU-15 live with one parent, who usually is the mother. The number of single-parents increased during the period 1990–2002 from 6 % to 10 %, and the increasing tendency seems to continue (European Commission, Directorate-General for Employment and Social Affairs 2003: 114–115). The single parenthood and especially single mothers have been studied in the frame of law sciences, too, as recipients of social assistance (e.g. Wennberg 2004). The single parenthood makes visible and even underlines the obligations of a modern citizen expected to fulfil both the role of an autonomous earner on the labour market and a caregiver in a nuclear family. At the same time, the single parents have gained special rights for the maintaining of children (e.g. extra child benefits for single parents). The single parents as an addressed group of certain benefits also emphasise the justified claim to both parents, and in the case where it is not gained, compensation is organised by family policy measures.

In most countries, the responsibility of both parents after the divorce has been regulated by the maintenance obligation since decades, as well as in the case of a non-marital child, too. The ensuring of maintenance and care for children today are interpreted as a gender-neutral duty of both parents despite the quality of their partnership. Obviously, the discussion about sin-

---

[61] By Total Fertility Rate usually the cross-sectional fertility is meant, which appears during a year; Cohort Total Fertility Rate is used to describe the fertility of the female cohorts who have already passed the fertile age, defined to be at 49 years.

[62] Detailed information, data and analyses of the population development, fertility included, can be found on the websites www.demographic-research.org and www.unfpa.org.

gle parenthood is a topic increasing its importance in the European Union where the participation at the labour market is highly emphasised as a factor for economic dynamics and social welfare. In general, the European societies have developed special measures (day-care services) in order to enable the single parents' opportunities to participate at the labour market.

In the EU-15 in 1988, the number of couples with dependent children was 52% of all couples; 12 years later, in 2000, the proportion of those couples was 46% (European Commission, Directorate-General for Employment and Social Affairs 2003: 114–115). The analyses of the mentioned tendency confirm the difficulties in reconciling work and family life (IPROSEC 2000–20043[63]); Leira 2002; Miettinen and Paajanen 2003). The emphasis of family policy put on the reconciling of work and family is intensely argued by several empirical and comparative studies. Today, most of the European Union member states have launched parental leave systems instead of the former maternal leaves; fatherhood has been taken on the agenda in order to balance the family life in the two-earner based societies.

The main trends of family development in the European Union contain of the following issues (European Commission, Directorate-General for Employment and Social Affairs 2003: 114–115):

- fewer and later marriages and more marital breakdowns;
- fewer children and later in life-course;
- a marked increase in non-marital unions;
- emergence of the 'step-family' as an accepted family formation;
- a rise in births outside marriage;
- smaller households and more people living alone, young people of fertility age included;
- a striking rise in the number of children living with one adult;
- a decrease in the number of couples with children.

In the current social situation within the European Union, the family is debated in the context of changes. The postponing of the family formation is discussed as well as the decreasing fertility, which as such is a sensitive issue from the point of view of global ethics.[64] The reconciling of the family life and the labour market participation is, however, the leading context for conceptualising the family policies, and has therefore been chosen as a frame-

[63] "Improving Policy Responses and Outcomes to Socio-Economic Challenges: Changing Family Structures, Policy and Practice". Into the study, eleven European countries were included, and during 2000–2003 the study was funded by the fifth frame programme of the European Union; the leader of the study was Linda Hantrais. There is an internet access to the design and results of the study: http://www.iprosec.org.uk

[64] We are aware of the ethical connotations of the fertility problematic – a topic we pass over in this study, where the focus will be on the reconciling of work and family. However, the empirical study makes visible the strong emphasis on fertility policy in Estonia.

work in this study, too. Additionally, family issues are currently considered in the context of poverty; the children living in poverty are interpreted as a vulnerable group and as one crucial target group in combating poverty in the European Union (Heikkilä and Kuivalainen 2002: 84–86). There are empirical evidence and theoretical considerations according to which the recent rapid changes of family structures have increased the poverty of children, and as an outcome, some decrease of the willingness of birth-giving can be expected (Case et al. 2003).

In all industrialised countries, social policy measures are arranged for newborn children and babies under one year of age, i.e. the parenthood is supported in the very beginning of the child's life. To a certain extend, the family policy measures offered for the newborn child's family can be understood as a classical example of application of the capabilities approach: through the public policies, the parenthood is confirmed as a capability worth to be supported by the state in order to strengthen both the family as a nuclear institution of the society, and the childhood as a special phase of the life course of the citizens. The second group of benefits and services consists of the measures organised for care of the under school age children, and finally, the support in general for the children and youth who are dependent on their parents' support until the age of 18 years.

3.4.1.2 Countries in Comparison: Estonia and Finland

In the entirety of the European Union, the two countries belong to the small member states: Estonia with 1.36 million and Finland with 5.21 million inhabitants. Estonia was among the EU-acceding states in May 2004, and Finland is a member state since 1995. Geographically, Estonia is one among the three Baltic countries; the other ones being Latvia and Lithuania. Today, those three countries do not share any strong self-identity as the Baltic states, although the countries have similarities concerning the history of the regained independence at the beginning of the 1990s, after the collapse of the Soviet Union. It is important to mention that the Baltic countries did not join the union of countries, which declared themselves sovereign from the Soviet Union/Russia. The Baltic countries decided to enter the NATO and the European Union, and in the current analyses, the social development of the countries since the 1990s is described as a return to the Western World (Lauristin and Vihalemm 1997).

Estonia and Finland have long and strong cultural ties, mostly based on a close kinship between the peculiar Finno-Ugric languages, Estonian and Finnish. Politically, the historical development of the countries is different even though both have been under Russian rule in the 19th century and in the beginning of the 20th century. Finland belonged to the Scandinavian welfare states since the beginning of the 1970s, and empirical studies confirm the similarity of the social (policy) development of Finland to that of the other Nordic societies after World War II. Among the Scandinavian societies,

Finland is different according to its long border with the former Soviet Union and currently with Russia, and according to its language which is definitely different compared to other Scandinavian languages which all are in a quite close kinship to each other. Finland has a minority of 6% of Swedish speaking citizens who have full Constitution based linguistic autonomy: all the official documents are published in Swedish, too, and all the educational and other institutions and services in Finland are organised in both official languages.[65]

The comparison of family policies of Estonia and of Finland will be conducted in a setting of two current European Union countries. As a background issue, it is important to know that after the regained independence, Estonia has launched an ultra-liberalistic economy, personally advised by Milton Friedman (Wrobel 2000). On the other hand, in the political speeches, expectations of developing the Estonian society towards the Scandinavian welfare model have been expressed. A certain gap between the institutional solutions and the political intentions still remains. In Finland, after the recession in the beginning of the 1990s, the social budgets have been decreased, and in 2001, the expenditure on social protection was 2% lower than the EU-15 average (European Commission, Directorate-General for Employment and Social Affairs 2003, 186). Nevertheless, the comparative analyses give evidence for arguing that Finland still belongs to the peculiar group of the Scandinavian welfare states (Kautto 2001).

The family policy can be interpreted as a field of social policy with connections to labour market, housing, education and health care related policies, too. In this analysis, the leading idea is to compare how the idea of reconciling the working life and the family life is advanced by the family policies in Estonia and in Finland and how the family policies facilitate the everyday living of the citizens with children under 18 years old. It will be asked if the concept of the enabling welfare state can be applied to the family policies of the two countries. Secondly, the focus of the analysis will be put on the child: how the childhood is taken into consideration and supported by the public policies, and how the capabilities approach can be interpreted in the frame of the family policies. Childhood is understood as a special phase of the life-course; from the viewpoint of family policy the age of a child is a meaningful criterion for organising benefits and services.

---

[65] After the regained Independence, Estonia launched a restricted language policy according to which the proficiency of Estonian language was a prerequisite for the citizenship of Estonia. At the same time, the country had a minority of Russian speaking inhabitants of more than one third of its population, half of which remained without the right to vote in the state elections. The Russian-speakers were also excluded from the higher positions of the administration. The ethnic and language relationships and the affiliated policies are very complex in Estonia, as Lauristin and Heidmets (2002) have profoundly analysed in their recent study.

Table 3.8: General societal data of Estonia and Finland 2002–2004

		Estonia	Finland	EU-15
1.	Population, in thousands	1 356	5 206	379 449
2.	Population aged 65+, %	15.9	15.3	16.1
3.	Population aged 0-14, %	16.6	17.8	16.7
4.	Natural population increase rate, %	-3.9	1.4	1.0
5.	Life expectancy at birth, females, years	76.0	81.0	81.2
6.	Life expectancy at birth, males, years	65.1	74.1	74.9
7.	GDP per capita in PPS, EUR	5 900	26 900	24 400
8.	Monthly gross earnings, EUR	453	2 210	1 118
9.	Unemployment rate 2004, %	9.2	8.9	7.7
10.	Inflation, %	1.9	1.5	2.0
11.	Tax rate, all taxes, % of GDP	36	46	~40
12.	Social expenditure, % of GDP	18	25	27

Source: For figures 1-6, 12: European Commission, Directorate General for Employment and Social Affairs 2003, 2004; STAKES 2004; Ministry of Social Affairs of Estonia 2003. For figures 7–11: Statistical Office of Estonia 2002, 2004; European Central Bank 2005.

Note: There occurs a high variation of the figures in different statistical sources, and some uncertainty remains, especially in respect to the monthly gross earnings, the unemployment rates, the inflation and tax rate. The national statistics give different figures than the European Central Bank.

While comparing the family policies of different countries it is useful to start by introducing the leading ideas and goals which the concrete family policy measures have been established on and are developed along; those are the ideals or expectations which can be in a tensional relation to the concrete measures. The comparison of the real benefits and services has to be related to the explicit family policy goals, although implicit family policies are practised, too, by means of housing and educational policies as well as by endorsing an ideological atmosphere which can be interpreted to be more or less family-friendly. The comparison will be done within the framework applied in the entire study: the approaches of the enabling welfare state, the viewpoint of capabilities, and the life-course perspective will be discussed while comparing and interpreting the differences and similarities of family policies in Estonia and in Finland. The ethical foundations are discussed, too.

3.4.2 The Case of Estonia

In the family development of Estonia, both the government and scientific notes have reported a backlash during and after the transition period since

1987, the regained independence in the year 1991 included, until the beginning of the 21st century. By backlash the decrease in numbers of families with children under 18-years age is meant. The former socialist regime and its welfare system came to its end, and radical reforms were needed in all sectors of the society as well as a reorganisation of family policies.

As a background, the main periods of transition are introduced in order to give a general picture of how the Estonian society changed from a province of the Soviet Union to a sovereign Western state which has had only a short period of independence in its history (1918–1940) before World War II. The first stage of the post-communist transition is called the period of political breakthrough (1987–1991); the next stage is described as a period of laying foundations of the Estonian state by establishing the new constitutional order and launching radical economic and political reforms (1991–1994); and the third is referred to as the period of stabilisation of society (1995–1999) (Lauristin and Vihalemm 2002). Since 1999 the transition period is interpreted to be passed, and the return to the Western world has been fulfilled by the membership of the NATO in 2003 and by acceding the European Union in May 2004. During the second period, 1991–1994, the emphasis was put on the radical constitutional and economic reforms, often called a shock therapy, through which the socialist regime was replaced by a parliamentary democracy and by an extremely liberal market economy with very scarce state regulation and low and not progressive taxation.

During the period of 1991–1994, the social policy, family policy included, was not in the focus of the Estonian politics. In the next stage, during the years 1995–1999, social policy was taken on the agenda, but as the analyses confirm, not very prominently. High poverty rates and high income differences are visible and obvious while compared to other European Union countries and especially to the Scandinavian countries (Kutsar 2002). The Ministry of Social Affairs of Estonia is monitoring the poverty in the country: 23 % of all households lived under the relative poverty rate in 2002, and the poverty was very serious among children. In Estonia, 34 % of all children under 18-years age lived in poverty (Ministry of Social Affairs of Estonia 2003: 18), whereas in the EU-15, 19 % of children lived in poor or low-income families in the beginning of the 2000s (European Commission, Directorate-General for Employment and Social Affairs 2003: 151). The social policy has been in a residual position, and the compensating role can be said to be the dominating feature of the measures in Estonia. It can be stated clearly that the enabling dimension is not emphasised in the Estonian social policy.

In the Estonian society, the drastic decrease of the number of children is interpreted as the most urgent issue of the family change and of the policies affiliated to it (The Government of Estonia 1999). As such, the change of the number of children in Estonia is an interesting case of the European family development during the 20th century. The leading demographer of Estonia,

Katus (2000) has demonstrated that there are reliable and credible statistics of fertility in the region of Estonia since the 1910s. The most prominent time frame is that during the years 1920–1990 – the period which includes the first independence era and the Soviet time until its collapse and the regained independence in 1991: during the mentioned decades the Cohort Total Fertility Rate was very stable and remained on the level of 2.0. In the history of the European societies, the rate is one of the most stable and on a remarkably low level earlier than in many other societies.

During the period of the stable fertility, since the end of the 1910s, the women's participation in the waged employment has been on a high level. During the Soviet regime, it was a normative rule for all adult male and female citizens to participate at the labour market (Kandolin 1997). Ideologically, the family-life was not respected in the form of favouring the home care for children, but the public institutional day care was launched. Despite the historically early and large entering into the labour market under the circumstances of the Soviet regime, the Estonian women on average gave birth to two children during seven decades. In Estonia, the decrease of fertility from two to close to one child happened as rapidly as the change of the whole society, in one decade during the 1990s.

It is a fact that all the current Estonian family policies are heavily tinged by the present situation, which due the collapse of the fertility rate is unavoidably interpreted as an emergency. In the current right-wing headed government, there is a Minister for Population Affairs, to whom the advancing of the population increase and the family policies are delegated. In 2003, the Estonian government created a special concept of the policies for children and families. There are three explicit principles according to which the family policies of Estonia are oriented: improving the quality of life of children and that of the families with children; support for combining family and work life; and attachment of value to raising children (www.vm.ee/estonia). As measures of implementing this family policy concept, the ministry for population affairs was created, and the other state agencies, local governments, associations of labour market partners and non-profit and other civic society actors are called to a mutually cooperate in all spheres of life essential for children and families. However, no concrete incentives have been organised for co-operation.

The radical family policy reform launched by the government since the beginning of 2004 is the full compensation of one year's salary for the mother or the father of a newborn baby (to the father after the child is six months old). The argument for the reform is to help parents to cope with the increased living costs with the child; the amount of the benefit is based on the previous earnings of the parent, or in the case of an unemployed parent, a minimum has been enacted (EUR 141; the level of the minimum salary). As an income-redistribution element, the ceiling was set at 2.5 times the average 2002 salary or EUR 1003. The first evidence of the implemented reform indicates a low increase of fertility, and a clear gender division in taking the benefit – the mothers are the ones who take the benefit, although the men could get more money on the basis of their higher salaries.

Five types of benefits have been implemented in the Estonian family policies: maternity benefit, parental benefit, universal family benefits, tax credits and holiday benefits. The full compensation of the salary during the first year of the baby dominates and is just in another category than all other benefits. Additionally, a birth grant of EUR 240 for the first child and EUR 192 for the next children is given, which emphasises the focus put on the newborn child. Relatively, the benefits for a newborn child are of high level. On the contrary, the other family benefits are on a lower level; a family with two children of whom one is under school age and one going to school gets in a whole EUR 97 universal family benefits per month, and an additional school allowance of EUR 29 once a year for the school age child. While adjusted to the average salary, the share of family benefits is 22 % of one salary. The single parent allowance is EUR 19 per month.

Three tax credits are applied to families with children: one as an incentive to give birth to more than two children, the other two for families whose children are attending the university or vocational education. As a special Estonian benefit, the paid breaks of 30 minutes every three hours for the breastfeeding working mothers with a child younger than 1.5 years have been launched. The state is compensating the employer for the time used in the breaks, which can be organised also as a reduced daily working time. There is a special childcare leave of 14 days for fathers during the pregnancy leave or maternity leave of the mother within two months after the birth of the child. In principle, the family benefits, some minor holiday measures included, are gender-neutral despite the specific maternity leave for the first six months of the newborn child and the childcare leave of 14 days for the father. The tax credits in a country with a non-progressive taxation are gender-neutral, too.

In Estonia, the family policies have gone through a huge transformation during the move from the Soviet system to sovereign national policies after the regained independence in 1991. The share of social expenditure for families and children in Estonia in 2003 was 15 %, but the figure is difficult to compare with the relatively low social expenditure in general, in terms of the GDP share of 18 % the lowest in the European Union (Ministry of Social Affairs of Estonia 2003). As in many other European Union countries, also in Estonia the risk of poverty is high among the (long-term) unemployed people. The second groups in the highest risk of poverty include the pensioners and the families with many children and single mothers (Einasto 2002).

The surveys of the family benefits during the first decade of the regained Independence confirm the fact that the Estonians evaluate the family benefits to be very low (Ainsaar 2003). The day care in Estonia is organised on a level which the families evaluate to be good, however, many parents must organise the care in a private kindergarten setting or in a private network. In 1999, a pre-primary education was launched for the age group of 3–6 year-old children in Estonia and in the same year, 74% of the children attended the pre-primary school education; in the urban areas, the enrolment ratio was close to 100%. In the rural areas, 54% of children participated in the pre-school education (Statistical Office of Estonia 2001: 21–23).

The analyses done in the context of the IPROSEC-study (Kutsar 2003) give arguments for the conclusion that the family policies implemented during the regained Independence have failed. In the last UNDP Human Development Report of Estonia the data on children and families given by the Estonian experts is focused on the poverty of families and children, on the psycho-social problems of children and on street-children problems (Human Development Report, Estonia 2003: 32–42). Not a single positive issue is mentioned in the report. Many unpublished theses done in social policy and social work have reported difficulties of parents living with handicapped children and with many children, especially in the country-side.

Making a synthesis of the repertoire of the family benefits in Estonia, demonstrates that the most prominent feature is the clear pro-natal orientation of the launched benefits. This is also an expressed goal of the state politics: the attempt is to increase the fertility in the country where population decreases by 5.000 persons every year. The emphasis put on the newborn children leaves many other problems unresolved, which are obvious in the current child and family policies in Estonia. The government tries to advance the child-friendliness by a radical reform but at the same time, such a prominent problem like school-drop-out is not managed, although the problem is widely realised and visible. Although the figures of early school-leavers (15%) in Estonia still are a little lower than in the EU-15 (19%) (the Southern European countries contribute remarkably on the EU-average; European Commission, Directorate-General for Employment and Social Affairs 2003: 195), the phenomenon is increasing rapidly. At present, there is no comprehensive policy to buffer the school-drop-out, which happens already on the secondary level. Some imbalance of the family policies remains, as the main emphasis is put on the newborn children, but the families with many children and problems are not considered as a target group.

The Estonian scholars (Kutsar 2003) have criticised the family policies, which emphasises the first year of the child, when at the same time the surveys give information that parents are most worried how to cope with children at the school age and how to ensure the education opportunities for the child. The housing policies are criticised to be not very favourable for families with children. There is also empirical evidence that a family with two children and one breadwinner is in a remarkably high poverty risk if compared to a family with two parents participating in the labour market. In reality, the Estonian couples postpone not only the birth giving for the first child but especially for the second child.

In the IPROSEC-study, the comparative analysis confirms the day-care opportunities to be on a good level in Estonia. Nonetheless, the Estonian parents reported uncertainty concerning the future of their children, and they also reported to feel themselves to have only scarce opportunities, if any, for exerting some influence on the family policies (Oras 2003). As an important result from the viewpoint of reconciling work and family, the role of the employers was evaluated as not active or favourable for developing family policies. In the conclusion of the Estonian IPROSEC-study (Reinomägi 2003), the coordination of family policies is emphasised and the active participation of all social partners is asked for in order to launch the goals of reconciling family and work.

3.4.3 The Case of Finland

After the end of the World War II, the family policies of Finland have been developed according to two principles: to facilitate the parents in managing the living costs with children, and, especially since the 1960s, to facilitate women's participation at the labour market. For the former goal, cash benefits and tax credits have been launched, and for the latter, the organisation of a reliable and high-level day care has been in the focus. Like in the other Scandinavian societies, the public support in the form of benefits and day-care services is conceptualised as an entitlement, a social right for working mothers and fathers (Leira 2002: 1–14). In Scandinavia, a special ideal model of Nordic family policy can be recognised, and the implemented family policy reforms are in line with the goals of the model (Hiilamo 2004).

The explicit ideal model of the Nordic family policy includes six aims (Hiilamo 2004: 124):

1. universalism of benefits, i.e. emphasis on the non-means-tested allowances in order to support the capabilities of all social classes' children and their parents;[66]
2. a wide range of state regulated and financed measures for ensuring the well-being of families with children, i.e. the orientation towards the enabling welfare state;

[66] Universalism as a ideological basis of social policy in the Scandinavian societies results in the fact that the middle-class actually profit most from the income transfers and public services channelled by the social policy arrangements. This policy is intended: the idea is that all citizens contribute in the form of relatively high taxation but all citizens benefit, too. The Scandinavian social democratic model can be described as an institutional solution, which is intended to increase and maintain the highly appreciated social equality among citizens. To this kind of orientation also belongs that the social reforms only seldom have been argued for by referring to improvements of the situation of the poor. However, after the economic depression in the beginning of the 1990s, the poverty has increased and affiliated to that also the arguments for a more selective social policy instead of the universal system have been expressed.

3. horizontal distribution of income in family policy, i.e. families with children are entitled to life-course adjusted benefits and services;
4. vertical distribution of income in family policy, i.e. supporting the life-phases in which the income level of the family is low like in the case of young families or single parents;
5. gender equality, i.e. advancing the reconciling of work and family for both parents, and during the recent years, especially, supporting the men's role as fathers;
6. weak pro-natalism, which means that the voluntary parenthood has been supported – instead of population policy intentions – as well as all parents' right to a high quality of life and well-being with so many children as they themselves want to have.

Currently, the reconciling of work and family and the strengthening of the father's role are the main targets of the family policies in Finland. The Ministry of Social Affairs launched a programme for advancing the reconciling of work and family in 1998, especially intended to encourage the men to use the parental leave and care allowance paid for a parent who cares for children at home. Basically, the goal has been to develop gender-neutral family policies. In practice, today the role of men as carers is highly emphasised and supported by special programmes and measures, too.

As a consequence of the deep depression at the beginning of the 1990s, the poverty has increased in the Finnish society, and today, 10 % of families with children live under the poverty line; the single parenthood increases the risk of poverty. An obvious deterioration of the living circumstances of families with (small) children has been confirmed by empirical studies (Sauli et al. 2004): per child the whole family benefit level is 5 % less than ten years ago, and the level of the universal child benefit is 14 % lower than ten years ago. On the other hand, the enlargement of the day-care right and preschool system has increased the opportunities of families. In an international comparison, Finland still offers day care at low cost on a universal basis, and the family and child benefits are clearly above the European average level (Bradshaw and Finch 2002).

In Finland, the repertoire of services and cash benefits is wide and comprehensive. The proportion of family and children was 12.5 % of all social expenditure in the year 2000; the same figure in the EU-15 countries was 8.2 % (European Commission, Directorate-General for Employment and Social Affairs 2003: 186). Ireland and Luxemburg had higher proportion than Finland. In the year 1994 in Finland, all the tax credits based on the number of children were given up and compensated by increasing the tax-free child benefits in cash. The reasons for giving up the tax credits was the (sharp) progressive taxation because of which the deductions favoured the families with higher income and did not impact similarly on the families with lower income. In Finland, all families with children get an equal child benefit, which is staggered according to the order of the child in the family, and is highest by the fifth child after whom it is the same for the next children (from EUR 100 to EUR 172 per month, and in the case of single parents, an additional EUR 36.60 for every child per month) (www.kela.fi).

Leira (2002: 75–105) has described the development of the parental leave system in Scandinavia by the concept of 're-familialised childcare'; she refers to the decades long evolvement from a short maternal leave to one year parental leave. In Finland, the outlined development can be seen very clearly. In the beginning of the 1970s, the maternal leave consisted of a couple of weeks. Today, the home care allowance system has even opened an opportunity to stay at home for several years. In Finland, the maternal leave consists of 105 calendar days, also for adoptive mothers, and an additional 158-days parental leave is offered to the mother or father of the newborn child. For fathers, an 18-days leave is available during the maternal leave. The salary compensation level for parental leave is 60 %, and the allowance is a taxable income. The employed parents are entitled to a care leave until the child is three years old; during this period, the salary is not paid but the employees keep their working place (www.kela.fi; Rostgard and Fridberg 1998).

In Finland, the repertoire of child and family benefits and allowances consists of ten cash transfers with life-cycle and situational variation, and some extra benefits for single parents. Additionally, allowances and services for families with handicapped and chronic ill children are given as well as means-tested housing allowances, which support the families with children. In the family policies of Finland, the early childhood is emphasised like in other societies. The explicit intention of family policy is that the parents really can choose if they want to stay at home and take care of the small child(ren). Since the beginning of the year 2005, the time spent with children at home will be taken into consideration in accounting the personal cumulative pension – the reform can be interpreted as an incentive to encourage younger generations to birth-giving as well as a measure to advance the reconciling of work and family life as equal spheres.

Finland belongs to the very unique countries which pays a relatively high compensation for caring of the own healthy child at home. The background for the mentioned specific solution is the consensus based political decision-making processes around the child day-care. The home care allowance has been developed as a compensation for the families who do not use the child day-care services.[67] The enacting of the day-care and home allowance laws was highly influenced by the regional division and the rural-urban interests. However, the care allowances and child benefits cannot reach the level of the

[67] In the year 1972, the child day-care law was enacted in a very affective political atmosphere, and the agrarian parties were ready to join to the enactment with the resolution that besides the day care also the home care had to be developed. In the year 1985, the law of the home care allowance was enacted (Rauhala 1996).

full salary. The agrarian parties originally had the idea of a mother's salary on their agenda but never succeeded in reaching that.

Although there is a wide repertoire of home care allowances in Finland, maternal and parental leave included, it is easily shown that the emphasis in developing the public support for the families with small children has been in the day-care. With regard to day-care, the intention has been to ensure women's entering into the labour market, and the Finnish women have participated in the labour market already since 1960 to a degree similar to that of some Central and Southern European countries today. Currently, half of all children under school age attend the day care centres or the family day care arranged on the local level, and 90 % of all 6-years-old children attend the half-day free-of-charge preschool (nursery education) which the municipalities are obliged to establish whit state financial subsidy since August 2001. Municipalities charge for day care an income-adjusted payment, varying between EUR 0 to EUR 200 per month, including breakfast, a lunch and an afternoon snack (STAKES 2004; Välimäki and Rauhala 2000).

Since 1996, all children under school age have a statutory subjective right to day care in a municipal day-care centre or in a family day-care. Municipalities can organise the care both in day-care institutions and in a family day-care, which is a special form of day care: the municipality pays the salary for a caregiver (usually a woman) who takes care of children in her own home and can care her own children in the same group with the care-children. The family caregivers get regular and professional advice by the municipality, as well as a monetary refund for meals, toys and some other equipment. According to the studies, the parents prefer the family day-care for younger children but the day-care centre for the 4-6 years old children. Academic parents clearly prefer the day-care centres in which academically educated, and since the beginning of the 1990s master-level kindergarten teachers are responsible for the care and education of the children (Takala 2000).

Today, the number of children per one caregiver is on average four children both in institutions and in family day-care. Additionally, there is a municipal arrangement of three-family day care, in which the care-givers rotate between the families' homes every week, and a group-family day-care where 2–3 care-givers jointly care for a group of 12 children in a house which the municipality equips. Because of the increasing flexibility on the labor market, the number of children requiring shift-care is increasing: today 7 % of under school age children need shift-care because of the working hours of their parents (Sauli et al. 2004: 32).

In Finland, half of the children under school age attend the municipal day care, and the number has to be considered quite low; the explanation is the long parental leave, and the home care opportunities. There are only 7 % of all Finnish children under school age who are in no day care or whose parents do not apply for any allowance (Takala 2000). Besides the municipal day-care arrangements, the Lutheran Church of Finland (85 % of the Finns belong to the church which has the taxation right) organises day clubs for children under school age in 3 hours meeting 3–5 times per week; 55 % of 600 parishes run day clubs. The clubs are free of charge but the parents pay for the snack. Municipalities and non-governmental organisations offer a guided playground activity with meals offered for children who can be school aged, too. Many municipalities organise the open day-care centres for children and parents to have an opportunity to meet other children and parents.

The comprehensive school system is offered free-of-charge to all 7–16 years old children who live in Finland (there are special services for children with learning and other difficulties). Finland is among the Scandinavian societies the only country, which provides the free-of-charge lunch to all schoolchildren of comprehensive, secondary and vocational schools. The families appreciate the school lunch, and according to the surveys on of the quality of the food, most of the pupils evaluate the meals to be of high quality.

As a conclusion, the Finnish family policy is highly focused on the small children, especially on the phase of family formation and on the support of the start of parenthood. On the macro-policy level, a clear intention is to enable and facilitate women's entering into the waged employment. During the last five years, the role of fathers has been raised in family programmes, and it is evident that Finland by concrete implementation measures has adopted the intention to enhance the reconciling of work and family. The chosen policies are in line with the efforts of the European Union, although the idea of dual breadwinners was adopted in the Scandinavian countries already in the 1960s. The studies made on family values emphasise that the Finns appreciate family life as such, and the situational issues like the current economic resources do not affect the willingness to have children. However, the timing of the life phase with children is considered very carefully (Miettinen and Paajanen 2003).

3.4.4 Comparison

The dominating difference between the family policies in Estonia and Finland concerns the evolvement of family policies. Finland has developed a structural, deep-rooted family policy over a long-span period of time. Estonia has had the transition period in the 1990s, and social policy, family policy included, was not on the political agenda. During the transition period, the dramatic decrease of fertility has happened, and currently, the family policies are tinged with a strong population policy approach. At the same time, in Finland the fathers' role is emphasised in the context of reconciling work and family, but Estonia challenges both the emergency situation of the fertility decrease and the demands of the dual bread-winner model in the post socialist situation. In a historically oriented approach, the family policies of these two countries have developed in a very different way, and as an implication of that, they are not comparable as such.

On the other hand, in a current cross-sectional consideration, the political aims of family policies can be stated to be similar in Estonia and in Finland: facilitating the parents to carry on the expenditure of children; supporting the family formation in the very beginning of the newborn child's life; developing measures to advance the reconciling of work and family. In a very general level, the chosen strategies both in Estonia and in Finland can be described as enabling in the meaning that they support the sensitive phases of the life-course of children and (young) parents in their efforts to fulfil the roles as employees in the labour market and as carers in the family sphere. As a whole, the chosen strategies, at least ideologically, can be said to be capability-oriented both in Estonia and in Finland. In the case of Estonia, the family policy is after the regained independence the first field with explicitly programmed goals for supporting the choices of citizens and for increasing the personal autonomy by social policy measures. The other side of the coin is that the new strategy has been launched in a situation, which is interpreted as a national emergency because of the extreme decrease of the population.

Methodologically, the comparison of a post-socialist country and a Scandinavian country is a challenge as such, and there are also several problems to be encountered. The family policy was not among the first issues on the agenda of Estonia after regained independence, and in fact, as the empirical studies emphasise, the families with many children and living in the countryside belong to the looser group of the transition. The other issue, which surprised the new Estonia, was that the generation born during the 1970s has postponed or even refused to form a family and to give birth. The reactive population policy does not improve the situation of the families with other than newborn children; the level of family benefits is low, and much of the responsibility of family support is left on the municipal level to be done on a voluntary basis in a public or private setting.

For the Estonian family policy, it is a prominent feature that the municipalities differ a lot according to the services and benefits they offer to the families. The capital and medium-size dynamic cities can offer a good repertoire of services and extra-benefits to families, and these cities are, too, the ones, which have organised the school social workers for buffering the school-drop-out. The differences of rural and urban areas are high in Estonia. In Finland, the state-level social policy has been and still is an efficient tool for equalising the regional and rural-urban differences.

As a whole, the goal of equality (on the basis of gender, class, region) is dominating the family policies of Finland more than any other factor, and in the Scandinavian context, the capability approach means first and foremost efforts to activate the available human resources by equality policies. While compared to other European Union countries, in Scandinavia, there is a strong continuum of equality policies, and it seems that the Scandinavian societies adjust themselves to the European practices by calibrating their equality policies instead of rejecting the original goal. While taking the capability approach as a starting point, it is not easy to say, what kind of policies should be launched in order to enhance diversity and difference as source of social dynamics. However, it is discussed to take the development of 'equality in diversity'-oriented policies on the agenda.

The Estonian scholars summarise that the decrease of fertility in Estonia is connected with three main issues: the postponing of the first birth giving; the economic restrictions both on state and personal level; and the diminishing of the value of children in the Estonian society (Ainsaar 2003; Kutsar 2003: 102). As a conclusion, the family with children as such can be interpreted among the groups left in the shadow during the transition period. It is reasonable to doubt, whether the very strong pro-natal policies can be sustainable in a long-run perspective. The Scandinavian development of the family policies has had the opportunity to be much more sustainable. Leira (2002) has concluded in her detailed empirical analysis that the Scandinavian societies have developed the social benefits for families as rights of working fathers and mothers from the beginning of the modern era, and that population policy as such has never been in the focus of the Scandinavian social policy. The mentioned spirit is written in the family laws of Scandinavian societies. As an outcome, a stable fertility rate has been reached.

In the Scandinavian countries, especially in Finland and Sweden, the tendency of re-familialisation of the childcare has been launched since the 1980s, and the economic depression in the beginning of the 1990s has accelerated the tendency (Leira 2002). Affiliated to that, the fathers are given the same responsibilities for care-giving to small children as mothers. In Estonia, during the Soviet time both women and men were obliged to participate in the labour market, and the system of public day-care was established. It is not yet possible to say if the currently launched family policies succeed in re-familialising the childcare on the basis of gender equality, or if the traditional male and female roles remain dominant. During the transition period, the women have postponed birth-giving, and, as the Estonian scholars have reported, at the same time the family as an attractive life-chance has lost a lot of its meaning. From this point of view, it might be questioned, if the enabling family policies are possible in a social atmosphere where the family is not appreciated but interpreted as a reproductive machinery of the population policies.

Discussing the question what can be learned from the cases of Estonian and Finnish family policies for the European Union policies, some promising ideas may be mentioned. The structural family policies of Finland can be interpreted as a prerequisite for reconciling work and family. The family policies with benefits and services form an infrastructure for the labour market participation, and the current challenge is to draw the men as fathers to share the family and care tasks in the private sphere. This goal has already been adopted by political decisions. The fathers' rights for the care of chil-

Table 3.9: Characteristics of the family policies

	Estonia	Finland
Pro-natal orientation	very strong	low, if at all
Population policy orientation	strong	medium
Diversity of family types	neutral	favouring of diversity
Child-friendliness in society	low	medium, high
Favouring the role of men as father	low	high, emphasised
Level of family benefits and services	low, medium	medium, high
Favouring women's waged work	neutral	emphasised
Life-course orientation	biased (on the newborn)	needs-tested
Capabilities approach	biased, not explicit	emphasised
Enabling welfare state	in progress	structural, institutional
Tendencies		
Fertility 1988-2001	-0.81	+0.14
Fertility 2002	1.2	1.7
Single parents 2002	7%	16%
Population growth since 1990	decreasing rapidly	stable

dren has been in the focus, and accordingly, the labour market partners have been informed by a governmental 5-years programme, in which the balance of family and work has been interpreted as a source for increased productivity in the knowledge society. It is reasonable to interpret the Finnish family policies as a capability-oriented approach: all the resources are intended to be used effectively and the focus is on equal opportunities for men and women.[68] Up to now, however, it is too early to say if some benchmarking of the Finnish experience could be useful or beneficial in the context of the European Union.

In Estonia, an exceptional reform has been launched in the frame of population policy. The case of Estonia can be interpreted as a social experiment by which the aim is to test if a high cash benefit can influence the fertility behaviour. Equally interesting is the question, if the role of men can be

[68] While discussing the results of the PISA-study and the observed high capabilities of the Finnish children, the high education of Finnish women has been debated as one possible explanatory factor. The mothers with high education understand the meaning of studying, and are also able to motivate and supervise the independent conducting of the school tasks of their children.

changed in direction of the ideal of "working fathers", who are ready to participate in the family tasks, too. In Estonia, the gender roles outside of the labour market are traditional. In the current situation, the subsidiarity in conducting the social policy, family policy included, has been the only way to cope the transition and its consequences. The family networks were important during the Soviet time, and the confidence is based on personal relations, not on the institutional actors. The Estonian social policy and family policy are still open for the choices, and it is too early to say how the increasing economic growth will affect social policy.

3.4.5 European Dimensions and Elements

The roots of the modern European family policy can be found in the population transition, which started during the second half of the 19th century. The infant mortality decreased rapidly due to industrialisation and urbanisation, and the development of the modern nuclear family started. As Cohen and Hanagan (1991) have analysed, the children made themselves visible as a target group of social policy by decreasing their number.[69] At the same time, the industrialised work and labour market began to require new individual skills which demanded years-long basic and occupational education. The modern child-pedadogy was launched by Friedrich Fröbel and Heinrich Pestalozzi who understood the child as a unique person on her own. The modern childhood as a special phase of the life course was discovered, and as such the childhood is a recent social phenomenon, still under its formation, as the variety of the family conceptualisations underline. The changes of family and childhood, and the decrease of the number of children as a consequence, have had a multilevel and complex impact on the formation of women's waged employment (e.g. Nave-Herz 2003).

The history of the European social policy after Word War II indicates that diverse family policies were launched in the European countries (Hantrais and Letablier 1996). However, the goals of the family policies were quite similar in all European societies. The aim of the family laws and of the implemented reforms has been to facilitate the maintenance and everyday caring of children by the state and municipal benefits in cash and kind. In some countries, the taxation of the bread-winner was adjusted in accordance with the family size, e.g. Germany and in the Scandinavian countries until the 1970s. In other countries, high cash benefits were favoured, e.g in France and since the 1980s in the Scandinavian countries. In the former socialist countries, and in the Scandinavian countries, too, the public child day care was offered to the families in order to draw both parents into the labour market. In the Southern European countries and in Ireland, the extended family was expected to be the main care-giver for the children. The situation is rapidly

[69] The elderly people, vice versa, make themselves visible by increasing their proportion of population.

changing in the European Union while the labour force of both genders is expected to be available for the needs of the labour market. This can be interpreted as one dimension of activation policies, although usually it is not addressed as such.

The child benefit schemes and tax credits on the basis of family ties can be seen as measures of familialisation. The idea has been to improve the household economy through (direct) income transfers in cash. From the very beginning of the family policies, there has been the explicit intention to practice distributive justice in organising the concrete measures. The gender based division of care work has come later into the repertoire. The state has given the final decision-making of the income sharing to be done within the family, and that is why a special "mother's wage" has not been launched in any society as an opportunity for a real choice between waged employment and domestic work. This feature of family policies refers to an enabling approach: the utmost decision-making has to be done by the family members themselves, whereby the personal autonomy in the meaning of right to privacy is respected.

The social services as measures of family policy represent the de-familialisation, which commodify women and decrease their dependency on men (Esping-Andersen 2000, 46 et seqq.). The services for family underline the life-phase of under school age. According to that, day-care and pre-primary schools have been developed. It seems evident that the adopted labour market policies of the European Union can function and succeed, if an infrastructure of high-level day-care arrangements is provided. The Scandinavian experience could be useful for developing the everyday-life sensitive care; there is an empirical study, which confirms the interpretation of day care as a prerequisite for dynamic labour market policies. The other issue to be taken under consideration is if and how the care outside of the nuclear family supports the personal autonomy and development of a child.

Currently, a double strategy of the family policies can be recognised; first, by the parental leave, the care opportunities of both mothers and fathers are strengthened in the very beginning of the newborn child's life; second, while the parents of small children are participating at the labor market, day-care is needed. The leader of the IPROSEC-study of the family and family policies in eleven EU-countries, Linda Hantrais (2003) has given a commentary that in the politics (decision making) and administrative settings, the family policies have a minor role while compared to how much the gender equality and both women's and men's participation at the labour market are emphasised. Hantrais draws the conclusion that the character of family policies as an infrastructure for a well-functioning labour market and high productivity of both genders has not yet been fully understood.

The Scandinavian experience supports Hantrais' commentary (Leira 2002). The enabling approach could be applied in studying the adopted double-strategy; what is ideological and political idealism, and what are the real opportunities to reconcile family and work in a successful way. In Europe, the single parenthood is a risk for poverty, but there is also evidence that a working single-parent with support of benefits and services can manage the maintenance of children (Leira 2002; Sipilä 1997). The European experience gives empirical evidence of the fact that the divorce rate as such is not dependant on the practiced family policies.

The European family policy is taking shape in a context which is highly emphasised by the dual-earner model, or, in other words, by equality policies which tries to enable the reconciling of family and work for both genders. How this goal will reached in the practical family policies of different countries, is still an open issue. The rights of children should be taken under (re)consideration in a setting where labour market policies are organised on a flexibility basis and where migration within the Union is favoured.

3.4.6 Concluding Remarks and Recommendations

In the current European situation, the conflicting interests of the family policies are obvious. The ageing societies need children in order to ensure the generation continuum and the mutual cultural exchange. In the contemporary discussions, the population policy interests are also introduced, and like in the case of Estonia, explicit family policy is established in order to advance the population increase. At the same time, the both parents' participation at the labor market starts to be a self-evident fact. In addition, it is obvious that the dual-earner model is also needed in order to buffer the social exclusion and poverty; according to several comparative studies, the single parenthood is among the evident factors, which increase the risk of social exclusion, in the Scandinavian societies, too. (Heikkilä and Kuivalainen 2002.)

In his study of the European societies David Coleman (1996) has pointed out that the wish of the number of children is on average a little bit more than two children but the realised fertility is two or less children, as the fertility rates confirm. The wished and realised fertilities were analysed in the IPROSEC-study. Quite in conformity, the European men and women would like to have two or three children. In Spain and in Germany, the life without children was expressed as an ideal more often than in other countries, and the attitude was sustained by environmental and sustainable development reasons (Kasearu and Kutsar 2003: 92). In the current situation the young adults have real opportunities to decide if they want to have children or not. The Scandinavian experience gives evidence that the family policy as such does not favour nor buffer the willingness of birth-giving. The same concerns women's participation at the labour market: the waged employment does not hinder women to give birth, and here the Estonian experience with very stable fertility rates during 1920–1990 is a prominent example.

The current European family policy orientations, intentions, and discussions concerning their modernisation are, surprisingly, quite similar to those in the beginning of the modern family policy a good one hundred years ago. The adequate care and education of children since their early years were taken on the agenda then, and the arguments for a (public) organisation of kindergarten und primary school for children of all social classes were based on the new challenges concerning the urban and industrial life. The necessary civic skills and the basic intellectual capabilities could be achieved only by professional care and education; also the parenthood was interpreted as a "profession" (Myrdal 1941). Today, there are discussions concerning the education of small children in the ICT-society, and it is thought to be 'rational' to give the basic skills, international skills included, already in the nursery schools. Pre-primary school systems include the goals of training besides nurturing and care – in fact, the very original idea of Fröbel.

Additionally, the "professional" parenthood should be supported by tests and other measures for monitoring the difficulties of the development of (small) children, since the intention is to diagnose the possible learning difficulties early and to ensure the needed treatment and support for the child. These aims can also be thought of in the frame of ethics: what kind of screening, for instance, can be implemented in order to avoid labelling? From the viewpoint of social inclusion, the monitoring measures of difficulties and deficiency can be rational, but ethical justification is needed where the personal autonomy of parents and children is touched.

The source for similar developments of family policies in the European Union and its member states, is the family structure itself: the nuclear family is dominating, the single parenthood is an increasing trend, the step-families are a reality, and the couples without children are increasing their share, too. While looking at the development of the family, the trends seem to be convergent despite very different family policies. We can hardly speak of the impact of policies on the family, but the issue seems more vice versa: how the change of family has influenced and is going to influence on the policies. The societies have to adjust the social practices according to the change of the family; in the individualistic and economically rich societies citizens themselves decide if they want children and how many children they want, and, finally, when they want to live with (small) children. The personal autonomy in the family formation issues has started to be of great importance in the era of developed contraception technology and in societies which emphasise individual values.

In the European Union labour policies, the reconciling of work and family is mainly discussed in the frame of employment policies, not (yet) within family policies. The goals of the increasing economic power and competition capacity can be based only on an effective use of educated labour force. The young women are highly educated and their occupational and professional contribution is expected from the society. The classical issue of taking care of children is no longer interpreted in a maternal frame, but in a parental where both parents should have opportunities to conduct family and working life. Compared to the beginning of the modern family policies, the emphasis put on man's role as educator and care-giver for children is the biggest change in the policy-discussions.

The gender equality can be seen as an increase of personal autonomy and justice in the gender relations. Nevertheless in practice, the care of small children still is interpreted primarily as a task of women, which the majority of the female single-parents indicates. One crucial change has happened in the attitude towards the children born out of wedlock and their mothers; the earlier disapproval in many societies has disappeared, and the children born out of wedlock are considered equal to the children of married couples. The focus of family policies has moved from the marriage onto the child. Besides the family policies, the childhood policies are in need for development. From the viewpoint of social inclusion, the critical phases of the life-course are not known; there is still a strong family ideology according to which the "family-childhood" is interpreted as favourable for developing a coherent personality who can manage well in the society. The current family policies try to combine the family ideology and the gender-neutral labour market participation by enhancing the role of the fathers. In several countries, the main tool chosen for the mentioned purpose has been the launching of parental leaves instead of the former exclusively maternal leaves.

It seems evident that the future for a successful reconciling of family and work depends on how the fatherhood is supported. Important in this regard are the signals given to the parent(s) by the society that both women and men while nursing, rearing and educating children are doing a socially appreciated work. In the countries of the European Union, fathers' rights and opportunities should been taken on the agenda in order to equalise women and men in their family tasks. It is evident that there will also be more differentiation in childhood (and consequently in the future, people with more variation in their skills, too) if fathers are taking more practical responsibility of the childcare and education from the early childhood on. There is not yet enough historical experience on how the men are doing as carers of small children. Nevertheless, it seems that in spite of strong incentives, the changing of traditional family roles is not any easy task. The Scandinavian experience is promising but the final breakthrough in equalising the parents as similarly competent persons to take care of the children and the household work has not yet happened.

The child-friendliness is one dimension in the social climate of a society. There are no clear indicators for describing the child-friendliness of society, although the implemented family policy measures give opportunities to make comparisons and draw conclusions on how the children are appreciated. According to the social policy expenditure used in the family policy, the European Union (EU-15) countries pay much attention to children (European Commission, Directorate-General for Employment and Social Affairs 2003: 186, table 5). Both in cash and kind, in the forms of tax deductions, child allowances and day-care services, the European Union countries offer the best possible compensation of the costs of having children. The development of indicators for measuring the child-friendliness in different countries should be taken on the agenda of the European Union, and as a target it could be managed by OMC.

The current emphasis of the family policies in the European Union countries is prominently put on the care of children under school age. According to the studies concerning the welfare of children, the school age and the adolescence should be emphasised more. The capabilities approach actually gives a tool to interpret all phases of childhood as crucial for the future development of a personality. The school-drop-out problem as a social problem is an increasing phenomenon in all European countries, and it should be combated by family policy measures, too. This issue is not yet in the common awareness. The lost generation consists of persons who have not been supported enough in getting the basic education, and probably they will be a target group for social policy in a society based on ICT; the experiences of the USA give some advice for avoiding similar consequences in the European Union. The social inclusion as an ethical foundation for the European social policy gives a reason to enlarge the family policy towards the later life-phases of children and youth.

The family policy needs extension towards a broader life-course perspective, too. The OMC could be applied in seeking for methods to buffer the school-drop-out. From the viewpoint of developing capabilities, the prevention of school-drop-out is among the target issues. A proposal could be an empirical comparative study of the school-drop-out phenomenon in the European Union in order to avoid the tendencies, which have occurred in the USA. The low educated persons are cheap labour force but they are also the persons in need for sensitive support in later phases of their life course. The OMC could be used for developing Europe wide programmes in the area of family policy, which would also be a prerequisite for the mobility of labour force in the EU. The parents of small children are the best educated persons, and they are also the group most likely to move in Europe before their children start the elementary school.

As a final conclusion, it has to be stressed that, although on the rhetorical level there exist common family policy goals in the EU, the practices and measures actually implemented present a large scope of variation, which is not yet studied enough in order to understand how the employment, education and family policies are intertwined into each other. The life-course perspective within the family policies refers to emphasising not only the very beginning of the family formation – as the repertoire of the benefits and services focused on a newborn and under school age children indicate – but on the family and family policies as a life-long process where childhood, adulthood and ageing are taking shape both personally and socially.

3.5 Poverty Prevention: Belgium/Denmark

3.5.1 Starting Point

Recently, promoting social inclusion and fighting poverty has become one of the key-objectives of the European Union. While the European Commission has had anti-poverty programmes in process since the 1970s (cf. Room 1995), the issue was brought to centre stage at the Lisbon Summit of the European Council in March 2001. The summit conclusions declared the number of individuals living in poverty and social exclusion in Europe to be unacceptable and called for decisive steps to tackle the problem, beginning with the setting of targets for particular indicators. The Lisbon Council also agreed that member states should coordinate their policies for combating poverty and social exclusion on the basis of the OMC. By now two rounds have been successfully completed of National Plans for Social Inclusion; there have been two Joint Reports on Social Inclusion; and the new member states have drawn up Joint Inclusion Memoranda.

This chapter aims to contribute to a clearer understanding of the social inclusion and poverty policy in the EU. We begin with elucidating our starting points and then focus on two EU member states, Belgium and Denmark. Our goal is to compare policy tendencies, goals and instruments in the field of social inclusion and poverty policy. The next section discusses the case for action at the European level to fight poverty and social exclusion. We conclude with some final remarks and policy recommendations.

3.5.1.1 Distinguishing Poverty and Social Exclusion

During the 1970s and the1980s the socio-political dialogue, both on national and European level, focused on income inequalities and their impact on the phenomenon of poverty. During the 1980s however, a new concept was introduced in the debate on European social policy: the notion of social exclusion. Slowly, but steadily 'exclusion' superseded 'poverty' as the key term in the political debate. In 1989 the European Council of Ministers passed a Resolution calling for action to combat social exclusion and in 1990, the Commission set up a research network, a so-called 'Observatory' to monitor national trends and policies in this field.

Although the vocabulary of social exclusion is nowadays widespread throughout the European Union, its precise meaning is not always clear. As originally coined by the French, social exclusion refers to persons who are excluded from the mainstream social protection – social insurance schemes – such as the unemployed youth in the banlieues. Its meaning nowadays, however, is far broader and may encompass many variant forms, ranging from financial poverty through labour market marginality to the denial of social, political and civil rights (Silver 1994; Room 1995; mayes et al. 2001; Leisering 2004).

What is the exact difference between poverty and social exclusion? Although the concept of 'social exclusion' builds on multidimensional and dynamic notions of poverty, it is a broader and more theoretical notion. While poverty focuses on the economic and financial situation and – if understood as multidimensional – on the question how poverty impacts a person's life situation, the concept of social exclusion is more interested in the institutional and sociological context of poverty. It focuses more particularly on the conditions for participation in society and studies the deficiencies or the factors hindering the participation or life-chances of individuals (Silver 1994; Paugam 1998; Mayes et al. 2001; Leisering 2004). The concept has thus both economic and social dimensions. Being excluded implies that someone's opportunities to earn an income, participate in the labour market or have access to assets are substantially curtailed. But it can also mean that people are excluded from public services, community and family support, and cannot even participate in shaping the decisions that affect their own lives. Thus, social exclusion always has a communal or societal dimension, since it can be caused not only by lack of personal resources but also by insufficient community resources (Room 1995).

Poverty does not necessarily imply social exclusion. Vice versa, social exclusion does not necessarily or not only refer to poverty (Barry 1998). Some forms of social exclusion mainly derive from physical handicaps, severe diseases, political and social discrimination, cultural stigmatisation and lack of education. Yet, in reality financial poverty and social exclusion tend to go together. An important reason for this is that poor people are often unable to participate in the customary consumption activities of the society in which they live. The most evident example is that of housing, but also significant are access to durables, health expenditure, food expenditure (nutritional content), and expenditure relating to recreational, cultural and leisure activities. A good example of exclusion in consumption is the telephone: "a person unable to afford a telephone finds it difficult to participate in a society where the majority have telephones. Children are not invited out to play, because neighbours no longer call round – they call up" (Atkinson 1998: 88). Another reason why poverty and social exclusion tend to go hand in hand is that they are often related to unemployment. Unemployed people not only have a higher risk of poverty, but are also more vulnerable to social exclusion. The rationale behind this is that unemployment and long-term unemployment in particular, damages individual autonomy, harms self-respect, and might lead to social isolation and other disadvantages in the realisation of future life chances (Gallie 1999; Paugam and Russel 2000). In this fashion, the EU Joint Report on Social Inclusion states that "long-term unemployment is very closely associated with social distress, as people who have been jobless for a long time tend to loose the skills and the self-esteem necessary to regain a foothold in the labour market, unless appropriate and timely support is provided" (European Commission, Joint Report on Social Inclusion 2004: 25).

3.5.1.2 Social Indicators for Poverty and Social Exclusion

How are poverty and social exclusion defined and measured? Until the late 1980s it was commonplace in poverty research – in particular, cross-national studies within Europe – to focus on disposable income (or expenditure) of an individual or a household at a given moment in time.[70] This was the basis for most of the estimates of the overall poverty rate in the EU that have appeared during the 1980s and 1990s. in the last decennium there was a growing acknowledgment that this focus on financial indicators as a proxy for poverty is too narrow (Berghman 1995; Saraceno 1997; Atkinson et al. 2002).

Although financial resources are very important for the whole range of life chances which a person can enjoy, a concern for the capacity of people to participate in diverse social activities and networks and share in the general quality of life, necessitates a more multi-dimensional approach. To disentangle the different elements of social exclusion and identify their interrelationships, one needs a set of multidimensional indicators, focusing not only on financial resources or disposable income, but also on housing, educational attainment, poor health, poor housing, etc. During the 1990s, the European Commission has shown great interest in the development of such a multidimensional set of indicators (Atkinson et al. 2002). Recognising the role played by quantitative indicators and targets in the implementation of the EMU, it was argued many times that there was a case for setting targets in terms of poverty and social exclusion similar to those that had been achieved earlier in the macro-economic field as part of the Maastricht Process (Commission of the European Community 1996, 2000). A solid basis for monitoring progress and assessing the effectiveness of policy efforts was established with the endorsement, at the European Council in Laeken in 2001, of 18 commonly agreed indicators to approach the measurement of poverty and social exclusion (Atkinson 2002; Atkinson et al. 2002). A variety of domains is covered – not only financial poverty and income equality, but also regional variation in employment rates, long-term unemployment, joblessness, low educational qualifications, low life expectancy and poor health – reflecting the widespread perception that poverty and social exclusion in Europe have a multidimensional nature and cannot be reduced to one single variable. These indicators should serve as a basis for the EU and each individual member state to assess objectively progress of the multi-annual process on the basis of verified outcomes.

---

[70] As concerns measurement of poverty, there is a diversity of possible judgements about the specification of the poverty line and choice of poverty measure (cf. Atkinson 1998; Van den Bosch 2001; Atkinson et al. 2002). However, the most common approach is to use relative income poverty lines, derived as proportions of mean or median household income. For example, the proportion of individuals living in households where equivalised income is below the threshold of 60 % of the national equivalised median income has been taken as an common indicator of relative poverty in the NAPs.

3.5.1.3 Preventing Entrapment in Deprivation

The shift from 'poverty prevention' to 'combating social exclusion' entails a growing concern not only with multidimensional deprivation but also with long-term poverty and deprivation. Various panel studies show that most poverty is short-term: there is plenty of income mobility, although many of those who escape from poverty remain on its margins and may subsequently descend in it once more. Often poverty is associated with particular life events, and especially transitions such as divorce, separation and leaving home among the young (Leisering and Walker 1998; Leisering and Leibfried 1999).

Individuals' life chances, however, are not per se affected by a short period of low pay or unemployment. But they are if this period prolongs and especially, if it bundles with other risks in the form of a multidimensional, multiple deprivation. In case citizens are indeed 'trapped' in inferior life chances, we will use the term social exclusion, denoting a situation of cumulative disadvantage, in which different forms of disadvantage reinforce each other (cf. Silver 1994; Paugam 1996; Layte and Whelan 2002). People will feel excluded not just because of their current situation, but also because they have little prospect for the future.

From the point of view of combating social exclusion, it is therefore important to distinguish between temporary and persistent poverty. It is not enough to count the numbers and describe the characteristics of those who are poor or disadvantaged; it is also necessary to identify and to focus on the factors, which can trigger entry or exit from this situation, and explain how individuals get caught in a situation of persistent deprivation or social exclusion (Tsakloglou and Papadopoulos 2002). Panel studies have well identified those most at risk of falling into poverty and staying there: people who are poorly educated and lowly-skilled, unemployed and sick people, and lone mothers. By now it's also well documented that a crucial mediator for entrapment in social exclusion is the household context (Mejer and Siermann 2000; Esping-Andersen 1999; 2002a). A person may be unemployed or a low-wage worker. But it is one thing if the only earner is low paid, and another if the low paid worker is but one of several income recipients. Hence, it is no surprise that full-time employment of the main breadwinner remains one of the most important barriers against social exclusion (Tsakloglou and Papadopoulos 2002).

The Commission has therefore underlined the importance of dynamic indicators of poverty (the share of the population below the poverty line for three consecutive year) and indicators of persistent poverty risk, taking into account the household context (such as the percentage of persons living in households who have been below the poverty line two of the previous three years). These provide a valuable way of focusing attention on those most likely to be at risk of social exclusion. We should note, however, that the implementation of such a dynamic approach requires that special attention is given to the collection of appropriate data (administrative data or panel data) (cf. Atkinson et al. 2002: 110). This new European emphasis on the dynamics of social exclusion might lead to a more innovative approach to social policy, which has the life course as a perspective and focuses on improving individual capabilities (Begg and Berghman 2002; Muffels et al. 2002). From a life-course perspective, welfare states should aim to intervene at the most efficient and effective time in people's lives. An example is the prevention of child poverty, which could be a very efficient and effective way of preventing the poverty and social exclusion of adults. Making sure that children receive a good education and grow up as healthy productive individuals will contain social expenditures at a later stage (Vleminckx and Smeeding 2001; Esping-Andersen 2002).

3.5.1.4 The Normative Framework: towards Full Citizenship

The focus on the fore-mentioned 'capabilities' is part and parcel of a wider attempt to embed the idea of social exclusion within a wider conceptual and normative framework (cf. e.g. Giddens 1998; Vandenbroucke 1999; Sen 1999; Giddens, Esping-Andersen 2002). Guideline here is the idea that full participation in society is one of the basic opportunities that should be the right of everyone. It is, to use the terminology of Sen, an essential 'functioning' or a basic condition for personal well-being. In this view, social exclusion often leads to 'low intensity citizenship', for one particular trait of it is that it not only entails inequality of income, wealth and resources, but also that it creates a real danger that the poor no longer have access to the full set of political, civil and social rights (Silver 1994; Merkel 2002; Amitsis et al. 2003).

Broadening the idea of poverty to a concept of social exclusion thus implies a stress on the capacity of citizens to fully exercise their civil, political and social rights, as advanced in Marshall's (1950) account of citizenship. For Marshall, the different categories of rights are not self-contained but are only meaningful when taken as parts of a common programme of social inclusion: civil and political rights (rights to political participation, free speech, legal equality, etc.) require some measure of welfare if they are to be more than formal guarantees. For example, providing high-standard education to everyone promotes people's abilities to know, use, and enjoy their liberties, due process rights, and rights of political participation. Ignorance is a barrier to the realisation of civil and political rights because uneducated people often do not know what rights they have and what they can do to use and defend them. This inclusive aspect of welfare rights is also easy to see in the area of democratic participation. Education and a minimum income make it easier for people economically at the bottom to distinguish between electoral campaigns and carnival, to follow politics, participate in political campaigns, and to spend the time and money needed to go to the polls and vote.

As indicated, high priority should be given to education and training, for these play a crucial role in empowering individuals to pursue a self-determined life and participate in the different (social, economic, political) spheres of society. Instead of compensating individuals for exclusion, one opts for the wiser and more just option of 'social investment' (Giddens), of enhancing individuals' capabilities (Sen) by preventing them from becoming obsolete or redundant, so they can work on their own future. This fits into a preventive approach to social protection, which is believed to be more fruitful than the curative one, based on the reparation of damages caused by the realisation of social and economic risks. Education and training are seen as investments in human capacities and considered to be an integral part of the European social policy framework. In the USA, the view that the education system is at the core of the welfare state is much older and some have even considered the education system as a substitute for social protection in the USA (Begg and Berghman 2002; Flora and Heidenheimer 1981).

The renewed stress on education and 'lifelong learning' is inextricably linked to the contemporary policy vocabulary of 'activation', 'employability', 'make-work-pay' and 'welfare to work'. Activation policies were advocated by the 'Third Way' and the 'active welfare state' paradigms, which became prominent in European social policy debates in recent years (Giddens 1998; Vandenbroucke 1999). The emphasis here is clearly on rehabilitation, training and insertion of the benefit population, with a view to preventing long-term dependency on income support, which may contain social expenditure levels and make the labour force more responsive to changing economic conditions.

But the ambition to get people 'back to work' is not only an economic necessity. Improving people's productive capabilities (relating to the economic domain) and social capabilities (relating to the social domain) will serve both economic efficiency and equity. The current policy discourse emphasises gainful employment more in general as the axial principle of full citizenship (Vandenbroucke 1999; Gallie 1999; Esping-Andersen 2002). Rationale behind this is the idea that unemployment is not primarily an economic problem that can be solved simply through extensive transfer payments. As already mentioned, unemployment and long-term unemployment in particular, also imply a social and ethical challenge because they easily lead to social isolation and other disadvantages in the realisation of future life chances. As long as not only income, but also status, respect and self-respect and social standing in European societies are primarily distributed via paid employment, inclusion in the labour market has to be a high political priority.

This is even more so in contemporary labour markets which often imply more stringent requirements in terms of knowledge, skills and intellectual flexibility, combined with lower demand for low-skilled labour. "The threat of job polarisation is real because investment in new skills is concentrated in higher occupational classes, while low-skilled jobs offer few opportunities for training, skill enhancement, or personal development. High employment levels may nourish a low-end labour market with a class of workers locked into inferior jobs and poor life chances" (Esping-Andersen 2002a: 22). It is thus considered to be of utmost importance that investments in people's capabilities and skills focus first on the lowest end of the labour market.

Employment may play an important role for enhancing social inclusion and reducing the risk of financial poverty. It should be noted, however, that the latter is only valid if 'work pays' and people in a job are not at risk of poverty. A job may not pay sufficiently to raise a family above the EU agreed poverty line. The same concern is evidenced in the 'unemployment trap'. If the jobs on offer pay no more than unemployment assistance and this is below the poverty line as defined at EU level, then employment does not provide a route out of poverty. Cutting unemployment assistance will not help, since it still leaves people below the poverty line.

3.5.1.5 Poverty and Social Exclusion: the EU Situation

Poverty and social exclusion policies in European member states vary significantly along several dimensions: methods of financing, organisation, level and duration of benefits, eligibility rules, scope of benefits, etc. Each national system has its own set of institutional, administrative and political characteristics (an overview of the main characteristics of different regimes of protection against poverty is given in table A.2 in the annex). Yet, what is common to almost all countries,[71] is that they combine some form of social security schemes which guarantee social and economic status through selective labour-income related benefits, with social assistance schemes, aimed towards those who cannot support themselves or their children by other means, such as salaries or income from the social security. Recent analyses of poverty policies show that both components are important. If a country wants to achieve low levels of poverty it needs an important social redistribution and an adequate minimum protection for those in work and those out of work (Cantillon et al. 2003; Nicaise et al. 2004). While work provides the best protection against social exclusion and poverty, it is no guarantee of social protection for all groups in all circumstances. Therefore, adequate minimum wages and minimum income protection in social security should be complemented by social assistance as a last safety net.

How effective and strong are the existing safety nets of guaranteed minimum income and, more generally, social security systems in the EU? When looking at the total population, around 15 % of citizens in EU-25 were at risk of poverty in 2001 (European Commission, Directorate-General for

[71] Not all EU member states have a fully implemented, nation-wide system of minimum income. For example, until now, Greece has no universal minimum income guarantee. For more information, see MISSOC 2004.

Employment and Social Affairs 2003). This figure represents around 68 million people. Across the Union there are considerable differences in the severity of the problem. Using 60 % of the national median as a cut-off threshold, the proportion of people at risk of poverty was relatively higher in Ireland (21 %), the Southern countries (19–20 %), the Baltic states (17–18 %) and the United Kingdom (17 %). In Southern countries, as well as in Ireland, poor people not only benefit comparatively less from the overall prosperity of their respective countries, but also are more likely to be subject to more persistent forms of poverty and deprivation. Figures were lower in Benelux countries, Germany and Austria, the Nordic member states and most Central and Eastern European countries (11–13 %). Amongst old member states the lowest poverty rate was in Sweden (10 %) and was even lower in certain new countries such as Slovakia (5 %) and the Czech Republic (8 %). In this context, it should be remembered that we are analysing relative poverty within each country and not absolute poverty by reference to an independent cut-off threshold. Social benefits (pensions and other transfers) reduce the proportion of people at risk of poverty in all European countries, but to varying degrees: the reduction ranging from 50 % or less in Greece, Spain, Ireland, Portugal, Cyprus and Malta to more than 75 % in Sweden, the Czech Republic, Hungary and Slovakia (European Commission, Directorate-General for Employment and Social Affairs 2003).

Within the various member countries there are substantial regional differences. In Italy, for instance, poverty rates in the North are among the lowest in Europe, while those in Sicily are among the highest. Spain, Belgium, the UK and Germany also show significant regional disparity in terms of poverty and social exclusion (Heidenreich 2003; Stewart 2003). In the context of the big differences in prosperity and poverty between countries and regions, the importance of the EU structural funds should be underlined. The regions with an average per capita GDP of less than 75 % of the EU average (Objective 1 regions) are the ones which mainly profit from these funds. The structural funds are thus an important instrument in the redistribution of financial support from richer to poorer countries and regions and to a certain extent constitute a functional equivalent for the lack of a genuine social policy at the EU level.

The structural funds are a good instrument to address 'place poverty', which emerges when regional circumstances, such as the quality of public services, contribute to poverty outcomes. Yet, besides regional differences in poverty indicators, the EU also knows important social differentiation in the poverty risk. In 2001, the risk of poverty in the EU tended to be significantly higher for particular groups such as the unemployed, single parents (mainly women), older people living alone (also women mainly) and families with numerous children. Women are generally at higher risk than men. A particular risk of poverty and social exclusion is faced by young people deprived of sufficiently solid skills to get a firm grip on the labour market. In 2002, almost 19 % of the people aged between 18 and 24 had exited the school system too early and were not following any training. Also children are in a vulnerable situation. They tend to experience levels of income poverty that are higher than those of adults (19 % in 2001), and material deprivation in the early years may affect negatively their development and future opportunities. A particular concern arises when children are living in jobless households, without almost any links to the world of work (10 % of all children in the Union, in 2002). The severity of the problem of child poverty has led the 2004 Joint Report to call for action to end "child poverty as a key step to combat the intergenerational inheritance of poverty with a particular focus on early intervention and early education initiatives which identify and support children and young families" (Joint Report on Social Inclusion 2004: 105).

The risk factors associated with poverty and social exclusion which were identified in 2001 are confirmed in the 2003 NAPs. These are: long-term dependence on low/inadequate income, long-term unemployment, low quality or absence of employment record, low level of education or training and illiteracy, growing up in a vulnerable family, disability, health problems and difficult living conditions, living in an area of multiple disadvantage, housing problems and homelessness, immigration, ethnicity, racism and discrimination. However, while the range of risks and barriers remains constant, the 2003 NAPs paint a more nuanced and complex picture and some situations emerge more strongly than before as particularly associated to social exclusion: living in a jobless household, inadequate income, over indebtedness, mental illness, alcohol or drug misusing, disability, depending on long term care, asylum seekers, refugees and migrants, living in urban and rural disadvantaged areas.

In relation to the link between unemployment, poverty and social exclusion four aspects are particularly noteworthy (Joint Report on Social Inclusion 2004: 27–28). First, in the EU as a whole, in 2002, around one in ten individuals aged 18 to 59 were living in a jobless household. At the same time, employment does not ensure escape from poverty: around a quarter of the people in the EU aged 16 and over at risk of poverty are in employment. Secondly, there is an increased recognition that the unemployed are not a homogeneous group and the reasons people have difficulties in accessing the labour market vary significantly from person to person. This is especially the case for the long-term unemployed, who tend to suffer from a cumulative series of disadvantages which may include functional illiteracy, outdated competences, lack of linguistic competences, disability or poor health, addiction, too long absence from the world of work and other factors. These, if not countered by adequate and timely support, may lead to permanent exclusion from the labour market. In addition, such disadvantages can be exacerbated by negative objective factors such as discrimination (against, for instance, ethnic minorities, women, etc.), lack of jobs, lack of child care, lack of transport, living in a region of high unemployment, lack of housing, etc. Thirdly, there is more emphasis on the very vulnerable situation of jobless households but also the greater poverty risks for households where one rather than two people are working. Fourthly, vulnerability of three particular groups is repeatedly stressed: older male and female workers whose skills became redundant, young men and women in the 16–25 age groups without formal competences or accredited qualifications and immigrants and ethnic minorities.

3.5.2 The Case of Belgium

3.5.2.1 Institutional Characteristics

Belgium is characterised by generous income transfers within a social insurance system that is primarily work-related and financed through social contributions on labour, by a high-wage (and consequently high-productivity) strategy, and strongly developed labour relations. From a macro institutional point of view, the Belgian welfare state can be regarded as an example of a Bismarckian system based on two pillars (insurance and assistance) with a broad access to social insurance and a limited social assistance pillar compared with other EU countries. This broad (expensive) but passive social security system has succeeded quite well in addressing the income-related consequences of unemployment and non-employment. The downside is that benefit dependency is high and employment rates are relatively low. Social expenditure levels are relatively elevated, but poverty and inequality are low.

3.5.2.2 Poverty and Social Exclusion

Belgium has been quite successful (more than other Continental European welfare states) in combating poverty and social exclusion. It ranks among the countries with low (financial) poverty rates and low degrees of income inequality. Poverty is low in Belgium, in part because of high wages, but it is, first and foremost, among the non-working that poverty levels are kept comparatively low thanks to a broad social security system (De Lathouwer 2005). Because of the very high coverage rate and the unlimited duration of entitlement to unemployment benefits the risk of poverty remains quite low. In 2002, 13 % of the population was at risk of poverty, the EU average being 15 %. However, without all transfers, Belgium would see a risk-of-poverty rate of 38 % (Joint Report on Social Inclusion 2004: 151). Yet, recent analyses also point to serious holes in the Belgian safety nets. One of the main problems seems to be a relatively high level of non-take-up of social assistance, i.e. many people do not claim their rights because they are afraid, ashamed, or they simply lack information (Groenez and Nicaise 2002). Another problem is the significant regional disparity in Belgium, a federal state that has decentralised significant elements of policy to the regional level. There are clear differences in poverty and social exclusion indicators between the different regions, with poverty and unemployment in Flanders several percentage points lower than in Brussels and Wallonia.

3.5.2.3 Promoting Social Inclusion through Gainful Employment

As already mentioned, income-related consequences of unemployment and non-employment in Belgium are compensated through a broad but passive social security system. The downside is that benefit dependency is high and employment rates are relatively low (see table 3.10). The capacity to fully include people into society is insufficient, due to the bad performance on the labour market.

Table 3.10: Employment rate (persons in work between 25–64 years), 2003

	2003
Belgium	**59.6**
Germany	64.8
Netherlands	73.5
UK	71.8
Denmark	**75.1**
Italy	56.1
EU-15	**64.3**

Source: Commission of the European Communities 2004: 27

Employment rates in Belgium are still below the EU average especially for older workers and non-EU foreigners and to some extent also for women. When we look at the employment rates of these three categories, Belgium stands out with extremely low figures. There is an exceptionally low rate of employment for the age group 55 to 64 years, 28.1 % compared to the EU-15 average of 40.2 % (2003). The gap in employment rates between the sexes is also significant: the Belgian employment rate for women is less than 75 % of that of men (51.8 % for women vs. 68.3 % for men). This may be related to the fact that unemployed women enjoy long entitlement periods under the Belgian system, unlike under foreign regimes (De Lathouwer 2003). In many countries unemployment insurance benefits are limited in time, followed by an (often income-tested) welfare allowance. Under such schemes, more unemployed persons, particularly married women, lose their benefit rights more quickly. Due to the length of the entitlement period, the (financial) disincentive for seeking employment may persist for a long time in Belgium.

Several studies also show that the employment rate of non-EU foreigners in Belgium is very low. While 61 % of the Belgian citizens were employed in

Table 3.11: Employment rate of older workers (employed person aged 55–64 as a share of the total population of the same age group, 2002) and women (15–64 years, 2004)

	Employment rate of older workers (2003)	Female employment rate (2003)
Belgium	**28.1**	**51.8**
Germany	39.9	58.8
Netherlands	48.8	65.8
UK	55.5	65.3
Denmark	60.2	70.5
Italy	30.3	42.7
EU-15	**41.7**	**56.0**

Source: Commission of the European Communities 2004: 27

2000, only 33% of non-EU citizens were included in the labour market (Koopmans 2003; Verhoeven 2004). The gap in unemployment rate is even more dramatic: only 6% of Belgian citizens were unemployed in 2000, but more than 30% of the non-EU foreigners were without a job. The latest national report on social inclusion states clearly that unemployment among immigrants may not simply be the result of low education and that efforts are being made through the employment services to prevent discrimination by employers. There have been educational priority measures favouring immigrant children and there is also a wide-ranging new anti-discrimination law.

3.5.2.4 Activation Measures

One of the important challenges for Belgian social policy is thus to escape from very low employment levels, from a 'welfare without work' trap. This is not only a financial and economic challenge – the broad and expensive social security system is only sustainable under the condition of an increase in the employment rate – it is also a social challenge. Because work provides the best protection against precariousness and poverty, labour market reinsertion of youth, women, older workers and ethnic minorities should be an urgent priority.

As everywhere in Europe, the Belgian welfare state stepped in the policy shift towards activation. An important element of the last NAP in Belgium was the great political effort to combat inactivity traps at the same time as raising minimum incomes to alleviate rising poverty. Benefits are increasingly linked to activation, not always without controversy, and targeted fiscal and para-fiscal measures have been taken to make work more attractive. In the period 1999–2002, the number of guaranteed minimum income recipients decreased by 13.4 % while the number of recipients in activation measures rose by 57 %, a real achievement in adverse economic circumstances.

The problem of lagging employment is hard to solve, however. Job growth in the market economy is made difficult by high wage floors and contribution burdens, and in public services because of severe budgetary constraints (partly caused by the high public debt in Belgium). Hence, the scope for investing in measures which might help people out of the 'welfare without work' trap (such as women-friendly benefits and services) is rather limited. Belgium – just as other welfare states of the continental model – therefore has opted mainly for the creation of new job opportunities in the private sector. Despite the intensive efforts results have been modest due to the high labour costs and the fiscal burden attributed to the already great number of unemployed citizens. In the absence of jobs, the response has been to subsidise early retirement requiring additional increases in social contributions.

3.5.3 The Case of Denmark

3.5.3.1 Institutional Characteristics

Denmark has a generous redistributive welfare state which is based on universal, tax-financed social benefits and citizens' rights to free social services, health care and education. All residents are guaranteed fundamental rights if they experience social problems such as unemployment, sickness or dependency. Welfare benefits are largely 'de-familialised', i.e. targeted at individuals rather than families (Esping-Andersen 1999: 78–81). Income taxes are progressive, social transfer payments are generous, and the public service is large and decentralised.

As its Swedish and Norwegian counterparts, the Danish welfare state especially expanded in the 1960s and 1970s (Torfing 1999; Benner and Vad 2000). An important element in this expansion was the establishment of a large-scale public social service provision. As the welfare state 'de-familialised' many caring functions, it also fostered demand for more social services. The net result was near-maximum employment among men and women alike, and very little social exclusion due to poverty and long-term unemployment. In the late 1970s and the 1980s, however, the Danish government was confronted with growing fiscal problems and rising unemployment; counter-cyclical Keynesian demand-management was attempted but this only increased public-sector and balance-of-payment deficits, and worsened the unemployment problem. From the 1990s onwards, Denmark therefore tried to solve these problems by adopting an active and inclusive labour market policy, combining methods to reintegrate excluded people of working age in the labour market and sharpening obligations on unemployed (and disabled) citizens to undertake activation measures in order to qualify for benefits, as well as their obligations to find and accept jobs.

The Danish social system is highly decentralised. While it is legislation at national level that determines social policy, it is local tax-collecting authorities who actually implement a large share of the social activities. Hence, it is the municipalities who, out of their own budgets – with varying degrees of co-financing by the state – pay the benefits, develop and agree a personal social plan with the client and decide on the level of money set aside to implement the national government structures and administrative support. The Danish model is based to a large extent on a culture of 'partnership' characterised by the involvement of social partners, local authorities and other relevant organisations, including user organisations.

3.5.3.2 Poverty and Social Exclusion

Thanks to its extensive system of social protection and very high employment rates Denmark has been very successful in keeping poverty and social exclusion at low levels. Denmark is the member state with the highest expenditure on social protection and one of the lowest risk of monetary poverty. According to 2001 ECHP data, 11 % of the Danish population lived on an income of less than 60 % of the median income. Denmark has one of the lowest risks of persistent poverty in the EU at 5 % (Joint Report on Social Inclusion 2004: 151–152). This reflects the fact that Denmark has one of the EU's most even income distributions. Yet, also Denmark has its problems in the field of poverty and social exclusion. Immigrants and long-term unemployed are over-represented in the lower income brackets. The employment rate of people with another ethnic background than Danish is also below the general level for the population as a whole. The data on health and life expectancy situate Denmark somewhere at the bottom. Moreover, while there has been an improvement in recent years, the increase in life expectancy between 1960 and 2000 was the lowest in EU-15.

3.5.3.3 Activation Measures

A remarkable feature of the Danish social model is that has been able to keep poverty and social exclusion at very low levels, while at the same time coping with problems of unemployment, insider-outsider problems on labour markets, and fiscal limitations deriving from the EMU. Together with the Netherlands, Denmark in the late 1990s has emerged as something of an alternative to both neo-liberal deregulation and traditional 'social democracy' as well (Torfing 1999; Benner and Vad 2000; Clasen 2000).

From 1994 onwards, a policy shift towards activation programmes has been set up that aims to raise labour supply, by improving vocational training and education, and restricting and directing benefit policies towards the unemployed. Arguably, the motivation for this policy change can be linked to the 'discovery' of the structural nature of unemployment on the one hand and the growing awareness of the detrimental effect of long-term unemployment, particularly on young people, on the other. As well in public discourse as in social scientific literature, younger employed people were increasingly portrayed as a potentially 'lost generation' threatened by social exclusion (Clasen 2000: 102).

This policy shift, prudently introduced by conservative-led governments in the early 1990s and generalised by the social-democratic governments after 1993, consisted of a mix of policies which eased temporary labour exit, and began to put more emphasis on qualification and education. Leave-of-absence programmes were made less attractive and other exit schemes, such as early retirement options, were phased out. By the late 1990s, activation policies became paramount within both labour market and social inclusion policy (Torfing 1999; Benner and Vad 2000; Clasen 2000). Strategies aimed at promoting social inclusion place a great deal of emphasis on employment and active labour market policies (although complemented by health, education, and urban and housing policies) (Government of Denmark 2001; 2003). These activation approaches of Danish social policy aim not only aim at raising labour supply, but also at improving self-support, preventing marginalisation and social isolation, and ensuring reasonable financial support without damaging incentives.

Based on individual 'action plans' which try to match the need and circumstances of individual claimants with local labour market conditions, there are a host of different types of activation programmes, including education, training and work experience schemes (Torfing 1999; Clasen 2000). A striking characteristic of these reforms is that the level of unemployment benefits remained more or less unchanged, but restrictions were introduced with respect to duration and eligibility. Since 1998, those in receipt of social protection cash transfers have both the right and the obligation (after a year of receipt, in the case of those aged 30 and over) to take up active work or take part in activation programmes. At the same time special programmes for activating vulnerable groups have been initiated, so as to prevent long-term unemployment and reduce labour market bottle necks. Subsidised jobs for handicapped people have been introduced (Benner and Vad 2000: 451). Especially the young have been targeted by active labour market policies, which help absorb unemployed youth in either training schemes or sponsored employment. Since 1994 it is virtually impossible for any Danish youngster to receive passive unemployment aid for more than 3 months – after which he or she will receive either subsidised education or jobs. There is a lot of evidence that this combination accounts for very brief poverty durations among the youth (Esping-Andersen 1999: 159–160; Clasen 2000).

All in all, the Danish labour market is highly flexible and unemployment spells, often in the form of temporary layoffs, are fairly common, frequent and dispersed. To put it in a more friendly fashion: Danish unemployment insurance provides 'high protection during unemployment but low protection against unemployment' (Clasen 2000: 101). There seems to be a trade-off between generous compensation levels on the one hand and employers' discretion over making workers redundant on the other. In this way, Denmark delivers one of the more attractive ways of dealing with the 'flexicurity' thesis (cf. Muffels et al. 2002). The Danish system of activation performs very well in initiating (youth) and re-integrating people into the labour market. Yet, recent analyses also show that the strict activation rules and the conditionality of the minimum income system might have exclusionary effects for a small number of people who are not able or not willing to participate in flexible labour markets. Since the strict activation rules that include so many people also involve higher rates of non-compliance and sanctioning or 'opting out' behaviour on the part of beneficiaries, there remain a small number of socially excluded people who are not able to benefit from labour markets (Van Oorschot and Abrahamson 2003; Nicaise et al. 2004).

3.5.3.4 The Importance of the Service Economy

There is a strong case to be made for the beneficial economic and social effects on many aspects of large-scale public social service provision, as we find it in Denmark (Esping-Andersen 1999; Scharpf and Schmidt 2000). As indicated, much of this provision focuses on the development of human capital. Active labour market policy, training and education provide crucial support for skill development, which in turn has made a key contribution to the high standards of living in the social democratic countries. At least as significant, however, is the success of a large, public social service sector in reconciling and supporting the desire of women to combine caring responsibilities, including child rearing, with active participation on the labour market. As a recent study by Esping-Andersen has shown, Denmark is the only European country in which the incompatibility problem of motherhood and work seems to be largely eradicated. This becomes apparent, for example, in the fact that Denmark emerges as the only case where female employment does not seem to have a negative effects on family formation[72] (Esping-Andersen 2002: 83). There is a strong case to be made for very high coverage of day care as a crucial element for fighting incompatibilities of motherhood and work.

But there is a downside to this highly developed 'caring economy' as well. If the Danish welfare state enhances employment opportunities and efforts to balance work and child rearing, it also produces high levels of gender segregation within the labour market (although less than in Sweden, Norway or Finland). The creation of a protected or 'soft' welfare sector created a female labour market, providing good pay and the kind of job security and flexibility that makes careers compatible with having children, "but at the expense of a virtual female employment ghetto" (Esping-Andersen 2002: 83).

---

[72] Where at the pan-European level, the odds are roughly twice as low for employed as for inactive women to have two-plus children (five times as unlikely in Germany and three times as unlikely in the UK), in Denmark being employed has no effect whatsoever on being also a mother of two-plus children.

3.5.4 Comparison

Belgium and Denmark have implemented different strategies when it comes to poverty and social exclusion. Although both countries were quite successful in keeping financial poverty at low levels, Denmark has proven to be the more efficient one in solving problems of exclusion and unemployment by means of active labour policy. In Belgium exit options (early retirement), the considerable burden of taxation on labour, long entitlement periods and a high level of labour market regulation have maintained a rather inflexible labour market that is not able to fully include people into society and reinforces the exclusion of different social groups like the elderly, immigrants and young people.

The offensive Danish workfare strategy has been largely successful in reducing the level of unemployment while, at the same time, keeping poverty and social exclusion at very low levels. By combining successful employment strategies – especially for vulnerable categories such as lone women and youth – with a generous benefit system, inclusion and re-insertion in the labour market is enhanced. Yet, the strict activation rules seem to have exclusionary effects for a small number of people who are not able or not willing to participate in flexible labour markets. Denmark has given a lot of attention to life-course politics and the combination of flexible labour markets with family life. By ensuring almost free and universal access to social services, such as day care, it has strengthened female participation on the labour market, provided solutions for the 'incompatibility problem'. Denmark has also been a forerunner in a 'social investment' approach, using life-long learning and training as crucial elements of their labour market and social inclusion policies. By focussing on the weak groups on the labour market, it tries to prevent entrapment in poor life-chances and to further social integration.

Belgium as well has been quite successful (more than other Continental European welfare states) in combating poverty and social exclusion. Its capacity to fully include people into society is insufficient, however, due to the bad performance on the labour market: employment rates are too low. The considerable burden of taxation on labour, long entitlement periods and a high level of labour market regulation as well as other engrained habits make it rather difficult to increase the employment rate. One of the more promising avenues for policy change is therefore including certain activating elements integrated into a Scandinavian-style, tax-financed and more universalistic welfare state (Vandenbroucke 1999). The Danish experience suggests a way for a country like Belgium, to pursue social inclusion by putting more emphasis on activation programmes, to boost service employment, to complement social spending with social investment in human and social capital. To reach this goal however, several structural barriers to increasing labour participation will have to be cut down. In the Belgian context this means for instance phasing out early retirement options, subsidising low-skilled labour and reforming the current benefit entitlement and long benefit duration periods in such a way they 'make work pay' (De Lathouwer 2003).

3.5.5 European Dimensions and Elements

In the EU reducing poverty and combating social exclusion is a challenge primarily for individual member states. Under the principle of subsidiarity, social inclusion policy is the prerogative of the individual member state governments. The EU as a whole can however mainstream the poverty and social inclusion initiatives; at the Lisbon European Council it has established shared objectives, and has agreed on a common set of indicators by which progress is to be measured. The idea was that benchmarking on the basis of the OMC would encourage the dissemination and adoption of best practices in the various states.

In the current European social policy context, with a lot of emphasis on mutual learning and exchange of 'best practices', the poverty and social inclusion policies of Denmark stand out as an example to follow. Activation policies in Denmark (but also in Ireland and the Netherlands) have been viewed as a success story by the EU (and other trans-national institutions) as well as national governments. It is an open question, however, to which extent the Danish experience can serve as an example. As will be discussed in the last part of this book, the vision of promoting policy convergence at the European level by the 'soft' means of the OMC is highly ambitious indeed, given the very 'hard' facts of national differences and priorities. The cases of Belgium and Denmark show that the peculiarities, path dependencies and institutional feedbacks – which have become central notions in current theories about the new politics of welfare (Pierson 2001 a) – as well as structural and institutional conditions have generated vast differences in terms of labour market performance and social protection. Each country's room to manoeuvre is seriously limited by this dead weight of previous institutional choices and by the fiscal capacity to respond to various challenges. Thus, we cannot simply presuppose that 'best practices' (for example, in Denmark) can be imported wholesale (for example, in Belgium) without taking into consideration the (national) context upon which they are dependent. What works best in one context is not necessarily the best solution for another context. One should also assess to what extent some solutions can or cannot be transposed to another national context. Learning in this sense does not mean simply taking over elements of a foreign system. It rather means receiving stimulation and developing other ideas in one's own national context.

Is the 'soft' instrument of OMC a good basis for a European social policy? We believe that the OMC process provides policy makers with a sound basis on which the starting positions and progress over time in the different member states can be reliably compared. Yet, to make the OMC more effective than it currently is and to enhance the development of a more coherent policy towards social exclusion in the various member states, some further measures have to be taken.

3.5.6 Final Remarks and Recommendations

Firstly, the OMC process should work towards a real and concrete target setting on the basis of the set of commonly agreed indicators. While the adoption of an initial common set of indicators represents a major achievement, social inclusion may only be given the same weight as employment and the macro-economy in the EU and national decision making when such targets are in place (Atkinson 1998; Atkinson et al. 2004). One can strive at targets in the field of social inclusion and poverty, which are comparable to the goals set in the EES (for instance, the goal of having 70% of the population of working age in each member state in employment by 2010). This is in line with the principle of subsidiarity because member states will be allowed to choose different means and instruments to combat poverty and to obtain the targets. In a first phase, countries could focus their target setting on social outcomes that are seen as particularly important to their own situation (for instance, reducing child poverty in the UK or raise the employment rate of ethnic minorities in Belgium). Different countries may then have different targets, and these may or may not be directly linked to the common EU indicators. This would still represent a significant step forward, given the impact which an explicit adoption of targets can have (Atkinson et al. 2004: 68). However, in a next phase member states should be working towards a situation where targets are framed in terms of those commonly-agreed indicators, and could even set EU-wide targets. There are a number of possible forms that a European target in poverty prevention could take (Atkinson et al. 2004: 70), including:

- a common target for all member states (e.g. poverty risk down to x% in all countries);
- an overall target for the European Union, set in terms of the poverty rate for the EU as a whole (the proportion of the total EU population at risk of poverty);
- different targets for each member state (for instance, a poverty rate of 21% should be reduced to 15% and one of 7% be reduced to 5%);
- member states being asked to emulate the best performing member states.

A common poverty target would be unrealistic given the existing wide differences in performance and would only address part of the member states. For instance, if the poverty risk would be set at a maximum of 15%, this would be challenging for some member states and irrelevant for others. An EU-wide target, on the other hand, does not take into account the huge differences in population size, with the 80 millions of Germany counting for no more than the 450.000 citizens of Luxembourg. In this situation a small country with a poverty rate of 50 % might make little difference to the EU-wide statistic. Hence, the third or fourth method would be preferable (Atkinson et al. 2004: 70).

Secondly, the OMC process in the field of poverty and social exclusion can only be made more effective on the condition of a better balance between economic and social policy. The multi-dimensional nature of poverty and social exclusion – as reflected in the multi-dimensional indicators – necessitates a continuous interplay between economic and social policy. The overlap between the Employment and Social Inclusion NAPs is an obvious example (Atkinson et al. 2004: 65). Policy to reduce long-term unemployment and joblessness requires inter alia joint action by the various national ministries of employment and social affairs. This implies that social protection and labour market policies should be more integrated as they currently are. For instance, in order to avoid long-term entrapment in welfare dependence and poverty member states should be encouraged to ensure that social protection preserves the incentives to work. The fight against poverty should as well be accompanied by investments in human capital. The example of Denmark shows that social inclusion policies perform better if they are linked with education and training programmes. In a dynamic labour market, transition from social security dependence to work can be improved by investing in people's productive capabilities (relating to the economic domain) and social capabilities. The Commission should therefore encourage developments towards active and human capital-enhancing policies, since that would contribute towards making Europe both social and competitive.

4 Elements of a European Social Policy

4.1 *Preliminary Notes*

In this final past, we deal with the question whether there is need and room for a European social policy and if the answer is yes, what sort of role it should play in the near future. The 'should'-question has been dealt with starting from a few leading perspectives, which are applied and elaborated further in this study: the multidisciplinary approach, the normative principles, the 'enabling welfare state' paradigm and the regime type approach.

First, the aim of this book on social policy has been to apply a multidisciplinary perspective, starting from a philosophical and normative view, which has been supplemented with economic, social and juridical views and analyses about the evolution of European social policies and their impact on the economic and social faring of Europe. Second, the idea developed and elaborated throughout the book was that notwithstanding the strong debate in the history of European integration on the merits of European interference in the social domain, there is a sort of implicit normative consensus in Europe about how 'social' European policies in the socio-economic domain should be and, though less clear, how 'European' national social policy should be. The concept of the 'European Social Model' is viewed upon as reflecting this underlying normative, but implicit, consensus. Third, the basic idea put forward in this book has been to track and to define new routes for European as well as national social policy, and the relationship between the two, starting from the so-called 'paradigm of the enabling welfare state', which goes beyond the notion of the active welfare state into the idea of an activating, responsible and life-course oriented welfare state. In the course of time, the concept of social policy at the national level had multiple and over time changing interpretations (cf. Kaufmann 2001a, b). This is shown by the multifarious and sometimes divergent ways in which national welfare states defined in the course of their history their social policy goals and the ways to achieve them. Therefore, our fourth perspective deals with regime type differences and explanations.

In this final part the first issue pertains to the question how 'social' European policy should be and therefore concerns the relationship between macro-economic, employment and social policy at European level. In terms of substance, it goes back to the idea of social protection being conceived as a productive factor. This has been translated into the capability (investment in human capital) and life-course approach to policy. This issue also pertains to the question by which interventions and mechanisms of social policy the paradigm of the enabling welfare state can be put into practice in the four different policy areas: health care, old-age security, family policy and poverty prevention.

The second issue deals with the question how 'European' national social policy should be. The distribution of authority between the national and EU level is ruled by the principle of subsidiarity according to which there is no role to play for Europe in the social domain except when commonly agreed social goals are unattainable without interference by Europe. In its positive meaning subsidiarity also implies that European interference should entail positive effects for the member states as such and for Europe as a whole (Fourage 2004). In this sense, a European engagement in the social domain at national level appears to be part of the European Social Model. However, part of the issue is to examine whether the existing institutional structures at the European Union level are sufficiently equipped and capable of coping with the tasks pending on them. Eventually, after drawing conclusions from the analysis conducted in the previous parts, the resulting recommendations and strategies of actions are summarised.

4.2 Orientation of the Social Policy at the Paradigm of the Enabling Welfare State

4.2.1 Ethical Foundations and Basic Elements of the Enabling Welfare State

The ethical foundations of the paradigm of the enabling welfare state are the two normative principles, first, of personal autonomy (i.e. the freedom of choice for the individual person to live the life he or she wishes to live) and second, of distributive justice (equal opportunities and access to resources)[73]. These constitute the basic principles on which a just and good society might be built if added to it some form of compensation for natural inequalities and social disadvantages exists. Such a society will also be able to safeguard social cohesion and integration and to avoid social exclusion of particular groups. As the reality of social policy in Europe shows, the European welfare regimes irrespective of their diversity seem to be grounded on these basic principles. Nevertheless, due to the openness of these basic values to diverse interpretation, this common ground is still underdetermined with regard to the fundamental normative models of the relationship between the individual and the community. Albeit on the one hand it excludes libertarianism and anti-liberal communitarianism on the other hand it doesn't exclude or prioritise neither moderate liberalism nor liberal communitarianism (see chapter 1.2). However, since the paradigm of the enabling welfare state is compatible with both positions, there is no need for a narrower scope of determination at the level of principles. Social policy in Europe relies on the notion of positive freedom, according to which the state is obliged not only to safeguard individuals' negative freedom from interference, but since the exertion of autonomy is impossible without social inclusion, to provide a social setting wherein individuals can live their lives autonomously.

Given these two normative principles the idea of an activating or enabling welfare state based on the two pillars of the capabilities approach and the life course perspective provides the most plausible strategy to achieve normative consensus about which policies are best equipped to deal with the challenges a society is confronted with. The capability approach focuses on raising the personal autonomy and freedom to choose by offering a wider set of opportunities and a wider range of life course options and strategies, but supplemented with a greater appeal on the self-responsibility

---

[73] Unlike in chapter 1.2 on the ethical foundations, where – for the sake of a more differentiated exposition – three basic principles have been distinguished, in this part we suppose a comprehensive understanding of personal autonomy, which embraces the social embedding of the individual as a constitutive part of autonomy. Thus in the following, the third principle of social inclusion (through active integration into society, recognition and material resources) will not explicitly be referred to any longer.

of individuals for the development of their capacities and the achievement of their functionings. The life course perspective additionally stresses the need for a balanced and flexible achievement of these goals over the entire life course, i.e. not only for simple and isolated decisions but also for complex and life course associated decisions.

This is striven for by fostering the investment in the capabilities of people during their entire life course and by providing opportunities for a better reconciliation of work and other activities in the various life domains (education, care, leisure, etc.). Eventually, a life course oriented approach also implies a redistribution of resources over the life course across the various generations. This is required to maintain a decent level of social benefits in the long run when the burden of public expenditures caused by social, economic, technical and particularly demographic developments will expand rapidly. Accordingly, the role of national social policy is to enable and activate people to achieve these goals, to improve social integration through active participation, mutual recognition and to safeguard a decent level of material resources. Due to the emphasis on a wider range of social actors than the state only for promoting personal autonomy and rendering more free choice, the enabling welfare state is better understood as an enabling welfare society. The 'enabling welfare society' asks for a more balanced and widespread distribution of competences and responsibilities across various social actors and institutions (market, state, citizen, civil society including NGOs and charity organisations), a more preventive and managerial rather than protective and compensatory role for the government, and a reformulation of social welfare policy goals in terms of "freedom to act" instead of "freedom from want".

Such a model does not rule out the relief of handicapped people for which the notions of free choice and autonomous responsible behaviour have an entirely different meaning and interpretation. Furthermore, it does not mean that 'free choice' options are always organised or produced by the market. On the contrary, also the state retains the task to guarantee that also the weakest members of society should have real options for enduring participation and realisation of life course preferences without burdening these people beyond what they can afford given their limited personal financial means and individual capacity to pay. Yet, even the state is not the sole actor in society that might be held responsible for creating the necessary conditions for safeguarding the personal autonomy and free choice of individual citizens. The state might also delegate its responsibility to market or societal organisations for implementation of the tasks assigned to them, who might even perform the task more efficiently and effectively.

4.2.2 Acknowledgement and Perception at Union Level

While there is a broad general acknowledgement of the notion of the European Social Model, the perception of the political paradigm of the enabling welfare state is still quite limited. Provided that the notion of consensus is understood in a broad sense as synonymous with common acknowledgement referring not only to explicit, but to implicit general acceptance as well, there is a consensus in Europe on the normative principles of autonomy and distributive justice as well as on the weaker version of the active welfare state. In contrast, the paradigm of the enabling welfare state still requires a broader appraisal at the Union level.

Currently, the active welfare state as well as the idea of activation seems to acquire stronger support in the social policy debates at EU-level. The European Commission in its policy documents speaks now about the active and dynamic welfare state and refers to an initiating and active role of the government in the creation of opportunities for social integration of every citizen. However, to date its approach is still mainly oriented to the conventional poles of active labour market policies such as employment creation, training and education, and the promotion of flexible and part-time work. Nevertheless, the Commission's view at least seems to evolve into assigning a more active role to the citizen and to governments at various implementation levels for improving the opportunities for social integration.

Yet, the general question how 'social' European socio-economic policy should be is not a settled issue yet, because many politicians still want to see the EU particularly as an economic and financial union with the social part as being subservient to the other ones. The main arguments articulated in this debate deal with the presumed trade-off between social and economic policy, whereas from the perspective of the European Social Model and the Lisbon agenda, the social policy sector is perceived as being conducive to employment and economic growth because it should also be considered a productive factor. The immediate positive economic (counter-cyclical) effects of social expenditures on welfare and economic growth – reinforced by promoting human capital formation, by improving the job 'match' on the labour market and by activating people through insuring them against downside risks – are too often neglected and misunderstood.

4.2.3 Why an Enabling Welfare State?

The performance records of policy regimes are to a large extent effected by structural societal changes in the last 25 years, which occur at an increased pace and which stem from various factors. First of all the increased international competition between countries and regions which is caused by globalisation and which puts a pressure on the levels of taxation and public spending; second the wake and development of the knowledge economy with the stronger impact of the ICT on production and the resulting lower demand for low skills in favour of the demand for high skills; third the flexibilisation of labour markets endangering the quality of work and work security; fourth the process of individualisation and differentiation of life courses and the lower economic growth rates due to the gradual shift from an industrial economy to a service economy in nearly all member states. With these changes come along the challenges of very unbalanced labour markets aggravated by the higher supply of labour due to increasing female labour market participation, and participation of the elderly due to policy responses to the ageing of the population, the increasing migration and mobility flows as well as national and international tendencies for larger disparities between regions, cities and social groups leading to increased social polarisation.

The assumptions and foundations of the European welfare systems, developed during the post-war period, have been diluted by these structural changes in the economic and social context. The European welfare states are not sustainable if they remain as they are. Structural reforms and reconfiguration of the systems are called for. Yet, this does not mean a retreat to the classical but minimal welfare state. Rather, the development of European welfare states should enter a new stage. Social policy should shift from a fear that employment and competitiveness are harmed if social costs and hence, wage costs are too high and levels of regulation are too burdensome, to the goal of attracting mobile (financial and physical) capital by offering a high-skilled labour force, made available through active policies of investment in human capital. Active policies of human investment and skill development are increasingly considered as active instruments for both, economic growth and innovation, and for rendering opportunities to all. In addition, in order to better meet the challenges of the increased international competition, policies should try to create a better balance between labour market flexibility and income and employment security, by offering people more employment and labour market opportunities through facilitating transitions within and outside the own work organisation, but at the same time by safeguarding high standards of employment security (flexicurity policies). These 'flexicurity' policies offer room for innovating social policies at the national and European level particularly when they become linked to the life course approach as explained earlier. The complex relationship between the economic and social performance of an economy cannot be reduced to a trade-off between equity and efficiency, but has to be interpreted in a positive mutual association between economic and social goals for the purpose of enhancing people's well-being. Social protection policies have to be re-configured and re-engineered to support these shifts.

It may seem paradoxical to make a plea for a stronger interference of the European Union with national social policy in times when member states, partly due to the commonly agreed principle of subsidiarity and partly due to the pressure of dealing with the international economic changes, tend to view social policy as predominantly of national concern. And it seems even more paradoxical to ask for a stronger commitment of social actors, in particular the state, when most governments try to reduce or at least retain their current levels of social spending. However, both paradoxes can be understood from a broader perspective. Concerning the first aspect, it has to be stressed, that the international competition between the large economic regions in the world has become much more powerful than in the past. Therefore, the competition the EU faces is above all against the USA, China and other Asian economies. In order to compete successfully in the global economy Europe might gain from reducing the competition within its region and acting as a single economic block rather than as a multitude of economies competing with each other on wages or social costs and taxation levels. For this purpose, a stronger involvement of the Union in economic as well as social matters may well be asked for. With respect to the second aspect, it should be kept in mind that the paradigm of the enabling welfare society claims a more active role of social actors including the state, but not necessarily entails a more spending state. It is not the quantity of the public role that matters most, but the quality of the operation and services that the state provides. The emphasis in welfare state or social policy needs to be shifted from a compensatory logic to a preventive one, which places less emphasis on providing income support to people out of work and more weight on fostering active participation in the labour force, enhancing the quality of human capital, enabling more people to work and making citizens responsible for their own conduct. This becomes manifest in practices and ideas about rendering citizens more free choice to enable them to take up their own responsibility for managing their lives (personal autonomy), by increasing investments in the social and human capital of citizens (capabilities), and by defining new routes to social policy that take into account the needs of time, money and education for flexible and heterogeneous life plans (life course perspective).

However, the preventive role of the state should not be misunderstood as paternalism or even as a means of coercion, to enforce citizens to conduct according to the norms and values of the state. Instead an activating social policy focuses its efforts on the opening of opportunities and choice options leaving in turn sufficient room for the own conduct and self-responsibility of the individual citizens or their organisations in the civil society. When the role of the state to finance and deliver social protection diminishes, it also implies a shift from redistribution of financial resources to control, management and coordination of the services provided either by the state or by other private or non-state agencies. This shift presupposes a growing responsibility of the market and of societal organisations in the provision of social services: private initiatives on the market, civil organisations in society, charity institutions, non-governmental organisations, individual and informal networks, above all relatives and friends.

All these changes illustrate the need for a new paradigm of the welfare state that integrates the elements sketched before and which we have labelled the activating or enabling welfare state. This new stadium of the development of the welfare state has just been set in motion but it is clear that in some countries it already has inspired politicians to propose strong reforms.

4.2.4 Activating Elements in National Social Policies and Reform Efforts

Irrespective of the low awareness at Union level and at the level of the member states, some reform proposals of social policies in single states are already reflecting the enabling welfare state paradigm, while in other cases, though they are not explicitly oriented at it, the effects are more or less in line with it. The most prominent examples are probably the activation programmes in the area of employment and poverty launched in Denmark since the second half of the nineties.[74] Active labour policies on the one hand have led to a reduction and redirection of direct unemployment benefits and to a lowering of the protection against unemployment risks but on the other hand to enhanced efforts to initiate education, training and work experience schemes aimed at raising human capital formation and enforcing reintegration. Efforts are made to match the needs and circumstances of the individual claimants with the local labour market conditions, with special regard to young unemployed and handicapped people. Such policies denote a clear shift from assistance to activation by attributing more self-responsibility to individuals in combination with supporting the improvement of their capabilities. Additional measures implemented to reduce early retirement and to keep older workers employed as well as to improve the compatibility of child caring and work for women are in line with the requirements of life long learning and continuous integration of the life course perspective.

Similarly to the Danish example the family policies in Finland, which encourage the younger generation to birth giving by focusing on reconciling work and family life, especially also for fathers, on the strong support of single parents and on the equalising of differences in public child care facilities between rural and urban areas, demonstrate a clear orientation towards a widening of the options of free choice available to families. In addition it has to be stressed that public child care facilities are very effective 'transversal' policy measures in the sense that they have a lot of positive 'spill-over' effects, for example, on the labour market (increasing the employment rate by facilitating the reconciliation of care and work and thus enabling women to enter the labour market), on the prevention of poverty for single parents, especially lone mothers and their children, on the stability of families (children's guarantee to a decent childhood and to be raised, if possible, by both parents), and on the fertility rate (ageing problem) – although this last effect is rather controversial.

The pension reform in Poland is another example that reflects the basic elements of an activating approach to social policy. The fundamental reform

---

[74] The so-called Hartz IV, the latest reform of the German unemployment insurance and its linkage with minimum social protection, goes in the same direction. Yet its implementation is still too recent to allow judgments on its success.

of the previous system with the introduction of a new three pillar system with partial capital funding not only acknowledges the individuals' self responsibility for the provision of old age security but by aiming at both objectives of adequacy and sustainability of the system at the same time commits itself to the claims of the life course perspective as well.

Another example deals with the 'life course arrangement (levensloopregeling)' in the Netherlands, introduced in 2004, which is aimed at giving workers more free choice options in planning their lives and to improve the opportunities for reconciling work and family life during the life course. It might be seen for workers as a savings account that they build up, collectively funded by employers, employees and the government, for taking up sabbatical leaves or care leaves or to finance individualised early retirement plans when they need or want it.

Finally, the present reform plans of the health systems in Germany and the United Kingdom are worth mentioning. In the first case, the main issues are sustainable financing and cost transparency and efficiency, which should be realised through co-payments, budget and price controls and a broader application of the insurance principle. These goals will be attainable only through a higher recognition of patients' individual responsibility, which not necessarily means only more market but can also lead to more options in determining the personal risk profile by individual efforts of health care and maintenance. In the latter case, the main reform goals consist in enhancing patients' free choice and equal access through the reduction of waiting lists. Here the focus is on the guarantee of not only a better health provision and the attainment of the therewith connected basic capabilities, but also on the enlargement of real options of choice about when and by whom to get medical care.

4.2.5 Supporting Elements at Union Level

Apart from 'good practices' of enabling policies at national level, also at supra-national level, within the Union's Treaty and as part of European social policy, such examples can be found.

The most striking ones relate to the fundamental freedoms of the Treaty and the principles of the common market (e.g. the free movement of workers) which both have had a significant effect on the palette of choices available to citizens. The possibility to migrate without any limitations from one member state to another, to acquire a job in another country without, at least in formal sense, loss of social security protection for the individual, to take up a study in a country of one choice with the possibility to get scholarships or grants irrespective of one's country of origin and – in the near future – with reciprocal accreditation of qualifications, and, theoretically, having the possibility to move to the welfare regime of one's choice are features of the Union hardly found in other regions. But also the free movement of products and services has an enabling aspect from the consumers' point of view, for example by allowing them to get social benefits in kind regardless of one's country of residence or nationality. In the pension and the health care sector, it allows people to buy policies from foreign insurers. In the latter, it also enables citizens of countries with long waiting lists to obtain medical treatment elsewhere.

In addition to these effects of the Union's law, also the substance of European policies in the areas, where the Union has competence, exerts a significant effect on the 'enabling perspectives' of people. The Structural Fund, the Regional Development Fund, the Agricultural Guidance and Guarantee Fund, the Social Fund all exercise significant impact on the economic and social conditions of specific regions and productive branches, therewith improving people's chances in otherwise disadvantaged conditions. Obviously one could argue that these efforts in part are necessary to compensate for the negative effects of the common market, but regardless of their justification, the enabling effect particularly in terms of their contribution to regional development cannot be denied. However, the enabling perspective also implies, that these funding mechanisms should be redirected from regional development in general to a more targeted approach to improving the capabilities of people in the region (human capital investments), and to invest in creating options for life course planning (such as building up at the regional level savings accounts or drawing rights for educational or caring purposes) which enable citizens to improve their life course options and raising their personal autonomy simultaneously.

4.3 European Social Policy: Coordination and Orientation of National Social Policies

4.3.1 Reasons for a Social Policy at Union Level

The first question we have to deal with is whether there is a reason for European involvement in the social domain and if the answer is yes, what sort of involvement that might be. There are several good reasons for a positive answer to the first question. Firstly, as argued earlier, since the international competition between member states can have the adverse effect of weakening their position in the global competition, a better coordination at Union level can reduce competition within the Union and thereby enhance international competitiveness of the Union as a whole. Secondly, the provision of social services is at national level to a large extent a public task and the EU, since it has acquired competences from the member states and fulfils legal functions, already takes on public tasks in other political areas. Thirdly and connected to the former points, social policy can not be treated separately from monetary and economic policy in which the Union plays a crucial role through the EMU and the BEPG. Fourthly, social policy constitutes a core element in the process of building a common understanding and appraisal of the fundamental normative self-image of the Union, of identification with the community and of solidarity enhancement, which are all indispensable for the development of a general European citizenship. Fifthly – from a pragmatic point of view and by virtue of the factual situation – because the European Union is already active in the social domain as the above-mentioned supporting elements show.

4.3.2 Regulative and Enabling European Social Policy

If for various reasons it is accepted, that the Union has to play a role in the area of social policy, the second question is in what form it has to exercise its influence. Due to the legal and political structure of the European Union, it cannot take over any tasks of an institutional or productive welfare state, but it can play a purely regulative role, exerting an ensuring and enabling influence on the social policies at member state level. An enabling European social policy should aim at the coordination of national policies and strengthen their orientation towards the paradigm of the enabling welfare state by regulative interventions.[75]

[75] "The term "regulation" is not clearly defined, it has a variety of meanings. In a very broad sense any government interference can be referred to as regulation, like in phrases 'market or state regulation'. A narrow use of the term as preferred by legal scholars and economists only covers basic legal rules and controls with regard to private actors. While "regulative" is mostly used in the sense of regulating private markets (e.g. by imputing them with social goals), we use it in a much broader sense as a negative dissociation from "productive".

However, in practice the distinction between institutional or productive and regulative is not as clear-cut as one would wish. A purely regulative policy approach excludes any form of social security engagement of the Union as well as any form of redistribution of resources between member states. While the first aspect is not an issue at present, the second one is more questionable, since the European Union already exerts a re-distributive influence particularly through the Structural Fund and the Social Fund. The assignment of authority to the Union to collect financial contributions is based on the juridical competences assigned to the Union by the member states in the various domains. The European Union possesses such competences and the necessary financial resources to draw on in the areas of monetary, agricultural and partly in structural policy, the common market and the tariffs sphere, but not in the area of social policy. Yet, the Union disposes of budgetary authority with respect to the Structural and the Social Fund, which have some impact on the development of national social policies and for the funding of investments in regional infrastructures. Even though the size of these funds is very limited, they illustrate that from a theoretical point of view the option of a social policy operating at Union level with own financial means and re-distributive power cannot be excluded from the outset. It is rather for various pragmatic reasons of cost containment for the Union, the subsidiarity principle and the notion of fiscal federalism, that redistribution at Union level should be kept at a minimum level. Also within the single member states the relationship and division of competences and financial means between national and sub-national, regional and even local tiers of government, can be structured in ways, that leave most of the productive tasks to the lower level whereas the national level has a mere coordinating and ensuring role. Translated to the Union level, it implies that the distribution of authority between the Union and the member states might be structured according to a similar model based on 'soft law' principles at the Union level and implementation at national level providing much room for national diversity and flexibility. The OMC seems to be designed according to this model.

To what extend the enabling approach due to a more activating role of the Union requires more regulation will be dependent on the success of the 'soft coordination' or OMC processes in the various domains (OMC/employment, OMC/inclusion, OMC/pensions). We doubt whether coordination through 'soft law' is sufficient in the longer-term to attain levels of coordination, which are required to make the enabling perspective a success. The experiences with the OMC processes in the employment domain (see 2.2.2) give rise to concerns about the progress that can be made through 'soft-law' and a process of mutual learning through best practices. Therefore, we defend a different view than some commentators do, i.e. that additional regulation is unnecessary while most of what should be regulated is already captured by directive no. 1408. One might argue that the directives on the coordination of social protection systems are on the one hand too much focussed on the free movement of workers only, whereas the enabling approach covers a broader range of topics and instruments, and on the other hand merely deal with technical and administrative coordination of the various systems in their current form. Reflection on future developments, on needs for reform to modernise the system and on developing new ideas about welfare state objectives and tools as for instance the paradigm of the enabling welfare state would imply, are not part of the current discourse and also impossible to implement in the current context of European decision making. But room for innovative approaches and perspectives is exactly what an enabling European social policy perspective asks for already in the near future.

4.3.3 Prospects of a European Social Policy: towards Convergence or Divergence?

Although social policies at Union level should be an integrated part of the policy formation process at national level, it does not necessarily imply that in the end a convergence of the different systems has to be attained. Coordination can also mean – and according to the authors of this study should primarily mean – the management of diversity, meaning that the systems define similar tools to attain shared goals but that the way these tools are designed differ across countries. It can also mean that the tools differ or that the ways of implementation of particular tools differ but the outcomes in terms of the attained goals being more or less similar.

Further to this, the idea of managing diversity is also compatible with the setting of minimum social standards like in poverty (minimum income levels), employment (minimum employment rates) or pensions (minimum pension level) which might differ at the national level while they are linked to national standards or norms. To give an example: a minimum wage or a minimum income standard at European level might be set at 40 % or 50 % of the median wage or the median equivalent household income in a country. This implies that standards are considered relative and adjusted to the national conditions, which would comply with the subsidiarity rule in European social policy.

4.3.4 Instruments for a European Social Policy

Though the various social systems in Europe have a lot in common with view to their cultural, social and economic evolution, they also differ a lot. There exist large structural differences between the various systems or regimes, which are rooted in a country's historical background, its cultural heritage, its paths of social and economic development, and its institutional setting, which jointly will determine their divergent pathways for the future as well. The question arises yet, what the form of coordination at Union level should be. It is rather obvious that harmonisation of the national social systems should not be considered a viable option for the European Union to move on for the future. But also a weaker form of the close tuning of benefit rights and entitlement conditions, implying a sort of European standardisation of social benefit rights, can not be viewed as a feasible alternative. And even the proposals constituting the so-called "Maastricht-criteria" for European social policies, which are grounded already on the subsidiarity principle, are yet not considered viable options for the future. The reasons for that are not only the missing legal competences of the Union – due to the subsidiarity principle, social policy at least from a legal perspective is still a strictly national matter – but also the experience in the recent past that does not seem to justify a move into this direction.

The diversity among welfare regimes should rather be appreciated as a potential source for policy evaluation and mutual policy learning. Simultaneously, it should be considered a knowledge base of best practices for defining reform proposals for policy innovation. Even the proposals, explained earlier, for the setting of minimum standards in the end might turn out to be 'one bridge too far' because they go beyond what the Union is able to attain, given the mainstream views expressed in the current political debate. Instead, the focus of European social policy should realistically only deal with the process of goal setting and definition of the general framework in which the national member states operate, but leaving enough room at national level to use different instruments and pursue different ways of implementation. In order to guarantee this, goals have to be assessed in sufficiently broad terms, since otherwise the risk arises of specifying too narrowly the aims and implicitly also the ways to reach them.

4.3.4.1 Enhancement of the Institutional Awareness

For the recognition of the important role European social policy plays in Europe, first and foremost, it is necessary that the significance of the social domain is perceived and valued at the highest level within the EU, above all the Council of Ministers but of course, first of all, the Commission itself. It also presupposes that there exist sufficient agreement at these highest levels on the main social policy goals and the basic paradigms underlying them. This requires also a strong awareness of the important role social policy can and should play, particularly for the way it might affect, directly or indirectly, the attainment of the ambitious welfare goals, but also for what it can bring about in other areas especially in the economic domain. The required awareness of the significance of the social policy issues not only has to be reflected in the decisions and actions of the institutions and bodies of the Union but has to be clearly communicated to those who it concerns, too. At present intensified top-down communication from the Commission to the member states is needed to enhance awareness and transparency of the social dimension of the Union both at the level of the broader public as well as at the level of the political representatives across all tiers of government within the member states.

Obviously, instead of a top-down perspective, a further democratisation of the institutions of the European Union and a more direct democratic legitimisation of its actions through a stronger involvement of the European Parliament in the decision making on social policy issues appear desirable. However, following the current debate, it might not be taken for granted that there is strong and common support for a stronger parliament or for shifting the democratic right to appoint the EU-representatives to the European level; this remains a nation's authority. For these reasons it are the representatives participating in the institutions and Councils and the politicians and officials involved in the OMC process, all elected or appointed respectively at national level, who have to be aware of the crucial role the social dimension plays in the Union to attain the commonly shared targets. They also have to communicate its relevance and importance for their country and for Europe as a whole to gather stronger support at national level and not – as has been sometimes the case – raise doubts about the appropriateness and usefulness of the rules, which they full-fledged agreed and commonly decided on at EU-level.

4.3.4.2 Integration of Social Policy across Various Policy Domains

If the ultimate goal of the European social policy is the realisation of the paradigm of the enabling welfare state, then intensified coordination between the main policy areas at Union level especially between the economic (macro-economic and fiscal) and financial policy (EMU) is indispensable to improve the performance of the Union with respect to the Lisbon targets. But a better coordination or integration across the various policy fields also serves the Union with view to the rising importance of market solutions in the social domain, e.g. in health care and pension system, which poses new pressures on the Union and particularly challenges the Internal Market and Union's competition law. A full-fledged implementation and removal of barriers to the attainment of the aimed achievements of the Internal Market, the EMU, the SGP and the ECJ, all will exert a significant impact on the outcomes of European and national policies in the social domain. Although not all of these influences can be tuned, coordinated and integrated, there are opportunities to improve the fine-tuning of the economic, employment and social policies by restructuring, integrating or streamlining the governance process, i.e. the decision making process at the EU level.

One of the instruments not yet mentioned in the current political debate is to institutionalise some form of collaboration between the ECOFIN and the ESPHCA. This might be worked out in various manners but one proposal might be to invite the Ministers of both Councils to participate in the discussions and decisions on matters of concern for both. If one carefully examines the minutes of the meetings of both councils it is rather evident that there are quite some issues appearing on both agendas instead of on a single one. To give one example, within the ECOFIN meetings, employment issues are frequently on the agenda while they have a natural bearing on economic and financial policies, though employment at national level is mainly the responsibility of the Ministers of Social Affairs who meet in the other council.

The Commission recently launched a document (COM 2005: 141) in which the first so-called integrated guidelines for growth and jobs are defined and proposed. It deals with the formulation of integrated guidelines making linkages between the economic and social domain and with the streamlining of the 'cycle of governance'. The recommendations on the BEPG therefore have a bearing on social policies. To mention a few, we refer to the guideline to *safeguard economic sustainability* (no 2) that constitutes part of our life course perspective. That guideline deals with reducing government debts to reform pension and health care system to make them more financially viable while being socially adequate and accessible. A next one deals with the *coherence between macro-economic and structural policies* (no 4) that the Commission translates into promoting flexibility and mobility. This guideline is closely linked to the guideline on employment dealing with *promoting flexibility with employment security and reduce labour market segmentation (no 20)*. The link between flexibility and employment security constitutes part of our capability approach and what we have called the road to 'flexicurity' policies. Also the guidelines on jobs have a clear linkage with the ones on economic and social policies. Guideline 16 deals with the *quantity and quality of jobs* having a clear impact on the social inclusion of those who get employed in a job. The guideline on the *promotion of a life-cycle approach to work* (no 17) deals with a better reconciliation of work and private life and the modernisation of pension and healthcare systems to make them more financially sustainable and responding to changing needs (active ageing, longer working life). This guideline clearly associates with the second economic guideline on sustainability. The employment guideline on *improving investment in human capital* (no 22) has a clear impact on growth to and goes very well with our capability and life-course approach.

Thus the notion of integration and streamlining fits perfectly in our approach and we might see the endeavour of the Commission to integrated guidelines and recommendations as an application of the consequences of our conceptual and normative approach. However, the approach sketched in this book leads to a wider range of options and recommendations than the ones mentioned before. In chapter 4.4, we will go more into detail about the implications for national and European policies.

4.3.4.3 The Open Method of Coordination

The OMC is not a new instrument for policy coordination in the EU: it has been used since several years now in a growing number of areas in the social domain (i.e. employment, social inclusion, pensions) with different success. From the onset, the method has clearly been developed in response to the restrictions put by the principle of subsidiarity on European interference in the social domain that was considered for long a national prerogative. The concept of an enabling European social policy postulated in this study asks for an OMC process that will better meet the requirements of the enabling welfare state. The OMC process as it has become operational in the various domains serves only three purposes: the general goal setting for social policies, the mutual exchange of learning experiences between the member states, and the benchmarking of achievements by comparison with the benchmarks and by rankings in league tables. Consequently, the OMC process allows the member states much room to differentiate with respect to defining concrete policy tools in accordance with the agreed principles and goals and the way to implement these.

This does not mean that the OMC process is too 'soft' to lead to binding agreements on the issues at hand. In practice it might well be that it is more effective than other modes of communication in the past such as the White and Green Papers which were essentially directed to formulate very general ends without much thinking how to achieve them in the coordination process. However, the idea of mutual policy learning should not be understood as a learning strategy for policy transferability, simply transferring successful practises from one member state to another, without taking into account the heterogeneity, peculiarities and path dependencies of the different systems. The use of the method for benchmarking purposes might enhance the pressure on low performing countries to change their policies in accordance with the recommendations through the public ranking of 'good' and 'bad' practices, a strategy whose effectiveness is increasing in parallel with the strengthening of the public awareness of the European social dimension.

4.3.4.4 Structuring Competences and Interactions between Levels of Governance

As long as any form of supra-national social policy seems to be in contrast with the principle of subsidiarity and the sovereignty of nation states in this area, any plea for a European social policy appears an unrealistic scenario without much chance of success on achieving significant coordination. The limitations build in the theoretical principle on the one hand, and the practical requirements on the other, seem to provoke endless debates and disputes on competences. Since these conflicts pertain not only to the relationship between the supra-national and the national level, but also between the various tiers of government at the national level, e.g. the regional and communal level, there is need for a clear assignment and distribution of authority between the various levels of competence. That is the more necessary while the picture becomes even more complicated taking into account the interactions between the supra-national and the regional level, too, for instance through regional and interregional programmes of the EU. It seems however very difficult to build a commonly agreed governance structure that, at the same time, guarantees an efficient decision making process, allows rapid adjustments to change and embraces all of the involved tiers of government. Yet any progress made in this respect is of significant instrumental value not only for the attainment of EU policy goals but even more for the assessment of their scope and content.

4.3.4.5 Amendment: no Proliferation of Instruments

Although, at first sight it might appear that the proposals for improvement of the coordination process at the EU level are rather limited in scope and impact, further inspection shows that this might easily prove to be untrue. Firstly, the proposals for increased policy integration and the restructuring of the competence distribution involves a wide range of implications at the level of national and European policies that are not simply to implement. Especially not when they are seen in their joint impact on the governance processes at the various levels of authority. Secondly, the OMC process is still in development in some areas and for a final judgement of its success or failure it is therefore too early. Since we already discussed a few caveats of the approach in terms of its limited and too optimistic view on learning effects and transferability of 'best practices' and of the lack of willingness of member states without legal enforcement to abide to the rules set out in the cooperation process, it might lead to recommendations to make the process more compulsory or binding in the longer-term. Thirdly, the longer the OMC instrument exists the more experience will be build up and the more likely the method will be adapted to permit a better fine tuning and to improve the governance process itself. For social policy matters are intrinsically rather complex, there is no point in making their governance even more complicated by a continuous innovation and testing of new methods and instruments at the supra-national level well before the process has been given sufficient time and space to prove itself.

4.4 Conclusions and Recommendations for Social Policy in Europe

4.4.1 Implications of an Enabling Perspective: Conclusions

Which conclusions for national social policy and European social policy can be drawn from the results so far presented in this book? Below we want to present some examples of 'best practices' in a few European countries to show what the enabling perspective implies for national and European social policy. We also want to highlight some recommendations for changes in policies at the national and European level, which can be drawn from the inferences in part three of this book. Before we move to that, we sketch the major implications of a shift into the 'enabling perspective'.

If we examine the content of social policy from this perspective and the underlying normative principles developed in the beginning, the following five features of social policy have to be emphasised:

- Firstly, there should be a shift from compensating income or health risks to prevention, since it is better to act in time than to wait until the risk occurred, because the damage caused might then be larger and more difficult to repair. This is a central element of our notion of personal autonomy and the capabilities approach.
- Secondly, the capabilities approach implies a pro-active stance of social policies aimed at increasing the opportunities for people to act in an autonomous and responsible way.
- Thirdly, the life-course approach gives rise to a new understanding of justice and redistribution, not only viewing the distribution between current generations at one point in time but also between current and future generations over the entire life course because of which intergenerational justice comes into sight more easily.
- Fourthly, personal autonomy specified along the lines of the life-course perspective and the capabilities approach show that increased mobility and flexibility are not only unavoidable effects of economic developments in Europe but also articulations of the larger need for autonomy. Although often viewed upon as a source of uncertainty, we would like to emphasise that mobility and flexibility might also increase employment and income security and might strengthen social inclusion in the long run, provided that the process is accompanied and supported by appropriate policies aimed at smoothening and facilitating transitions on the labour market.
- Fifthly, the shift of social policy to endow people with the skills and capacities to prevent exclusion or to remain integrated by endorsing public and private investments in human capital is a direct consequence of our approach, which is backed normatively by the capabilities approach and the principle of personal autonomy.

For sure, these five key features of our overall approach do certainly overlap. Investments in human capital have a preventive character, and a proactive way of creating more freedom to choose will undoubtedly have a positive effect on personal autonomy. This means that they will strengthen each other in terms of the realisation of concrete policies, but that they can also be used to justify each other from a normative point of view.[76] These five implications of our enabling perspective explain how the normative principles and contents of policies are linked into new modes of policies which are better equipped to deal with the pressures put on the national systems and which provide a firm basis also for European interference in one way or another.

We have to be aware that social policies in Europe at the national and European level are already evolving and shifting in the course of time and in many countries reforms have been enacted which have to deal with these fundamental changes. Challenges derive from social and demographic developments (aging and migration), from the economic pressures (low growth and high unemployment rates, high wage and tax levels), from globalisation (international competition) and from the economic, social and political integration at the EU level itself. The competitive forces at the global level require more economic integration within the economic regions in order to be better equipped to compete successfully at the global level with other regions (e.g. Asia, Japan and the United States).

The high pressures combined with a loss of national political autonomy due to the growing interdependencies within the EU narrow the scope for action of the member states. In order to cope with the present situation, fundamental changes are necessary along the five lines set out before. The requirements for such shifts are already there given the fact that, as we argued extensively in the first part of the book, there is a sort of implicit common understanding of the ethical or normative foundations underlying the so-called European Social Model based on the principles of personal autonomy and distributive justice. The European Social Model seems to support the notions of an active welfare state safeguarding a high level of free choice, enabling people to take care of themselves, and of the positive role of Europe to improve the outcomes of national policies which are in accordance with the principle of subsidiarity. Nevertheless, it is fair to say that whereas the awareness of the need to shift to the paradigm of the enabling welfare state, based on the capabilities approach and the life course perspective, seems rising, particularly in academic circles, it is still in its

---

76 As stated before, some of these key elements promote but at the same time also endanger the personal autonomy (for example through compulsory participation in prevention programmes in health care). However, this is not necessarily the case; and, without a reorientation of social policy along the lines described here, personal autonomy will less easy be raised.

infancy at the political level, both nationally and supra-nationally, although there are significant differences between single states and specific areas of social policy.

In the sequel, we will discuss the practices at national and European level fitting into the perspective of the enabling welfare state. The findings should be identified for the various areas of social policy analysed in this study. With regard to the Union level we have to keep in mind that due to the subsidiarity principle most policies are implemented at the national level, although there is a significant impact of EU regulations on the social policies of the member states. At the EU level, the most important instruments for the coordination and orientation of national social policies towards an enabling welfare society deal with the OMC process in the employment and social domain; the increased consensus about the role the Union should play in the field of social policy; the 'streamlining' or integration debate in Europe focusing on the fine-tuning of macro-economic, employment and social policies in Europe and its implications for the structuring of competences and interrelations between supra-national, national and regional levels of governance.

4.4.2 Recommendations and Strategies of Action

On the basis of the conclusions sketched earlier we will identify strategies of action for defining concrete policy measures at national and European level, which fit into the framework and goals of the enabling policy perspective. Some of these measures are already enacted and mirror recently ongoing changes in some welfare states, others are still in debate and are part of the trend to modernising and innovating social systems to make them more activating.

The findings will be identified for the four areas of social policy analysed in part three of this study: health care, old-age security, family policy, and poverty prevention. In addition and in reference to part two, where we analysed employment policies and labour markets from a European perspective (see section 2.2.2), recommendations for this area will be given as well. Subsequently, separate and specific recommendations will be summarised for actions and instruments at the supra-national level.

4.4.2.1 Enabling Social Policy: Concrete Policy Measures in Different Areas

Health Care

- Based on the principles of equal access, fair financing and evidence-based medicine, nationally adjusted minimum standards should be defined for the provision of health care to raise the autonomy of especially the weak and vulnerable in society. The setting of European minimum standards closely tuned to the national specific conditions of the member states operates as a safety net to safeguard proper levels of health care also for sick people in the poorer regions of the Union.

- In order to reach a clear division of allocation and distribution in the financing structure of the health care systems at national level a shift towards the financing of re-distributive elements through general taxes is required. This in turn allows reducing pay roll taxes and contributions. It also contributes to reaching more equivalence between paid contributions and risk profiles with gains in efficiency and competitiveness of the systems.
- By introducing obligatory private insurance schemes for the whole population with public support for people with low income, more room should be given to market solutions.
- Additionally a shift to partially funded systems in public health insurance schemes leaves more space for private insurances and hence, might enhance the sustainability and efficiency of the whole system, thus responding to the demographic challenges according to the principle of intergenerational justice. This shift seems also to improve the transparency of the system and to raise therewith the autonomy of citizens with respect to health care decisions.
- Through the further implementation of the Treaty with respect to free movement of labour, capital and goods and services restrictions in cross-border health care could be reduced without having an adverse impact on the national social systems. To the extent that these freedoms indeed have a bearing on the provision of cross-border health care services, it raises the choices and opportunities people have to take up health care when- and wherever it is most appropriate. The national and international widening of the range of choices for consumers, to choose between providers, insurers (public and private) and treatments (in the form of products and services) will support the autonomy of patients. However, it might also endanger equal access and therefore the autonomy of the 'sick and disabled', because of its stronger reliance on market solutions.
- Finally, for the investment in capabilities and their maintenance over the life course of individuals a stronger emphasis on preventive health care matters is indispensable and should be supported as far as possible, but without restricting personal autonomy through obligatory participation in prevention programmes.

Old-age Security

- The enhancements of old age security systems should aim at their long-term budgetary sustainability while maintaining adequate levels of income replacement in old age. Safeguarding sustainability in old-age protection, in the framework of an ageing population, implies improving the intergenerational justice, the distributive justice between generations.
- The role of funded and private schemes should increase, not in the form of replacing public and pay-as-you-go schemes, but through reaching a better balance between various designs of old-age protection. This will

not only improve the sustainability of public schemes but it will also better account for new and more flexible life course patterns. However, private and funded schemes also create new income risks for which – just as it is now – also the state should take responsibility (by regulation, supervision or the assessment of a safety net).

- Modernised systems of old-age protection should aim at allowing much more individual freedom and choice and more opportunities to take up ones own responsibility. In quite a few national pension systems this has indeed be the way policies evolved over time, but in others it still seems extremely difficult to adapt the systems to meet the challenges of an ageing population. Generally, in Europe obligatory public pensions have become so important that to a large extent they have replaced personal responsibility and free choice with respect to old age. Instead, the state should allow people more free choice and responsibility in their later life with respect to both their participation in work and the therewith associated income as well as the extend they choose to rely on offers of the market, employers and the civil society.
- Old-age security has to be better adapted to individually defined life courses, allowing for more diversity and flexibility. This can be realised, inter alia, through the individualisation of public systems (individual saving accounts, notional defined contributions), through the move from defined benefit to defined contribution or the development of fully capital funded private systems, through flexible retirement age arrangements and the introduction of partial pensions.
- Activation has to remain one of the main objectives of old-age protection reforms. To allow people to work is most salient to safeguard adequate income protection in old age, as well as to maintain the sustainability of old-age systems in the long run. On the one hand, it is especially important to prolong working life, to postpone the factual retirement age, especially through restricting early retirement options. On the other hand, flexible retirement age arrangements or partial pensions are options not only for enhancing diversity and flexibility on the individual life course – as mentioned above, but for improving the flexibility and sustainability of pension systems as well. In addition, stricter earnings- (and contribution-) tests to increase the equivalence between premiums and benefits also belong clearly to the entire range of measures aimed at activation and increasing personal responsibility.
- Other policy domains, apart from income support, clearly also affect living standards in old age. Hence, it is important to have a balanced policy-mix of measures, which are well tuned and coordinated, especially with view to services as primary health care, long-term care, social services and housing.
- Albeit old-age security should remain within the responsibility of the single member states, the OMC in the EU applies to pensions and might be

a very useful instrument to remove barriers to mobility and a helpful tool – through mutual learning – for reconciling both parts of old-age policies: to increase sustainability while maintaining appropriate levels of protection. The OMC on pensions should be developed without changing its nature and objectives, through further work on indicators, better streamlining of various areas important for security in old age and a more unified structure.

- The coordination mechanisms of social security systems which have fulfilled their role properly should be further adjusted to the growing numbers of member states and the diversification of solutions. For instance, portability of pension rights under the statutory public schemes is not an issue for people migrating within the EU, thanks to coordination of social security systems; in contrast, cross-border portability of occupational pension rights should be further improved also at the European level.

Family Policy

- The promotion of the reconcilableness of work and family life over the life course is central not only for experiencing personal autonomy and for having the opportunity to invest in one's capabilities, but even more so for the well-being of the children. The crucial goal of social policy can be attained through various instruments: by extending the options for taking up parental leaves, by strengthening the fathers' role in care, by allowing both parents to act as breadwinners and child carers simultaneously, by offering day care facilities, which are very flexible in terms of the choices they offer to parents, respecting thereby the core value of the privacy of the concerned families. Although the reconciliation of work and family life and the improvement of the opportunities for both parents to care may have the effect that it also contributes to halt the trend of declining fertility rates (though we know that this trend is extremely hard to turn), this certainly is not the main argument for it. In any case, it can be attained only through a close tuning of family and employment policies. It may therefore be feasible only if it rests on a European wide enforceable agreement either through the setting of minimum standards or by extending the OMC process to the area of family policy. This is not at stake yet, but might be recommendable for the future.
- The emphasis of family policies on young children should be broadened to families with children at school age too, especially for the purpose of dealing with the rising problem of early school drop-out. Appropriate measures to deal with it would be flexible part-time and job-sharing arrangements combined with parental leaves, more flexible school times and better after-school caring facilities, possibly provided by private organisations or the 'civil society' (e.g. non-governmental organisations). The focus here is mainly on the building of capabilities, the social inclusion and the life course of children, but secondarily on the autonomy of

parents too, both of which are not sufficiently taken into account by present family policies centred almost exclusively on the first months or year of the children's life.

- Although it is a specific feature of European family policies to take the child as the central locus, the particular policies often fail to meet the associated requirements. More emphasis has to be put on the rights of children, for instance the right to a decent childhood or, if possible, to be raised by both parents. At the European level, this implies the setting of European minimum standards and rights. Furthermore, it implies at national and European level measures to combat child poverty, to support single mothers or fathers, and to offer equal opportunities for education to all children. For an international evaluation and comparison of policies it may be useful to develop European indicators for the measurement of the degree of child-friendliness of different policy measures.
- The ageing problem is a challenge for social policy because there is a growing proportion of older people dependent for extended periods on formal and professional care, whereas at the same time the supply of care does not rise proportionally neither in budgetary terms nor in terms of personnel. Hence, family policies should give support to middle-aged women and men acting as volunteers to take care of their parents and to facilitate them with sufficient guidance and assistance. The responsibility should not entirely be left to the market or the family but to safeguard 'freedom of choice' the co-responsibility of public institutions and the state is needed for the relief of the elderly. Otherwise, the informal support could easily and unintentionally lead to an obligatory requirement to children or relatives to fully take care of the older people living at home and therewith endangering the quality of the care services as well as restraining the personal autonomy and free choice options of the persons concerned.

Minimum Protection and Poverty Prevention

- The shift of policies of minimum protection and poverty prevention from a compensatory into an enabling and activating approach implies, first of all, a shift in the applied definition of poverty from an income-based measure into a multidimensional concept. Secondly, it means that instead of using a static indicator of e.g. low income at one point in time, also longitudinal indicators have to be considered, which take into account housing conditions, health, education and social and cultural participation over time. Such a measure allows us to give a comprehensive description of social exclusion patterns over time. Only in such a perspective, it is possible to determine the most critical long-term risks of poverty and to identify the proper measures for enhancing social inclusion over the life course of individuals and families.
- As the most vulnerable groups are the long-term unemployed, lone mothers and children the main focus of poverty policies has to be on the

prevention of long term unemployment, on the reinsertion into the labour market, on the reconciliation of work and care particularly for lone mothers and fathers with particular attention to promoting female labour participation, and on safeguarding sufficient levels of minimum protection. Most of the recommendations given with view to family policies come also into play with respect to child poverty prevention. From the life course perspective the childhood is the most critical phase for the investment in capabilities because the remaining life course is sufficiently long to let the investment pay off.

- In any case minimum income protection policies will remain indispensable to tackle poverty. For that aim, national and European minimum income and/or poverty standards might be defined. In general, national income standards are already applied in many welfare states, but a European standard that might be even set relative to the specific situation or conditions of the single member state is rather new.
- Social inclusion and poverty prevention policies have to be linked to active labour market policies, particularly when the entitlement to social assistance is made dependent on the willingness to work. Although this condition alone would be very pressing for the most disadvantaged, in combination with an enabling employment policy it may be commonly supported and considered beneficial for capabilities building and social inclusion, supporting thus in the end also the autonomy of the single individual.

Employment and Labour Market Policies

- Enabling labour market policies should focus on two aims: first, the activation of the unemployed and second, the flexibilisation of the labour market. Before we have denoted such an approach as 'flexicurity' policies according to which it is believed that flexibility can be promoted without endangering employment security. Policies should therefore aim at the improvement of the employability of people, because of which they can move more easily from one job into another either within the firm or outside the firm or organisation. Policies should generally aim at facilitating transitions on the labour market through reducing risks connected to transitions. This can be pursued in many ways: one is to support investments in human capital formation through life-long learning, education, training and skill formation, providing better opportunities to upward moves and preventing people from downward moves, of being excluded from the labour market due to obsolete skills.
- Public interventions on the labour market are aimed at improving the flexibility and efficiency on the labour market. This might indeed imply a less regulated labour market, with lower levels of employment protection and less influence of the unions, but in connection with policies to invest in the capabilities of people it might in reality increase employment chances instead of reducing them. An activating labour market policy

however also encompasses a wide number of other measures salient to raise the employability and flexibility of the labour market such as: decentralisation and regionalisation of labour market interventions, a system of tax credits on employment rather than wage subsidies, more transitional arrangements and opportunities for retirement leaves, education or study leaves and leaves for caring duties and family obligations. In facilitating transitions, the final outcome in terms of better matches on the labour market and shorter non-work spells should lead to more employment and a better performance of the labour market.

- Both, poverty prevention as well as employment policies, have a genuine European dimension, because they are inextricably intertwined with the economic policies of the Union and the member states. For this reason poverty and employment targets, adequate minimum protection standards for the unemployed and minimum wages for those in work should be determined and coordinated at the European level taking into account the national and regional differences of the single member states. Indeed in both fields, poverty and employment, the OMC has already been applied and may serve the purpose in the future, too.

4.4.2.2 European Social Policy: Streamlining and Integration

The implications of an enabling strategy for the institutions operating at European level with regard to Social Policy allow giving the following recommendations.

Enhancement of the Institutional Awareness and the Policy Communication

In order to coordinate and direct the European social systems towards the paradigm of the enabling and activating welfare state the awareness of the meaning and impact of European social policy on the entire policy of the Union need to be improved. This pertains first to the awareness of the social dimension of policies at supra-national level. It has to be acknowledged explicitly that generally the political decisions and measures of the European Union, irrespective of the particular area concerned, the basic freedoms, the internal market, the monetary and macro-economic policy, or the structural and social policies, all have a social bearing and impact. The awareness of the importance of social issues might gain from clear and ambitious quantitative target setting of the aims to achieve. This is especially the case when these targets are to be announced and seriously evaluated. Many member states are already engaged in elaborating this quantitative target setting. The institutional awareness seems to be raised mainly through a strong top-down communication process from the highest organisational and hierarchical levels of the institutions of the Union, spreading consequently over political representatives of all tiers of government at national level and eventually the broader public in general.

On the other hand, the paradigm of the enabling welfare state has to be emphasised and reinforced at all tiers of political decision making. In this respect it is not the social dimension itself that has to be recognised, but the explicit role the enabling welfare state perspective might play when it comes to elaborate and implement the required reforms. The paradigm of the activating welfare state is debated not only in the academic field, but to a growing extent also in the statements of the involved political institutions of the Union. At member state level there is an increasing number of countries, which already formulated or enacted new reform proposals and innovative social arrangements apparently reflecting the activating or enabling approach. In addition, it can be argued that some aspects of existing social and related programmes have some bearing as well on the enabling approach. The appraisal of the approach should better not stem from a top-down communication process only, but should be the outcome of an intensified dialogue between all parties and actors involved in the process of political decision making.

Coordination and Integration of Policies and Processes

The integration and tuning of policies across the various domains of European socio-economic policy require a stronger interaction and dialogue between the institutions involved in the decision making processes across the various policy areas. It is clear that political measures taken in one domain of policies will exert a significant effect on the other. What has been decided upon in monetary policy is likely to have clear effects on the room for economic and social policies as explained in the first part. Tight budget constraints as agreed on in the framework of the SGP are likely to have a substantial impact on the way national governments try to reduce spending in the social domain, especially in economic bust periods. It also implies that austerity measures in the social sector contribute to the way budget policies might fulfil the requirements of the SGP. Since in theory this interplay leaves open the question at what level of government spending and economic growth a budgetary balance can be attained, only a joint treatment of both policy domains allows to determine which overall target levels might be achieved – given the conflicting goals and priorities. The same holds for areas such as the structural and agricultural policies, the internal market, the basic constitutional freedoms and employment policies, which are all closely connected to social policy. These interrelations must be analysed more properly in order to detect potential sources of conflict or reinforcement between the various goals and instruments, which should be taken into account in the various domain-specific and more general decision processes.

With reference to the institutional integration at Union level as mentioned earlier, the preferable place where things might come together, at least in theory, would be the European Parliament with its closer connection to the electorate and strong democratic legitimisation. Indeed, in the longer run a stronger role of the European Parliament is surely desirable and probably feasible too. At present, however, it is actually the institution with the weakest linkage to the European voter who has the final saying: the ECJ. Yet, before it comes to the involvement of the ECJ the most powerful institutions are the Councils, above all the Council of Prime Ministers and the ECOFIN. Since the decisions of these Councils on the one hand have the strongest political impact and on the other reflect the will of the member states, it is obvious that the interaction of social policy with the other political domains takes place at this hierarchical level. The tuning and integration between the two domains might take place in many different forms. The occasional or permanent broadening of the ECOFIN or other Councils, either by joint sessions or by the participation of the Ministers of Social Affairs and Employment in the ECOFIN and the Ministers of Finance and Economic Affairs in the ESPHCA, might be another option to consider. Additionally, a rescheduling of the entire decision process by taking better account of the sequence and timing of the decisions in the other domains, as well as a better interlocking of procedural times and agendas, could improve the possibilities of mutual recognition and tuning of each other's decisions.

An option with a stronger and more direct participation of citizens would be the broader involvement of social partners and trade unions in the decision process at Union level. However, one may cast doubt whether the international representation of national unions at the European level is sufficiently matured to gather common support for such a stronger form of participation.

Definition of Competences and Interactions between Levels of Governance

Apart from the integration across the various policy domains, the distribution of authority and competences across the various levels of governance is important, especially for the implementation of social policies. The relationship between different tiers of government at national and sub-national level, i.e. regional and communal levels, has to be organised with a clear distribution of competences and responsibilities as well as structures of cooperation and coordination taking into account the subsidiarity principle.

Except for particular circumstances, a further transfer of competences on the European level is not a solution to the disputes on competences. Instead, the interplay between the different tiers of governance has to be improved through shifting the focus from the national and sub-national levels, which are already tightly regulated, to the interaction between these levels and the Union.

The Open Method of Coordination

The OMC should serve three purposes: the general goal setting, the mutual exchange of learning experiences between the member states and the benchmarking of achievements.

The experiences with the OMC in the various domains are not sufficient yet to allow a final judgement, with respect to its effectiveness in achieving its purposes. In order to avoid that the method is perceived as a hidden instrument for harmonisation and as a threat to national autonomy and to the principle of subsidiarity, its application should be limited to the formulation of commonly shared goals and to the exchange of experiences. If through this interaction process, some convergence arises it will not be in conflict with the principles of national autonomy and subsidiarity. However, the chances for convergence are rather small. Although most of the European member states face similar challenges and pressures on their policies to innovate and modernise their social systems, the concrete measures to be taken might be very dissimilar due to the different national conditions, institutional settings and historical roots. For instance, the challenge of guaranteeing the sustainability of a benefit system means something quite different dependent on whether in the particular country unemployment rates are high or low. The limitation of the OMC process to the setting of and the agreement on goals is appropriate for the purpose of leaving to the member states plenty of room for choosing and adapting nation-specific tools for the achievement of these goals. The OMC therefore remains an excellent method to manage diversity.

However, one could object that because of its focus on goal setting rather than on its achievements, the OMC will be reduced to some form of 'beauty contest', allowing the member states to apparently commit themselves to ambitious targets without being afterwards evaluated on their achievements in attaining the goals. Although admittedly there are no formal sanctions in case of insufficient results, public campaigns of naming, shaming and blaming may have a stronger effect, the stronger the awareness for benchmarking of social policies among the public, the electorate as well as politicians, are. Possibly this could lead to a cycle of self-reinforcement. A higher awareness boosts the effects of rankings and public campaigns, which in turn will be used more intensely. The public attention on social topics from an international perspective will grow further, thus strengthening in the end the bindingness of goals stated through the OMC. Under the condition that the pressure deriving from such a form of competition between social systems is sufficiently high, even the setting of continuously more ambitious goals becomes feasible. Moreover, it seems reasonable to expect that the more market elements will be introduced into the welfare systems the more transparent comparisons will get, increasing thereby the pressure on the worse performers.

In matters of policy learning and mutual exchange of experiences, a possible objection could be that up to now, there are only few examples of cross-national learning and policy transfer, demonstrating either an under-utilisation of the learning potential or the limited value of the potential itself. Nevertheless, two examples proof the contrary. First, the Scandinavian countries exhibit a long tradition of mutual learning in the social and other domains. Admittedly, the similarities between the Scandinavian countries facilitate the exchange of experiences, but these similarities exist within other groups of member states, too. On the other hand, also such heterogeneous systems like the transition countries have shown in the recent past a strong capacity to reconfiguration and policy learning, not only adopting existing models from other countries but partly even anticipating innovation needs not yet realised also. Hence, the conclusion may be drawn, that by appropriately taking into account the differences and peculiarities of national policies and by putting pressure on a country's achievement trough benchmarking exercises, an effective process of mutual learning can be initiated and implemented through the OMC process.

The current experiences with the OMC as well as the prospects of its application in the near future illustrate, that there is still some effort needed to make it an effective instrument for the purposes postulated here. In the areas of social inclusion and employment, shared objectives and common sets of indicators have been accomplished and put on record in Joint Reports and NAPs. In the field of pensions, the agreed goals are very fundamental and not yet binding in the very sense of the word and the learning process has progressed not further than to the identification of best practises and innovative approaches. In health care at present, there is a plea for a rather slim OMC, serving only as a platform for the exchange of information and the determination of comparison indicators. An application of the OMC to family policies is not yet in sight.

4.5 Final Remarks on the Perspectives of Innovation

The overall approach developed and presented in this book not only allows to formulate innovative recommendations or to justify claims that certain ongoing or beginning political processes should be strengthened. There are two more features of our study we would like to emphasise here, which are innovative on a general level.

Firstly, the trans-disciplinary approach unfolded in this book allows us to deal with the complexities of the phenomena under consideration without being in danger to reduce complexity due to constraints rising from one single methodology or from the 'logic' of one single discipline. As we experienced ourselves in writing this book learning from each other made possible to discover features of social policy, which do not come into view if one concentrates on the perspective related to one's own scientific background. We are convinced that one of the most innovative aspects of our study is to demonstrate that the complexities of the ongoing processes in social policy call for such trans-disciplinary studies.

Secondly, the overall normative stance we have taken in this book – besides the normative principles and specifications we rely on – has been characterised as a reconstructive perspective. This allows us to keep in touch with ongoing social processes, political developments and normative discussions (the descriptive dimension of our approach) on the one hand. But it also allows us to formulate recommendations backed up by normative principles and values (the normative dimension of our approach) on the other. This stance not only has the advantage to rely on specified normative principles, which are rich and flexible enough to deal with our topic. It also guarantees that our normative claims have the chance to be connected to political and social realities.

In this book, which has been produced as a collective effort of scholars coming from various disciplines and applying a variety of approaches to the issues dealt with, we have taken the view that a more coherent knowledge of the complex field of European social policy demands greater collaboration and understanding across academic approaches and disciplines than currently exists. Following this, we have sought to develop an understanding of the complementarity of a range of theoretical and empirical models and have drawn upon tools from a number of interrelated disciplines. All in all, our multidisciplinary linkage of a normative approach to European social policy with Sen's theoretical model of capabilities and the 'enabling perspective', and with the empirical 'state of the art' policy analyses of national and European interference in four policy domains has provided a rich account of both the potentialities and the limitations of European social policy. It also made clear some important challenges that European social policy will confront in the next decade.

We believe that the different elements of our approach together constitute a coherent analytical framework for dealing with the shifting and uncertain patterns and trends within European social policy. We started from a regime type analysis of welfare states, complemented with detailed pairwisely organised comparative analyses of social policies of some countries, and contended that this approach provides a suitable methodology to bring common trends and various policy responses to the fore. The welfare regime approach showed how policies in various countries tackle common pressures and challenges in different and sometimes adverse ways. The 'activating' or 'enabling perspective' as elaborated and developed in this book complemented this idea of welfare diversity with a broader theoretical and normative view on the goals of European welfare states. The principles of personal autonomy, social inclusion and distributive justice were specified along the lines of the life-course perspective and the capabilities approach. This enabled us to develop a positive understanding of the increased mobility and flexibility in the European Union. In our view, the combination of a life-course perspective and the idea of flexicurity provide a better alternative to current policies than the 'active and dynamic' approach the EU is moving at. Rather than limiting ourselves to conventional ideas of social policy and employment policy ('making work pay'), a full-fledged shift to prevention and investment in human capabilities is needed. After all, human capital is the most important growth factor. At the institutional level, we made a plea for a better tuning and integration of policies at various levels of governance, down to the lowest levels within the member states. The latter is particularly important if the aim is to pass on social policies to a lower level of competence at the national and regional level.

The 'enabling' perspective provides us with a rich number of new proposals and a wide range of innovative approaches and tools for national and European social policy, which are discussed throughout this book. We are convinced that this book in providing new analyses and new material will constitute an important contribution to the ongoing debate within European policy making on the integration and tuning of economic and social policies. We sincerely hope that our work is to the benefit of all people in Europe who are concerned with the issue how a highly productive and efficient society can be at the same time a fair and just society. The ideas and evidence presented here show that it is not futile to think that such a society might be attained for the majority of welfare states considered in this book.

Appendix

Table A.1: Main regulations (new systems) of obligatory general state (social) old-age security systems in EU member states (EU-25), (Situation on 1 May 2004)

Country	Type of the system	Field of application	Pension contribution rate (1) (%)		Legal retirement age	Minimum period of membership	Conditions for drawing full pension	Pension formula	Automatic adjustment-base
			employee	employer					
Austria	Social insurance	employees	10.25	12.55	W: 60 M: 65 (2)	180 months	40 years	DB	Wages
Belgium	Social insurance	employees	SI13.07	SI24.87	W: 63 M: 65 (2)	–	W: 43 M: 45 years	DB	Prices
Cyprus	Social insurance	economically active	SI6.30	SI6.30	65	156 weeks	–	DB	Prices and wages
Czech Republic	Social insurance	economically active	6.50	21.50	W: 59 and 4 months M: 61 and 4 months (3)	15 years	25 years	DB with flat-rate component	Prices and wages
Denmark	1/ National pension 2/ Social insurance (ATP)	1/ residents 2/ employees	1/ taxes 2/ EUR 10.00	1/ taxes 2/ EUR 20.00	65	1/ 3 years residence 2/ -	1/ 40 years residence 2/ Contributions since 1964	1/ EUR 7,495 yearly 2/ EUR 2,948 yearly	1/ Wages 2/ No automaticadjustment
Estonia	Social insurance	residents	–	20.00 (4)	W: 59 M: 63(2)	15 years	–	DB with flat-rate component	Prices and social tax revenues

Country	Type of the system	Field of application	Pension contribution rate (1) (%)		Legal retirement age	Minimum period of membership	Conditions for drawing full pension	Pension formula	Automatic adjustment base
			employee	employer					
Finland	1/ National pension 2/ Social insurance	1/ residents 2/ All economically active	1/ – 2/ 4.60	1/ 1.35 – 4.45 2/ 16.80	65	1/ 3 years residence 2/ 1 month	1/ 40 years residence 2/ 40 years	1/ EUR 419.16–496.38/ month 2/ DB	1/ Prices 2/ Prices and wages
France	Social insurance	employees	6.55	8.20	60	3 months	40 years	DB	Prices
Germany	Social insurance	employees	9.75	9.75	65	60 months	–	DB	Wages
Greece	Social insurance	employees	6.67	13.33	65 (5)	4 ,500 working days	35 years	DB	Prices
Hungary	Social insurance: 1. DB 2. Private pension funds	economically active	0.5 08.00 (6)	18.00 –	62	1. 20 years 2. –	–	1. DB 2. DC	Prices and wages
Ireland	Social insurance with flatrate benefits	economically active	SI 4.00	SI 8.50 – 10.75	65	260 weeks	260 weeks	EUR 167.30 per week	Prices
Italy	Social insurance NDC	employees	8.89	23.81	57-65	5 years	40 years	NDC	Prices
Latvia	Social insurance: 1. NDC 2. Funded	economically active	SI 9.00	SI 24.09	W: 60 M: 62 (2)	10 years	–	1. NDC 2. DC	Prices and wages
Lithuania	Socialinsurance	economically active	2.50	23.40	W: 59M: 62.5(2)	15 years	30 years	DB	Prices and wages
Luxembourg	Social insurance	economically active	8.00	8.00	65	120 months	40 years	DB	Prices
Malta	Social insurance	economically active	SI EUR 6.66 – 12.98 per week	SI EUR 6.66 – 12.98 per week	W: 60 M: 61	10 years		DB	Prices and wages

Country	Type of the system	Field of application	Pension contribution rate (1) (%)		Legal retirement age	Minimum period of membership	Conditions for drawing full pension	Pension formula	Automatic adjustment-base
			employee	employer					
Netherlands	National pension financed by contributions	residents	19.15 (7)	–	65	–	50 years	EUR 921.28 per month (8)	Wages
Poland	Social insurance: 1. NDC 2. Private pension funds	economically active	Old age only: 9.76 1. 2.46 2. 7.30 (6)	Old age only: 9.76	W: 60 M: 65	–	–	1/ NDC 2/ DC	Prices and wages
Portugal	Social insurance	employees	SI 11.00	SI 23.25	65	15 years	40 years	DB	Prices
Slovakia	Socialinsurance	economically active	4.00 (9)	16.00 (9)	62	10 years	–	DB	Prices and wages
Slovenia	Socialinsurance	economically active	15.50	8.85	W: 63 M: 65 (10)	15 years	-	DB	Wages
Spain	Social insurance	employees	SI 4.70	SI 23.60	65	15 years	35 years	DB	Prices
Sweden	1/ Social insurance NDC 2/ Fully funded system 3/ Guarantee pension	1/ economically active 2/ economically active 3/ residents	1 + 2/ 7.00 3/ taxes	1 + 2/ 10.21 3/ taxes	65	1/ – 2/ – 3/ 3 years residence	1/ – 2/ – 3/ 40 years residence	1/ NDC 2/ DC 3/ up to EUR 762 per month	1/ Wages 2/ Wages 3/ Prices
United Kingdom	Social insurance: 1/ Flat rate Basic State Pension 2/ State Second Pension	economically active	SI 9.40 or 11.00	SI 9.30 or 11.80 or 12.80	W: 60 M: 65 (2)	1/ 11–12 years 2/ 1 year	1/ 90 % of W: 39 M: 44 years 2/ –	1/ EUR 118.00 per week 2/ DB	Prices

DB defined benefit

DC defined contribution

NDC notional defined contribution

W women, M – men

SI an overall contribution for social insurance, no separate pension contribution

(1) If not otherwise stated: contribution for old age, invalidity and survivors

(2) The retirement age for women is gradually raised to that for men (in Austria, Belgium, Estonia, Latvia) or to 60 (in Lithuania).

(3) In the Czech Republic, the retirement age is being gradually increased; for every child raised (until five), the retirement age for women is lowered by one year

(4) In Estonia, part of social tax, earmarked for pension insurance

(5) In Greece, for people insured before 31.12.1992 the legal retirement age is 60 years for women and 65 years for men

(6) In Hungary and Poland, if a person is only insured in First pillar the whole contribution is paid into it.

(7) In the Netherlands, contribution for old age and survivors, of which 17.90 % for the old-age scheme

(8) In the table – amount for single person. For persons sharing a household, both over 65, EUR 631.76 per month for each person.

(9) In Slovakia, the contribution rate shown in the table covers old age and survivors, and a separate contribution for invalidity is paid.

(10) In Slovenia, the retirement age is lower when the number of insurance years accedes 20 (61 women and 63 men) or 38 years (58 for women and men). Also, child-upbringing can lower the retirement age to 56 years form women and 58 years for men.

Source: Based on MISSOC 2005

Table A.2: Welfare regimes and protection against poverty in Europe

Protection from poverty risks	Liberal	Corporatist conservative	Social democratic	Familialistic
Through work	– High labour market participation – High (UK)/low (Ireland) level of female employment – High occurrence of low-paid labour	– High labour market participation – Medium/low level of female employment – Low occurrence of low-paid labour	– High labour market participation – High level of female employment – Low occurrence of low-paid labour	– Low labour market participation – Low level of female employment – High occurrence of low-paid labour
Through social security	– Medium(UK)-low (Ireland) social expenditure – Modest universal transfers – Means-tested – Flat rate benefits	– Medium-high social expenditure – Contribution related social insurance – Categorical insurance: related to class and status – Unequal levels of benefits	– High social expenditure – Universalistic insurance – High level of benefits	– Low social expenditure – Contribution related social insurance – Categorical insurance: related to class and status – Immature and fragmented insurance system – Low level of benefits
Through intra-family transfers		– Extended family obligations	– Dependence on the family is minimised	– Traditional family structures
Through minimum income	– Extensive system of SA – Medium-low levels of generosity	– Extensive system of SA – Generous benefits	– Residual system of SA – Medium -high levels of generosity	– No universal guaranteed minimum income system – Categorical schemes for the elderly
Countries	UK, Ireland	Austria, France, Germany, Belgium, Luxemburg	Denmark, Finland	Greece, Spain, Portugal, Italy

Source: Nicaise et al. 2004: 31

List of Abbreviations

BEPG	Broad Economic Policy Guidelines
CEEC	Central and Eastern European Countries
DRG	Diagnosis Related Group
ECB	European Central Bank
ECJ	European Court of Justice
ECOFIN	Economic and Financial Affairs Council
EEC	European Economic Community
EES	European Employment Strategy
EGL	Employment Guidelines
EMU	Economic and Monetary Union
ESPHCA	Employment, Social Policy, Health and Consumer Affairs Council
GDP	Gross Domestic Product
ICT	Information and Communication Technologies
ILO	International Labour Organisation
IPROSEC	Improving Policy Responses and Outcomes to Socio-Economic Challenges
KRUS	Agricultural Social Insurance Fund of Poland (Kasa Rolniczego Ubezpieczenia Spolecznego)
NAP	National Action Plan
NHS	National Health System
OMC	Open Method of Coordination
SGP	Stability and Growth Pact
SHI	Social Health Insurance
ZUS	Social Insurance Institution of Poland (Zaklad Ubezpieczen Spolecznych)

Index

References

Adam H, Henke KD (1994) Ökonomische Grundlagen der gesetzlichen Krankenversicherung. In: Schulin B (ed) Handbuch des Sozialversicherungsrechts, Band 1 Krankenversicherungsrecht, München, pp 113–144

Adam H, Henke KD (2003) Gesundheitsökonomie. In: Hurrelmann K, Laaser U (eds) Handbuch Gesundheitswissenschaften, 3. Auflage, Weinheim, München, pp 779–797

Advisory Board of the Ministry of Finance (2004) Nachhaltige Finanzierung der Renten und Krankenversicherung. Gutachten, Band 77, Berlin

Advisory Council for the Concerted Action in Health Care (1997) The Health Care System in Germany, Cost factor and branch of the future. Volume II, Progress and Growth Markets, Financing and Remuneration, Special Report, Summary

Adolph H, Heinemann H (2002) Zur Lebenssituation älterer Menschen in Deutschland. Ausgewählte Daten und Kerninformationen, Deutsches Zentrum für Altersfragen, DZA-Diskussionspapier Nr. 37

Ainsaar M (2003) Laste- ja perepoliitika Eestis ja Euroopas. [Child and family policies in Estonia and in Europe.] Rahvastikuministeriumi büroo, Tallinn (In Estonian)

Alber J (1982) Vom Armenhaus zum Wohlfahrtsstaat: Analysen zur Entwicklung der Sozialversicherung im Westeuropa. Campus Verlag, Frankfurt a. M.

Alber J, Bernardi-Schenkluhn B (1992) Westeuropäische Gesundheitssysteme im Vergleich. Campus Verlag, Frankfurt a. M.

Alter KJ (2001) Establishing the Supremacy of European Law. The Making of an International Rule of Law in Europe. Oxford University Press, Oxford

Amitsis G, Berghman J et al. (2003) Connecting Welfare Diversity within the European Social Model. Report Submitted to the Hellenic Presidency of the European Union, Athens

Arts W, Entzinger H, Muffels R (2004) Verzorgingsstaat vaar wel. Koninklijke Van Gorcum, Assen

Atkinson AB (1998) Poverty in Europe, Blackwell, Oxford

Atkinson AB (2002) 'Social Inclusion and the European Union', Journal of Common Market Studies 40 (4) 625–643

Atkinson AB, Cantillon B, Marlier E, Nolan B (2002) Social Indicators. The EU and Social Inclusion, Oxford University Press, Oxford

Atkinson AB, Marlier E, Nolan B (2004) 'Indicators and targets for Social Inclusion in the European Union', Journal of Common Market Studies 42 (1) 47–75

Baldwin P (1990) The Politics of Social Solidarity. Class Bases of the European Welfare State 1875–1975. Cambrdige University Press, Cambridge

Banting K (1995) The Welfare State as Statecraft: Territorial Politics and Canadian Social Policy. In: Leibfried S, Pierson P (eds) European Social Policy. Between Fragmentation and Integration. The Brookings Institution, Washington D.C., pp 269–300

Barbier J-C (2004) Research on open methods of coordination and national social policies: what sociological theories and methods?, Paper for the RC 19 international conference, Paris 2–4 September 2004

Barfuss KM (2002) Globale Migration: Triebkräfte, Wirkungen und Szenarien aus ökonomischer Sicht, IMIS-Beiträge (Institut für Migrationsforschung und Interkulturelle Studien der Universität Osnabrück), Heft 19/2002, August 2002, pp 43–64
Barr N (2001) The Welfare State as Piggy Bank. Information, Risk, Uncertainty, and the Role of the State. Oxford University Press, Oxford
Barr N (2002) Reforming Pensions: Myths, Truths, and Policy Choices. International Social Security Review, Vol. 55, No. 2: 3–36
Barr N (2004) The Economics of the Welfare State. 4th edn., Oxford University Press, Oxford
Barry B (1998) Social Exclusion, Social Isolation and the Distribution of Income, Centre for the analysis of social exclusion – LSE, CASE paper 12
Bauer TK (2002) Migration, Sozialstaat und Zuwanderungspolitik, Forschungsinstitut zur Zukunft der Arbeit, IZA Discussion Paper No. 505, March 2002
Baumol P (1967) The Macroeconomics of Unbalanced Growth, American Economic Review 57:415–426
Beauchamp T, Childress JF (2001) Principles of Biomedical Ethics. Oxford University Press, Oxford (5th edition)
Begg I, Berghman J, Chassard Y, Kosonen P, Madsen PK, Matsaganis M, Mayes D, Muffels R, Salais R, Tsakloglou P (2001) Social Exclusion and Social Protection in the European Union: Policy, Issues and Proposals For the Future Role of the EU. Southbank University, London
Begg I, Berghman J (2002) Introduction: EU social (exclusion) policy revisited? Journal of European Social Policy 12, 3:179–194
Begg I, Muffels R, Tsakloglou P (2002) Conclusions: Social Exclusion on the Crossroads of EU Employment and Inclusion Policies. In: Tsakloglou P (ed) Social Exclusion in European Welfare States, Edward Elgar, Cheltenham
Belke A, Hebler M (2000) EU Enlargement and Labour Markets in the CEECs, Intereconomics, 5:219–230
Benner M, Vad TB (2000) Sweden and Denmark: Defending the Welfare State. In: Scharpf F, Schmidt V, Welfare or Work in the Open Economy Vol. II, Cambridge University Press, pp 399–466
Benner M (2003) The Scandinavian Challenge. The Future of Advanced Welfare States in the Knowledge Economy. Acta Sociologica 46:132–149
Berghmann (1995) Social Exclusion in Europe: Policy Context and Analytical Framework. In: Room G (ed) Beyond the Threshold, the Measurement and Analysis of Social Exclusion. Policy Press, Bristol, pp 10–28
Berghman J (1997) The Resurgence of Poverty and Struggle against Exclusion: a New Challenge for Social Security in Europe? International Social Security Review, No. 1/1997
Berhanu S, Henke KD, Mackenthun B (2004) Die Zukunft der Gemeinnützigkeit von Krankenhäusern, Zeitschrift für öffentliche und gemeinwirtschaftliche Unternehmen ZögU
Beutler B, Bieber R et al. (eds) (2005) Das Recht der Europäischen Union. Loseblatt, Nomos
Björklund A (2000) 'Going Different Ways: Labour Market Policy in Denmark and Sweden'. In: Esping-Andersen G, Regini M (eds) Why Deregulate Labour Markets? Oxford University Press, Oxford, pp 148–180
Bonoli G (1997) Classifying welfare states: a two-dimension approach. Journal of Social Policy 26, 3:351–72
Bovenberg L (2003) Nieuwe levensloopbenadering. Discussion Paper OSA, 1, June 2003, OSA/Tilburg University, Tilburg
Bradshaw PL (2003) Modernizing the British National Health Service (NHS) – some ideological and policy considerations, Journal of Nursing Management, March 11 (2), pp 91–97

Bräuninger D (2002) Demographics and Pension Reforms in Large Central and Eastern European Countries, Frankfurter Voice, Demography Special, November 14, 2002, Deutsche Bank Research
Bradshaw J, Finch N (2002) A Comparison of Child Benefit Packages in 22 Countries. University of York, Huddersfield
Breyer F, Haufler A (2000) Health Care Reform: Separating Insurance from Income Redistribution, Discussion Paper No. 205, Deutsches Institut für Wirtschaftsforschung (DIW), Berlin
Briggs A (1961) The Welfare State in Historical Perspective. European Journal of Sociology/Archives Europeénnes de Sociologie, II:221–258
Brown L, Amelung V (1999) 'Manacled Competition' in Germany, Health Affairs May/June, Vol.18, no.3:76–92
Brücker H (2001) The Impact of Eastern Enlargement on EU-Labor Markets, Zentrum für Europäische Integrationsforschung, Rheinische Friedrich-Wilhelms-Universität Bonn, ZEI Policy Paper B 12/2001, pp 4–22
Brücker H, Boeri T (2000) The Impact of Eastern Enlargement on Employment and Labour Markets in the EU Member States, Final Report, Research carried out on behalf of the Employment and Social Affairs Directorate General of the European Commission, Berlin and Milano, European Integration Consortium 2000
Brücker H, Alvarez-Plata P, Siliverstovs B (2003) Potential Migration from Central and Eastern Europe into the EU-15 – An Update. Deutsches Institut für Wirtschaftsforschung, Berlin
Buchanan A et al. (2000) From Chance to Choice. Cambridge UP, Cambridge
Buchholz W, Edener B, Grabka MM, Henke KD, Huber M, Ribhegge H, Ryll A, Wagner HJ, Wagner GG (2001) Wettbewerb aller Krankenversicherungen kann Qualität verbessern und Kosten des Gesundheitswesens senken, Discussionpaper, No. 247, Deutsches Institut für Wirtschaftsforschung (DIW), Berlin
Bundesministerium für Arbeit und Sozialordnung u.a. (ed) (2000/2001) Soziale Grundrechte in der Europäischen Union. Baden-Baden
Bundesministerium für Familie, Senioren, Frauen und Jugend (BMFSFJ) (2000) Dritter Bericht zur Lage der älteren Generation in der Bundesrepublik Deutschland: Alter und Gesellschaft, Drucksache 14/5130, Berlin
Bundesministerium der Finanzen (2000) Freizügigkeit und soziale Sicherung in Europa, Gutachten erstattet vom Wissenschaftlichen Beirat beim Bundesministeriums der Finanzen, Schriftenreihe des Bundesministeriums der Finanzen, Heft 69, Bonn
Bundesministerium der Finanzen (2001) Nachhaltigkeit in der Finanzpolitik. Konzepte für eine nachhaltige Orientierung öffentlicher Haushalte, Heft 71, Bonn
Bundesministerium der Finanzen (2004) Nachhaltige Finanzierung der Renten- und Krankenversicherung, Gutachten erstattet vom Wissenschaftlichen Beirat beim Bundesministerium der Finanzen, Schriftenreihe des Bundesministeriums der Finanzen, Band 77, Bonn
Busse R, Henke KD, Schreyögg J (2004) Regulation of pharmaceutical markets in Germany – improving efficiency and controlling expenditure, forthcoming in: International Journal of Health Planning and Management
Busse R, Riesberg A (2004) Germany, Health Care Systems in Transition, European Observatory on Health Care Systems, World Health Organisation Regional Office for Europe, Copenhagen
Busse R, Schlette S (2003) Gesundheitspolitik in den Industrieländern – Trends und Analysen 1/2003, Verlag Bertelsmann Stiftung (ed), Gütersloh
Busse R (2002) Survey Report: Health Care Systems. Towards an Agenda for Policy Learning between Britain and Germany. Anglo-German Foundation, London
Busse R (2004) Disease Management Programs in Germany's Statutory Health Insurance System, Health Affairs, May/June, Vol. 23, No. 3, pp 56–67

Calliess C, Ruffert M (eds) (2002) Kommentar zum EU-Vertrag und EG-Vertrag. 2nd ed, Luchterhand
Cantillon B, Marx I, van den Bosch K (1997) The Challenge of Poverty and Social Exclusion. In: OECD, Family, Market and Community: Equity and Efficiency in Social Policy, OECD, Paris
Cantillon B, Marx I, van den Bosch K (2003) The Puzzle of Egalitarianism: Relationship between Employment, Wage Inequality, Social Expenditure and Poverty, European Journal of Social Security 5, 2:108–127
Case AC, Lin I-F, McLanahan SS (2003) Explaining Trends in Child Support: Economic, Demographic and Policy Effects. Demography 40:171–191
Casey BH, Gold M (2005) Peer review of labour market programmes in the European Union: what can countries really learn from one another?, Journal of European Public Policy, Vol 12 (1):23–44
Chassard Y (2001) European and Social Protection: from the Spaak Report to the Open Method of Coordination. In: Mayes D, Berghman J, Salais R, The Social Exclusion and European Policy, Edward Elgar, Cheltenham
Chłoń A, Góra M, Rutkowski M (1999) Shaping Pension Reform in Poland: Security through Diversity, Social Protection Discussion Paper No. 9923, The World Bank, August 1999
Chłoń A (2000) KRUS and the Challenges of Rural Development. Poland's Social Security System for Farmers, Draft Discussion Note, The World Bank
Clasen J (2000) 'Motives, Means and Opportunities: Reforming Unemployment compensation in the 1990s', West-European Politics
Cohen M, Hanagan M (1991) The politics of gender and the making of welfare state, 1900–1940: A comparative perspective, Journal of social history 24:469–484
Coleman D (1996) New patterns and trends in European fertility. In: Coleman D (ed) Europe's Population in the 1990's. Oxford University Press, Oxford, pp 1–61
Commission of the European Communities (1996) First Report on Economic and Social Cohesion (Brussels: CEC)
Commission of the European Communities (1999) Towards a Europe for all Ages – Promoting Prosperity and Intergenerational Solidarity. Communication from the Commission, Brussels http://europe.eu.int/comm/employment_social/social_situation/docs/com221_en.pdf
Commission of the European Communities (2000) 'Structural Indicators', Communication from the Commission, COM (2000), Luxembourg
Commission of the European Communities (2004) 'Joint report on social inclusion 2004', COM, Luxembourg
Commission of the European Communities (2004) Employment in Europe 2004. Recent Trends and Prospects, Luxembourg
Cornelissen VR (1996) The Principle of Territoriality and the Community Regulations on Social Security, Common Market Law Review 33:13–41
Council of the European Union (2003) Draft Joint Report by the Commission and the Council on Adequate and Sustainable Pensions, ECOFIN 51, SOC 72, 3.03.2003 (http://europa.eu.int/ comm/employment_social/soc-prot/pensions/2003jpr_en.pdf)
Crouch C (1999) Social Change in Western Europe. Oxford University Press, Oxford
Deacon B, Ollila E, Koivusalo M, Stubbs P (2003) Global Social Governance. Themes and Prospects. Globalism and Social Policy Program. Hakapaino, Helsinki
Dekker P, Ederveen S, Jehoel-Gijsbers G, de Mooij R, Soede A, Wildeboer-Schut J-M (2003) Social Europe. CPB/SCP, The Hague
International Labour Organisation (1994) Defending values, promoting change. Social Justice in a global economy. An ILO-Agenda. Geneva

De Lathouwer L (2005) Reforming the Passive Welfare State: Belgium's New Income Arrangements to Make Work Pay in International Perspective. In: Saunders P (ed) Welfare to work on practice: social security and participation in economic life, Ashgate, Aldershot, 133–158

De Swaan A (1988) In Care of the State. Health Care, Education and Welfare in Europe and the USA in the Modern Era. Polity Press, Oxford

Deutsche Krankenhausgesellschaft (2003) Krankenhausplanung und Investitionsfinanzierung in den Bundesländern. Köln

Devos K, Zaidi MA (1998) Poverty measurement in the EU: Country-specific or Union-wide Poverty lines?, Journal of Income Distribution 8 (1):77–92

Donzelot J (1984) L'invention du social. Essai sur le decline des passions politiques, Fayard, Paris

Dworkin R (2000) Sovereign Virtue: The theory and practice of equality. Harvard UP, Cambridge

Eichenhofer E (1994) Internationales Sozialrecht. München

Eichenhofer E (2001) Sozialrecht der Europäischen Union. Schmidt

Eichenhofer E (2002) Der aktuelle Stand europäischer Sozialpolitik, Deutsche Rentenversicherung Nr. 6, pp 322–331

Einasto M (2002) Income and deprivation poverty, 1994 and 1999. In: Kutsar D (ed) Living Conditions in Estonia Five Years Later. Norbalt II. University of Tartu and Fafo Institute for Applied Social Research Norway, Tartu, pp 109–130

Enthoven AC (2000) In Pursuit of an improving National Health Service. In: Health Affairs, Vol. 19, No.3, pp 102–119

Entrena F, Gómez-Mateos J (2000) Globalization and Socio-economic Restructuring in Andalusia. Challenges and Possible Alternatives, European Sociological Review 16:93–114

Esping-Andersen G (1990) The Three Worlds of Welfare Capitalism. Polity Press, Oxford

Esping-Andersen G (1996a) After the golden age? Welfare state dilemmas in a global economy. In: Esping-Andersen G (ed) Welfare states in transition. National adaptations in global economies. Sage Publications, London, pp 1–31

Esping-Andersen G (1996b) Welfare states without work: the impasse of labour shedding and familialism in continental European social policy. In: Esping-Andersen G (ed) Welfare states in transition. National adaptations in global economies. Sage Publications, London, pp 66–87

Esping-Andersen G (1999) Social Foundations of Postindustrial Economies. Oxford University Press, Oxford

Esping-Andersen G (2000) Social Foundations of Postindustrial Economies. Oxford

Esping-Andersen G (2002) Why we need a New Welfare State. Sage Publications, London

Esping-Andersen G (2002a) Towards the Good Society, Once Again? In: Esping-Andersen G (ed) Why we need a New Welfare State. Sage Publications, London, pp 1–25

Esping-Andersen G (2002b) A Child-Centred Social Investment Strategy. In: Esping-Andersen G (ed) Why we need a New Welfare State. Sage Publications, London, pp 26–95

Europäische Kommission (2002) Auf dem Weg zur erweiterten Union. Strategiepapier und Bericht der Europäischen Kommission über die Fortschritte jedes Bewerberlandes auf dem Weg zum Beitritt, 9. Oktober 2002

European Central Bank (2005) Statistics Pocket Book – June 2005, Frankfurt a. M.

European Commission (2000–2005) Biannual Update of the Scoreboard to Review Progress on the Creation of an Area of 'Freedom, Security and Justice' in the European Union, Luxembourg; biannually updated information of the Communications of the European Commission: www.migpolgroup.com/monitors (the most recent report updated on 28 February 2005)

European Commission (2001) The Free Movement of Workers in the Context of Enlargement, Information note, March 2001, Luxembourg
European Commission (2003a) Communication from the Commission to the Council, the European Parliament, the Economic and Social Committee and the Committee of the Regions: Strengthening the Social Dimension of the Lisbon Strategy: Streamlining Open Coordination in the Field of Social Protection, Luxembourg (http://europa.eu.int/comm/ employment_social/news/2003/may/lisbonstratIP280503_en.pdf)
European Commission (2003b) Employment in Europe 2003. Office for Official Publications of the European Communities, Luxembourg
European Commission (2004) MISSOC Mutual information system on social protection, Social protection in the Member states of the European Union of the European Economic Area and in Switzerland, January, Luxembourg
European Commission, Directorate General for Economic and Financial Affairs (2001) The Economic Impact of Enlargement, Enlargement Papers No. 4, June 2001, Luxembourg
European Commission, Directorate General for Employment and Social Affairs (2002a) Joint Report on Social Inclusion. Luxembourg
European Commission, Directorate General for Employment and Social Affairs (2002b) The Social Situation in the European Union 2002. Luxembourg
European Commission, Directorate General for Employment and Social Affairs (2003) The Social Situation in the European Union 2003. Luxembourg
European Commission, Directorate General for Employment and Social Affairs (2004) The Social Situation in the European Union 2004. Luxembourg
European Community (ed) (2004) Europa Social Report, 12/2004
European Observatory on Health Care Systems (1999) Health Care Systems in Transition – United Kingdom, Copenhagen
EUROSTAT (2003a) First Results of the Demographic Data Collection for 2002 in Europe, Joint Council of Europe/Eurostat Demographic Data Collection, Statistics in Focus: Population and Social Conditions, Theme 3 – 20/2003
EUROSTAT (2003b) Women and Men Migrating to and from the European Union, Statistics in Focus: Population and Social Conditions, Theme 3 – 2/2003
EUROSTAT (2004a) GDP per capita in Purchasing Power Standards for EU, Candidate Countries and EFTA, Nowcast 2003, Statistics in Focus: Economy and Finance, 27/2004
EUROSTAT (2004b) Pensions in Europe; Expenditure and Beneficiaries, Statistics in Focus: Population and Social Conditions, 8/2004
Ewald F (1986) L'État providence. Grasset, Paris
Ewert HA (1987) Der Beitrag des Gerichtshofs der Europäischen Gemeinschaften zur Entwicklung eines Europäischen Sozialrechts, Florentz
Fassmann H (2002) EU-Erweiterung und Arbeitsmigration nach Deutschland und Österreich. Quantitative Vorhersagen und aktuelle Entwicklungstendenzen, IMIS-Beiträge (Institut für Migrationsforschung und Interkulturelle Studien der Universität Osnabrück), Heft 19/2002, August 2002, pp 65–88
Fassmann H, Münz R (2003) Auswirkungen der EU-Erweiterung auf die Ost-West-Wanderung, WSI Mitteilungen, Heft 1/2003, pp 25–32
Ferrera M (1996) The "Southern Model" of welfare in social Europe, Journal of European Social Policy 6, 1:17–37
Ferrera M (2004) Social Citizenship in the European Union: Toward a Spatial Reconfiguration? In: Ansell CK, Di Palma G (eds) Restructuring Territoriality. Europe and the United States compared, Cambridge University Press, Cambridge, pp 90–122
Ferrera M, Hemerijck A (2000) Recasting the European Welfare States, Frank Cass, London

Ferrera M, Hemerijck A (2003) Recalibrating Europe's Welfare Regimes. In: Zeitlin J, Trubek DM (eds) Governing Work and Welfare in a New Economy. European and American Experiments, Oxford University Press, Oxford, pp 88–128

Ferrara M, Rhodes M (2000) Recasting the European Welfare State, Frank Cass, London

Fertig M, Schmidt C (2000) Aggregate Level Migration Studies as a Toll for Forecasting Future Migration Streams. In: Djadic S (ed) International Migration: Trends, Policy and Economic Impact, London: Routledge, pp 110–136

Fertig M, Schmidt CM (2002) Mobility within Europe – What do we (still not) know?, Forschungsinstitut zur Zukunft der Arbeit, IZA Discussion Paper No. 447, March 2002

Fetzer S, Moog S, Raffelhüschen B (2001) Zur Nachhaltigkeit der Generationenverträge: Eine Diagnose der Kranken- und Pflegeversicherung, Diskussionspapier 99/01, Freiburg, pp 279–302

Figueras J, McKee M, Cain J, Lessof S (2004) Health systems in transition: learning from experience, European Observatory on Health Care Systems and Policies, Copenhagen

Finch J (1989) Family Obligations and Social Change. Polity Press, Cambridge

Flora P, Heidenheimer A (1981) The Development of Welfare States in Europe and America, Transaction Books, Rutgers NJ

Flora P (1986–1987) Growth to Limits: The Western Welfare States Since World War II, Vol. 1-V, de Gruyter, Berlin

Forsander A (ed) (2002) Immigration and Economy in the Globalization Process. The Case of Finland. SITRA, Vantaa

Fourage D (2004) Poverty and Subsidiarity in Europe. Minimum protection from an Economic Perspective. Edward Elgar, Cheltenham

Frank L, Schulte B, Steinmeyer HD (1996) Internationales und Europäisches Sozialrecht, Luchterhand

Gallie D (1999) Unemployment and Social Exclusion in the European Union, European Societies 1:1

Gallie D, Paugam S (2000) The Experience of Unemployment in Europe: The Debate. In: Gallie D, Paugam S (eds) Welfare Regimes and the Experience of Unemployment in Europe, Oxford University Press, Oxford, pp 1–22

Gesellschaft für Versicherungswissenschaft und -gestaltung (GVG)/BMA (ed) (2003) Soziale Sicherheit der Grenzgänger in Europa

Gethmann CF, Gerok W, Helmchen H, Henke K-D, Mittelstraß J, Schmidt-Aßmann E, Stock G, Taupitz J, Thiele F (2004) Gesundheit nach Maß? Eine transdisziplinäre Studie zu den Grundlagen eines dauerhaften Gesundheitssystems. Akademie Verlag, Berlin

Giddens A (1990) The Consequences of Modernity. Polity Press, Cambridge

Giddens A (1991) Modernity and Self-Identity. Self and Society in the Late Modern Age. Polity Press, Cambridge

Giddens A (1998) The Third Way: the Renewal of Social Democracy. Polity Press, Cambridge

Giddens A (2000) The Third Way. Polity Press, Cambridge

Giesen R (1999) Die Vorgaben des EG-Vertrags für das Internationale Sozialrecht, Carl Heymanns

Giesen R (2004) Rationierung im bestehenden Gesundheitssystem, in: Zeitschrift für die gesamte Versicherungswissenschaft, Heft 4, pp 557–582

Göbel M (2002) Von der Konvergenzstrategie zur offenen Methode der Koordinierung, Nomos

Golini A (1999) Population ageing in developed countries: lesson learnt and to be learnt. In: Cliquet R, Nizamuddin M (eds) Population Ageing. UNFPA & CBGS, New York and Brussels, pp 49–84

Goodin RE (2003) Perverse principles of Welfare Reform. In: Pieters D (ed) European Social Security and Global Politics. Kluwer Law International, The Hague, pp 199–225
Goodin RE, Heady B, Muffels R, Dirven H-J (1999) The Real Worlds of Welfare Capitalism, Cambridge University Press, Cambridge
Government of Denmark (2001) Denmark's National Action Plan to Combat Poverty and Social Exclusion (NAPinl) 2001/2003, EU website, europa.eu.int/comm./employment social/news/2001/jun/napincl2001en.html
Government of Denmark (2003) Denmark's National Action Plan to Combat Poverty and SocialExclusion (NAPinl) 2003/2005, Euwebsite: europa.eu.int/comm/employment social/news/2001/jun/napincl03da_en.pdf
Graefe P (2001) Whose social economy? Debating new state practices in Quebec, Critical Social Policy 21:35–57
Grauwe P de (2002) Challenges for Monetary Policy in Euroland. Journal of Common Market Studies 40, 4:693–718
Grauwe P de (2003) The Euro at Stake? The Monetary Union in an Enlarged Europe. CESifo Economic Studies 49, 1/2003:103–121
Greß S, Axer P, Wasem J (2003) Europäisierung des Gesundheitswesens, Perspektiven für Deutschland, Studien der Bertelsmann Stiftung (ed) Gütersloh
Groenez S, Nicaise I (2002) Traps and Springboards in European minimum income systems. The Belgian Case. Hiva/KULeuven, Leuven
GUS (1992, 1994, 1995, 2001) Rocznik Statystyczny Rzeczypospolitej Polskiej 1992, 1994, 1995, 2001 [Statistical Yearbook of the Republic of Poland 1992, 1994, 1995, 2001], G∏ówny Urzàd Statystyczny, Warsaw
GVG (2003) Social Protection in the Candidate Countries: Country Studies Czech Republic, Slovak Republic, Poland, ed. Gesellschaft für Versicherungswissenschaft und -gestaltung e.V., Schriftenreihe der GVG, Band 41, Akademische Verlagsgesellschaft Aka GmbH, Berlin
Häberle P (2000) Gibt es eine europäische Öffentlichkeit? de Gruyter
Hailbronner K, Klein E, Magiera S, Müller-Graff P-C (1998) Handkommentar zum Vertrag über die Europäische Union, Loseblatt, Heymanns
Hall P (1952) The Social Services of Modern England. Routledge & Kegan Paul Ltd, London
Hanau P, Steinmeyer H-D, Wank R (2002) Handbuch des europäischen Arbeits- und Sozialrechts. Beck
Hantrais L (2003) Sissejuhatus [Introduction]. In: Kutsar D (ed) Millist perekonnapoliitikat me vajame? IPROSEC [What kind of family policies we need?]. Tartu Ülikooli Kirjastus, Tartu, pp 5–11 (In Estonian)
Hantrais L, Letablier M T (1996) Families and Family Policies in Europe. Longman, London & New York
Haverkate G et al. (eds) (1999) Casebook zum Arbeits- und Sozialrecht der EU. Nomos
Haverkate G, Huster S (1999) Europäisches Sozialrecht. Nomos
Headey B, Goodin RE, Muffels R, Dirven H-J (1997) Welfare over Time: Three Worlds of Welfare Capitalism in Panel Perspective. Journal of Public Policy 17,3:329–359
Healthcare Commission (2004) State of Health Report. London
Heclo H (1974) Modern Social Politics in Britain and Sweden: From Relief to Income Maintenance. Yale Univerity Press, New Haven
Heidenreich M (2003) 'Regional inequalities in the Enlarged Europe', Journal of European Social Policy 13/4, pp. 313–333

Heikkilä M, Kuivalainen S (2002) Using social benefits to combat poverty and social exclusion: opportunities and problems from a comparative perspective. European synthesis report. Council of Europe, Strasbourg (published in the same volume also in French: Utiliser des aides sociales pour combattre la pauverté et l'exclusion social: examen comparatif des opportunités et des problèmes)

Helne T (2001) Syrjäytymisen yhteiskunta. [The Society of Marginalization, in Finnish.] STAKES, Helsinki

Hemerijck A (2001) Prospects for Effective Social Citizenship in an Age of Structural Inactivity. In: Crouch C, Eder K, Tambini D (eds) Citizenship, Market and the State; Oxford University Press, Oxford

Hemerijck A (2002) The Self-Transformation of the European Social Model. In: Esping-Andersen G (ed) Why we need a New Welfare State. London: Sage Publications, pp 173–213

Hemerijck AC, Visser J (1999) Negociated Change: Towards a Theory of Institutional Learning in Tightly Coupled Welfare States. In: Public Policy and Political Ideas, edited by A. Busch. Edward Elgar, Cheltenham

Henke KD (1992) Financing a national health insurance, Health Policy, Vol. 20/3, pp 253–268

Henke KD (2001) The Allocation of National Resources in Health Care in Germany between competition and solidarity, Edition Dräger-Stiftung, Band 17

Henke KD (2002) Soft co-ordination versus hard rules in European economic policy – Does co-ordination lead to a new direction for a common European economic policy?, Diskussionspapiere zu Staat und Wirtschaft, Nr. 34, Berlin

Henke KD (2004) Financing and Purchasing in Health Services – A Book with Seven Seals. In: Henke KD, Rich RF, Stolte H (2004c) Integrierte Versorgung und neue Vergütungsformen in Deutschland, Lessons learned from comparison of other health care systems, Europäische Schriften zu Staat und Wirtschaft, Band 14, pp 11–21

Henke KD (2005) Was ist uns die Gesundheit wert? Probleme der nächsten Gesundheitsreform und ihre Lösungsansätze. In: Perspektiven der Wirtschaftspolitik, Schriften des Vereins für Socialpolitik, 6(1), 2005, pp 95–111

Henke KD, Friesdorf W, Marsolek I (2004a) Genossenschaften als Chance für die Entwicklung der Integrierten Versorgung im Gesundheitswesen, Deutscher Genossenschafts- und Raiffeisenverband e.V. (DGRV) (ed) Berlin

Henke KD, Mackenthun B, Schreyögg J (2004b) The Health care sector as economic driver: an economic analysis of the health care market in the city of Berlin, Journal of Public Health, Vol. 12, pp 339–345

Henke KD, Mühlbacher A (2004) Gesundheit, ein Privileg der Reichen? In: Schumpelick V, Vogel B (eds) Grenzen der Gesundheit, Herder Verlag, Freiburg, pp 405–428

Henke KD, Rich RF, Stolte H (2004c) Integrierte Versorgung und neue Vergütungsformen in Deutschland, Lessons learned from comparison of other health care systems, Band 14, Europäische Schriften zu Staat und Wirtschaft, Baden-Baden

Henke KD, Borchardt K (2003) Reform Proposals for Health-Care Systems – Capital funded versus Pay-as-you-go in Health-Care Financing reconsidered. In: CESifo DICE Report, Nr. 3, pp 3–8

Henke KD, Hesse M (1999) Gesundheitswesen. In: Korff W et al. (eds) Handbuch der Wirtschaftsethik, Band 4, Ausgewählte Handlungsfelder, Gütersloh, pp 249–290

Henke KD, Schreyögg J (2004) Towards sustainable health care systems. Strategies in health insurance schemes in France, Germany, Japan and the Netherlands – A comparative study –, vol 16 Nomos-Verlag Berlin

Hiilamo H (2004) Changing family policy in Sweden and Finland during the 1990s. In: Heikkilä M, Kautto M (eds) Welfare in Finland. STAKES, Helsinki, pp 123–144

Hill S, Garattini S, v. Loenhout J, de Joncheere K (2003) Technology Appraisal Programme of the National Institute for Clinical Excellence, A review by World Health Organization (WHO) Regional Office for Europe, Kopenhagen, p 3

Hinrichs K (2004) Active Citizens and Retirement Planning: Enlarging Freedom of Choice in the Course of Pension Reforms in Nordic Countries and Germany, Zentrum für Sozialpolitik, Universität Bremen, ZeS-Arbeitspapier Nr. 11/2004

Hinrichs K, Kangas O (2004) When is a Change Big Enough to be a System Shift? Small System-shifting Changes in German and Finnish Pension Policies. In: Taylor-Gooby P (ed) Making a European Welfare State? Convergences and Conflicts over European Social Policy, Blackwell, Oxford

Höhn C (1999) Die demographische Alterung – Bestimmungsgründe und wesentliche Entwicklungen. In: Grünheid E (ed) Demographische Alterung und Wirtschaftswachstum, Schriftenreihe des Bundesinstituts für Bevölkerungsforschung, Band 29, Opladen, pp 9–32

Hönekopp E, Werner H (2000) Is the EU's Labour Market Threatened by a Wave of Immigration?, Intereconomics, 1, pp 3–8

Holzmann R (2004) Toward a Reformed and Coordinated Pension System in Europe: Rationale and Potential Structure, Social Protection Discussion Paper Series No. 0407, The World Bank, Washington D.C.

Holzmann R, Hinz R (2005) Old-Age Income Support in the Twenty-first Century: An International Perspective on Pension Systems and Reform, The World Bank, Washington D.C.

Holzmann R, Münz R (2004) Challenges and Opportunities of International Migration for the EU, Its Member States, Neighbouring Countries and Regions: A Policy Note, Social Protection Discussion Paper Series No. 0411, The World Bank, Washington D.C.

Huber E, Stephens J (2001) Development and Crisis of the Welfare State, Chicago University Press, Chicago

Ipsen HP (1972) Europäisches Gemeinschaftsrecht, Mohr

Igl G (1988) Für eine integrierte Alterssicherung, Deutsche Rentenversicherung Nr. 10

Iliopoulos-Strangas J (2000) La protection des droits sociaux fondamentaux das les Etats membres de l'Union Europeenne. Athen

International Labour Organisation, ILO (1994) Defending Values, Promoting Change, Social Justice in a global economy: An ILO agenda, Report of the Director General to the 81st Session of the International Labour Conference. ILO, Geneva

IPROSEC (2000–2004) Improving Policy Responses and Outcomes to Socio-Economic Challenges: changing family structures, policy and practice. A European Commission Framework Programme 5 project, co-ordinated by Professor Linda Haintrais. Updated list of reports available in paper and electronic form in the homepage www.iprosec.org.uk

ISSA (1994) Migration: A Worldwide Challenge for Social Security, Studies and Research No. 35, International Social Security Association, Geneva

Iversen T (2001) The Dynamics of Welfare State Expansion: Trade Openness, De-industrialization, and Partisan Politics. In: Pierson P (ed) The New Politics of the Welfare State, Oxford University Press, Oxford, pp 45–79

Iversen T, Wren A (1998) Equality, Employment and Budgetary Restraint: The Trilemma of the Service Economy, World Politics 50:507–546

Jacobs K (2003) Die GKV als solidarische Bürgerversicherung, Wirtschaftsdienst, Heft 2, pp 88–90

Jaźwińska E, Okólski M (2001) Ludzie na huśtawce. Migracje miedzy peryferiami Polski i Zachodu [People on a Swing. Migrations between Peripheries of Poland and the West], Wydawnictwo Naukowe SCHOLAR, Warsaw

Johansson S (2001) Den sociala omsorgens akademisering. [The development of academic social care science, in Swedish.] Liber, Stockholm

Jorens Y (ed) (2003) Open Method of Coordination. Nomos

Jorens Y, Schulte B (eds) (1998) European Social Security and Third Country Nationals, de Keure

Jorgensen H (2004) The European Employment Strategy (EES). From discourse to beauty contest or implemented policy? Presentation at CARMA conference, 28 and 29 October 2004

Kainulainen S, Rintala T, Heikkilä M (2001) Hyvinvoinnin alueellinen erilaistuminen 1990-luvun Suomessa. [The regional differentiation of welfare in Finland during the 1990s, in Finnish.] STAKES, Helsinki

Kandolin I (1997) Gender, worklife and family responsibilities in Finland and Estonia – effects on economic and mental well-being. Hakapaino Oy, Helsinki

Karl B, von Maydell BB (2003) Das Angebot von Zusatzversicherungen. Dürfen gesetzliche Krankenversicherungen Zusatzversicherungen anbieten?, PKV-Dokumentation 28/2003, Köln

Kasearu K, Kutsar D (2003) Perekonna ideaalkujundid ja nende realiseerumise võimalikkus. [The family ideals and the opportunities to realise those, in Estonian] In: Kutsar D (ed) Millist perekonnapoliitikat me vajame? IPPROSEC [What kind of family policies we need?] Tartu Ülikooli Kirjastus, Tartu, pp 88–103

Katus K (2000) General patterns of post-transitional fertility in Estonia. TRAMES, Journal of the Humanities and Social Sciences 4(54/49)[77]:213–230

Kaufmann F-X (2001a) Towards a Theory of the Welfare State. In: Leibfried S (ed) Welfare State Futures, Cambridge University Press, Cambridge, pp 15–36

Kaufmann F-X (2001b) Social Security. in: Smelser NJ, Baltes PB (eds) International Encyclopedia of the Social and Behavioral Sciences, Vol. 21, Elsevier, Leiden, pp 14435–14439

Kaufmann F-X (2003a) Sozialpolitisches Denken. Die Deutsche Tradition Suhrkamp, Frankfurt am Main

Kaufmann F-X (2003b) Varianten des Wohlfahrtsstaats. Der Deutsche Sozialstaat im internationalen Vergleich. Suhrkamp, Frankfurt am Main

Kautto M (2001) Diversity among Welfare States. Comparative Studies on Welfare State Adjustment in the Nordic Countries. Research Report 118. STAKES, Helsinki

Kekes J (1996) The Morality of Pluralism. Princeton. Princeton University Press, New Jersey

Kleinman M (2002) A European Welfare State? European Union Social Policy in Context. Palgrave, Houndsmills

Klusen N (ed) (2003) Europäischer Binnenmarkt und Wettbewerb – Zukunftsszenarien für die GKV. Nomos

Kohli M (1987) Retirement and the Moral Economy: An Historical Interpretation of the German Case, "Journal of Aging Studies", No. 2, pp 125–144

Kohli M (1995) Möglichkeiten und Probleme einer Flexibilisierung des Übergangs in den Ruhestand. Freie Universität, Berlin

Kongshoi Madsen P (2004) The Danish model of 'flexicurity': experiences and lessons. Transfer 10,2:187–207

---

[77] Several volume numbers refer to the ruptures of the publishing history of the journal. Acta et Commentationes Universitatis Tartuensis (Dorpantensis) B, established 1893; Proceedings of the Estonian Academy of Sciences. Humanities and Social Sciences, established 1952; TRAMES: Journal of the Humanities and Social Sciences, established 1997.

Koopmans R (2003) Het Nederlandse integratiebeleid in internationaal vergelijkend perspectief: Etnische segregatie onder de multiculturele oppervlakte. In: Pellikaan H, Trappenburg M (red.) Politiek in de multiculturele samenleving, Boom, Meppel/Amsterdam

Kortteinen M, Vaattovaara M (2001) Why and how do urban spatial inequalities grow during the information age? A case study of the development of the Helsinki region. In: Urban Futures Anthology. [Proceedings of an expert meeting; no page numbering] European Commission, Södertälje

Krueger AB (2000) From Bismarck to Maastricht. The March to European Union and the Labor Compact. Labor Economics 7, 2000:117–134

Kuhnle S (ed) (2000) The Survival of the European Welfare State. Routledge, London

Kuhnle S (2001) The Nordic welfare state in a European context: dealing with new economic and ideological challenges in the 1990s. In: Leibfried S (ed) Welfare State Futures, Cambridge University Press, Cambridge, pp 103–122

Kutsar D (ed) (2002) Living Conditions in Estonia Five Years Later. Norbalt II. Tartu University Press, Tartu

Kvist J (2004) Does EU enlargement start a race to the bottom? Strategic interaction among EU member states in social policy. Journal of European Social Policy 14:301–318

Kymlicka W (1990) Contemporary Political Philosophy. Oxford UP, Oxford

Lauristin M, Heidmets M (eds) (2002) The Challenge of the Russian Minority. Emerging Multicultural Democracy in Estonia. Tartu University Press, Tartu

Lauristin M, Vihalemm P (2002) The Transformation of Estonian Society and Media: 1987 –2001. In: Vihalemm P (ed) Baltic Media in Transition. University of Tartu, Tartu, pp 17–63

Lauristin M, Vihalemm P, Rosengren KE, Weibull L (1997) Return to the Western World. Tartu University Press, Tartu

Layte R, Whelan CT (2002) 'Cumulative Disadvantage or Individualisation? A comparative analysis of poverty risk and incidence', European Societies, 4 (2):209–233

Le Grand J (1999) Competition, Cooperation, or Control? Tales from the British National Health Service, Health Affairs, Vol. 18, No.3:27–39

Leibfried S (1992) Towards a European welfare state? On integratinfg poverty regimes into the European Community. In: Kolberg J (ed) Social policy in a changing Europe. Campus Verlag, Frankfurt a. M.

Leibfried S (2000) National Welfare States, European Integration and Globalization: A Perspective for the Next Century. Social Policy and Administration 34, 44–63

Leibfried S, Pierson P (1995) Semisovereign Welfare States: Social Policy in a Multitiered Europe. In: Leibfried S, Pierson P (eds) European Social Policy. Between Fragmentation and Integration, The Brookings Institution, Washington D.C., pp 43–77

Leichsenring K, Alaszewski A M (eds) (2004) Providing Integrated Health and Social Care for Older Persons. A European Overview of Issues at Stake. European Centre Vienna & Ashgate, Vienna

Leidl R (1999) Was leisten ökonomische Methoden in der Gesundheitssystemforschung? In: König HH, Stillfried D (ed) Gesundheitssystemforschung in Wissenschaft und Praxis, Schattauer Verlag, Stuttgart, New York, pp 24–32

Leira A (2002) Working parents and the welfare state. Family change and policy reform in Scandinavia. Cambridge University Press, Cambridge

Leisering L (2003) Nation State and Welfare State: An Intellectual and Political History, Journal of European Social Policy, 13:175–186

Leisering L (2003) Government and the Life Course. In: Mortimer J T, Shanahan J (eds) Handbook of the Life Course. Kluwer Academic, pp 205–225

Leisering L, Walker S (eds) (1998) Dynamics of modern society: poverty, policy and welfare. The Policy Press, Bristol

Leisering L, Leibfried S (1999) Time and poverty in Western welfare states. Cambridge University Press, Cambridge
Lejour A (2003) Quantifying Four Scenarios for Europe. CPB Document, no. 38. CPB, The Hague
Loikkanen H, Saari J (2000) Suomalaisen sosiaalipolitiikan alueellinen rakenne. [The regional structure of the social policy in Finland, in Finnish.] Sosiaali- ja terveysturvan keskusliitto, Helsinki
Longo M (2003) European Integration: Between Micro-Regionalism and Globalism, Journal of Common Market Studies, Vol. 41, No. 3, pp 475–494
Luhmann N (1965) Grundrechte als Institution, Duncker & Humblot, Berlin
Luhmann N (1980) Gesellschaftsstruktur und Semantik. Studien zur Wissenssoziologie der modernen Gesellschaft. Band 1. Frankfurt a. M., Suhrkamp
Luhmann N (1990) Political Theory in the Welfare State. De Gruyter, Berlin-New York
Lyon-Caen G, Lyon-Caen A (1993) Droit social international et européen. 8th edn, Duncker und Humblot
Marhold F (2001) Auswirkungen des Europäischen Wirtschaftsrechtes auf die Sozialversicherung. In: Theurl E (ed) Der Sozialstaat an der Jahrtausendwende. Analysen und Perspektiven, Heidelberg
Marshall TH (1950) Citizenship and Social Class. Oxford University Press, Oxford
Marshall TH (1975) Social Policy. Hutchinson, London
Massey DS, Taylor JE (eds) (2004) International Migration. Prospects and Policies. Oxford University Press, Oxford
v. Maydell B (1967) Sach- und Kollisionsnormen im internationalen Sozialversicherungsrecht
v. Maydell B (ed) (1990) Soziale Rechte in der EG. Erich Schmidt
v. Maydell B (1994) Perspectives on the future of social security, International Labour Review, pp 501–510
v. Maydell B, Karl B (2003) Das Angebot von Zusatzkrankenversicherungen – Dürfen gesetzliche Krankenversicherungen Zusatzversicherungen anbieten? Verlag des Verbandes der privaten Krankenversicherung
v. Maydell B, Schulte B (eds) (1995) Zukunftsperspektiven des Europäischen Sozialrechts. Duncker und Humblot
v. Maydell B, Schnapp FE (eds) (1992) Die Auswirkungen des EG-Rechts auf das Arbeits- und Sozialrecht der Bundesrepublik. Erich Schmidt
Mayes D, Berghman J, Salais R (eds) (2001) Social Exclusion and European Policy. Edward Elgar, Cheltenham
Mayes D, Berghman J, Salais R (2002) Social Exclusion and European Policy. Edward Elgar, Cheltenham
Mayes D, Viren M (2002) Macroeconomic Factors, Policies and the Development of Social Exclusion. In: Tsakloglou P (ed) Social Exclusion in European Welfare States. Edward Elgar, Cheltenham
McKee M et al. (eds) (2002) The Impact of EU Law on Health Care Systems, Lang
Meeuwisse A, Swärd H (eds) (2002) Perspektiv på sociala problem. [Perspectives on social problems, in Swedish]. Natur och Kultur, Stockholm
Merkel W (2002) 'Social Justice and the Three Worlds of Welfare Capitalism', Archives Européennes de Sociologie, XLIII (1): 59–91
Merrien F-X (2000) L'État-providence. Presses Universitaires de France, Paris
Miettinen A, Paajanen P (2003) Value Orientations and Fertility Intentions of Finnish Men and Women. Yearbook of Population Research in Finland XXXIX: 201–226
Ministry of Social Affairs of Estonia (2003) Social Sector in Figures 2003. Tallinn
Ministerstwo Połityki Społennej (2005) Strategia połityki społennej 2007–2013. Warsaw

Mishler W, Rose R (2001) What Are the Origins of Political Trust? Comparative Political Studies 34:30–62

MISSOC (Mutual Information System on Social Protection in the EU member states and the EEA) (2005) Social Protection in the Member States of the European Union, of the European Economic Area and in Switzerland. Situation on 1 May 2004, European Commission. Directorate-General for Employment, Social Affairs and Equal Opportunities (http://europa.eu.int/comm/employment_social/social_protection/missoc_tables_en.htm)

MISSOC Info (2001) Old-age in Europe, MISSOC_Info 01/2001, European Commission. Directorate-General for Employment and Social Affairs

Mossialos E, McKee M (2002) The influence of EU Law on the social character of health care systems. Brussels

Mossialos E, Kanavos P (1996) The Methodology of International Comparisons of Health Care Expenditures – Any Lessons for Health Policy? LSE Health, The London School of Economics and Political Science, Discussion Paper, No. 3, London, 1996

Müller K (1999) The Political Economy of Pension Reform in Central-Eastern Europe. Edward Elgar, Cheltenham

Müller K (2003) Privatising Old-Age Security. Latin America and Eastern Europe Compared. Edward Elgar, Cheltenham

Muffels RJA, Luijkx R (2005) Job Mobility and Employment Patterns across European Welfare States. Is there a 'Trade-off' or a 'Double-Bind' between Flexibility and Security? TLM.NET 2005 Working Paper No. 2005-12. SISWO/Social Policy Research, Amsterdam

Muffels R, Tsakloglou (2001) Introduction: Empirical Approaches to Analysing Social Exclusion in European Welfare States. In: Muffels RJA, Tsakloglou P, Mayes DG, Social Exclusion in Welfare States, Edward Elgar, Cheltenham

Muffels RJA, Tsakloglou P, Mayes DG (eds) (2002) Social exclusion in European welfare states. Edward Elgar, Cheltenham

Muffels RJA, Fouarge DJAG (2004) Explaining Resources Deprivation in European Welfare States. Social Indicators Research 65, 3:299–330

Muffels RJA, Fouarge DJAG (2002) Working Profiles and Employment Regimes in Europe. Schmollers Jahrbuch 122,1:85–110

Munro N (2001) National Context or Individual Differences? Influences on Regime Support in Post-Communist Societies. University of Strathclyde, Glasgow

Myrdal A (1941) Nation and Family. The Swedish Experiment in Democratic Family and Population Policy. Harper & Brothers, London

Myrdal A, Myrdal G (1935) Kris i befolkningsfrågan. [Crisis in population issue, in Swedish] Bonniers Förlag, Stockholm

Nave-Herz R (2003) Familie zwischen Tradition und Moderne. BIS-Verlag, Oldenburg

Neumann MJM (1998) Ein Einstieg in die Kapitaldeckung der gesetzlichen Rente ist das Gebot der Stunde, Wirtschaftsdienst, pp 259–264

Nicaise I, Groenez S, Adelman L, Roberts S, Middleton S (2004) Gaps, traps and springboards in the floor of social protection systems. A comparative study of 13 EU countries. Paper presented at ESPAnet conference, Oxford

Nickell S (1997) Unemployment and Labor Market Rigidities: Europe versus North America. Journla of Economic Perspectives, Summer, Vol. 11, No. 3, pp. 55–74

Nink K, Schröder H (2004) Europa Märkte unter der Lupe, Gesundheit und Gesellschaft, 7. Jg., Nr. 2, pp 16–17

NSB (2002) Nationaler Strategiebericht Alterssicherung (NSB), Bundesrepublik Deutschland (http://europa.eu.int/comm/employment_social/soc-prot/pensions/de_pensionreport_ de.pdf)

Nußberger A (2003) Die Wirkungsweise internationaler und supranationaler Normen im Bereich des Arbeits- und Sozialrechts. In: Ekonomi M, von Maydell B, Hänlein A (eds) Der Einfluss internationalen Rechts auf das türkische und das deutsche Arbeits- und Sozialrecht, Nomos, pp 43–64

OECD (2003) OECD-Gesundheitsdaten 2003, Vergleichende Analyse von 30 Ländern, CD-Version

OECD (2004) Employment Outlook 2004. OECD, Paris

Oetker H, Preis U (eds) (2005) Europäisches Arbeits- und Sozialrecht (EAS), Rechtsvorschriften, systematische Darstellungen, Entscheidungssammlungen. Loseblatt, Forkel

Offe C (2003) The European Model of 'Social' Capitalism: can it survive European Integration? Journal of Political Philosophy 11:437–469

Office of the Government Plenipotentiary for Social Security Reform (1997) Security Through Diversity: Reform of the Pension System in Poland. Warsaw

Okólski M (2000) Polen – wachsende Vielfalt von Migration. In: Fassmann H, Münz R (eds) Ost-West-Wanderung in Europa, Böhlau Verlag, Wien/Köln/Weimar, pp. 141–162

Okólski M (2004) The Effects of Political and Economic Transition on International Migration in Central and Eastern Europe In: DS Massey, Taylor JE (eds) (2004) International Migration. Prospects and Policies, Oxford University Press, Oxford, pp 35–58

Oras K (2003) Perekonda soovitav laste arv ja ootused riigi perepolitiikale. [The wished number of children of the family and expectations concerning the state family policies] In: Kutsar D (ed) Millist perekonnapoliitikat me vajame? IPROSEC [What kind of family policies we need?]. Tartu Ülikooli Kirjastus, Tartu, pp 71–87 (In Estonian)

Otting A (1993) International labour standards: A framework for social security, International Labour Review 1993, pp 163–171

Palola E (2004) Euroopan sosiaalinen malli. [The European social model, in Finnish with an English summary, a referee article]. Yhteiskuntapolitiikka 69:569–582

Parfit D (1973) Later Selves and moral principles. In: Montefiore A (ed) Philosophy and Personal Relations. Routledge, London, pp 137–169

Parfit D (1991) Equality or Priority? In: Parfit D, The Lindley Lecture. University of Kansas, Kansas, pp 1–42

Parfit D (1997) Equality and Priority. In: Ratio (1997) pp 202–221

Parfit D (2002) Equality or Priority? In: Clayton M, Williams A (eds) The Ideal of Equality. Palgrave MacMillan, Basingstoke, ch 5

Palier B, Sykes R (2001) Challenges and Change: Issues and Perspectives in the Analysis of Globalisation and the European Welfare States. In: Palier B (ed) Globalization and European Welfare States. Challenges and Change. Palgrave, Houndsmills, pp 1–16

Parsons T (1977) Equality and Inequality in Modern Society, or Social Stratification Revisited. In: Parsons T, Social Systems and the Evolution of Action Theory, the Free Press, New York

Paugam S (1996) 'Poverty and social disqualification: a comparative analysis of cumulative disadvantage in Europe', Journal of European Social Policy 6 (4): 287–304

Paugam S (1998) 'Poverty and Social Exclusion: A Sociological View'. In: Meny Y, Rhodes M (eds), L'Exclusion. L'Etat des Savoirs. La Découverte, Paris, pp 41–62

Paugam S, Russel H (2000) The Effects of Employment Precarity and Unemployment on Social Isolation. In: Gallie D, Paugam S (eds) Welfare Regimes and the Experience of Unemployment in Europe, Oxford University Press, Oxford

Pestoff VA (1998) Beyond the Market and the State: Social Enterprise and Civil Democracy in a Welfare Society, Ashgate, Aldershot

Pickles J, Smith A (eds) (1998) Theorising Transition: The Political Economy of Post-Communist Transformations. Routledge, London/New York

Pierson P (2000) 'Three Worlds of Welfare State Research', Comparative Political Studies. Vol.33, No.6/7:791–821

Pierson P (eds) (2001a) The New Politics of the Welfare State. Oxford University Press, Oxford

Pierson P (2001b) Post-industrial Pressures on the Mature Welfare States. In: Pierson P (ed) The New Politics of the Welfare State, Oxford University Press, Oxford, pp 80–106

Pierson P (2001c) Coping with Permanent Austerity: Welfare State Restructuring in Affluent Democracies. In: Pierson P (ed) The New Politics of the Welfare State, Oxford University Press, Oxford, pp 410–456

Pierson P, Leibfried S (1995) The Dynamics of Social Policy Integration. In: Leibfried S, Pierson P (eds) European Social Policy. Between Fragmentation and Integration, The Brookings Institution, Washington D.C., pp 432–465

Pillinger J (2001) Quality in social public services. European Foundation for the Improvement of Living and Working Conditions, Dublin

Pitschas R (1999) Europäisches Wettbewerbsrecht und soziale Krankenversicherung, in: Vierteljahresschrift für Sozialrecht (VSSR), 3/1999, pp 221–237

Pitschas R (2001) Soziale Sicherungssysteme im 'europäisierten' Sozialstaat. In: Badura P, Dreier H (eds) Festschrift 50 Jahre Bundesverfassungsgericht, Bd. II, 2001, Mohr Siebeck, pp 827–871

Pitschas R (2002) Die Rolle des europäischen Wettbewerbsrechts für die Leistungserbringung im Gesundheitswesen. Berlin

Pitschas R (2003) Kommunale Sozialpolitik. In: v. Maydell B, Ruland F (eds) Sozialrechtshandbuch, 3rd edn., Nomos, C 24, para. 21

Polanyi K (1944) The Great Transformation, Rhinehart, New York

Popenoe D (1994) Scandinavian Welfare. Transaction Social Science and Modern Society 31:78–81

Preusker UK (1999) Warten auf der Warteliste, Gesundheit und Gesellschaft, Ausgabe 2, 2. Jg., pp 16–17

Prior PM, Sykes R (2001) Globalisation and the European Welfare States: Evaluating the Theories and Evidence. In: Prior PM (ed) Globalisation and the European Welfare States: Challenges and Change.Palgrave, Houndsmills, pp 195–210

Quante M (2000) Personal Identity as basis for autonomy. In: Becker GK (ed) The moral status of persons: Perspectives on Bioethics. Rodopi, Amsterdam, pp 57–75

Quante M (2002) Personales Leben und menschlicher Tod. Suhrkamp, Frankfurt a. M.

Quante M (2003) Einführung in die Allgemeine Ethik. Wissenschaftliche Buchgesellschaft, Darmstadt

Quante M, Vieth A (2002) In defense of principlism well understood, Journal of Medicine and Philosophy 27: 621–649

Raffauf P (2004) Gesundheitssysteme unter Druck: Ein lohnender Blick über den Telerand. In: Die Ersatzkasse, Oktober, p 379

Raffelhüschen B (2000) Aging and Intergenerational Equity: From PAVGO to Funded Pension Systems. In: Petersen H-G, Gallagher P (eds) Tax and Transfer Reform in Australia and Germany, Australia Centre Series, Vol 3, Berliner Debatte, Wissenschaftsverlag, pp 263–284

Riesberg A, Weinbrenner S, Busse R (2003) Gesundheitspolitik im europäischen Vergleich – Was kann Deutschland lernen?, Aus Politik und Zeitgeschichte, B 33-34, pp 29–38

Rauhala P-L (1991) Sosiaalialan työn kehittäminen. [Developing the social care work, in Finnish, an empirical study]. University of Tampere, Tampere

Rauhala P-L (1996) Miten sosiaalipalvelut tulivat suomalaiseen sosiaaliturvaan? [How the social care services emerged into the Finnish social policy? In Finnish; Habil. Diss.]. Acta Universitatis Tamperensis ser A vol 447, University of Tampere, Tampere

Rauhala P-L, Simpura J, Uusitalo H (2000) Hyvinvoinnin tutkimusperinteet, hyvinvointipolitiikka ja 1990-luvun uudet avaukset. [Traditions of welfare research, welfare politics and new openings of the 1990s; in Finnish with an English summary, a referee article] Yhteiskuntapolitiikka 65:191–207

Rawls J (2001) Collected Papers. Harvard UP, Cambridge

Reinomägi A (2003) Perede hinnangud eesti perepoliitikale ja võimalustele selle täiustamiseks. [How the families evaluate the Estonian family policies and what kind of opportunities they think to be for improving.] In: Kutsar D (ed) Millist perekonnapoliitikat me vajame? IPROSEC [What kind of family policies we need?]. Tartu Ülikooli Kirjastus, Tartu, pp 125–156 (In Estonian)

Richardson HS (1990) Specifying norms as a way to resolve concrete ethical problems. Philosophy and Public Affairs 19:279–310

Richardson HS (2000) Specifying, balancing and interpreting bioethical principles. Journal of Medicine and Philosophy 25: 285–307

Rieger E, Leibfried S (2003) Limits to Globalization. Welfare States and the World Economy. Polity Press, Oxford

Rimlinger G (1971) Welfare Policy and Industrialization in Europe, America and Russia, Wiley, New York

Rivža B, Stokmane I (2000) Economic and Social Analyses of the Baltic Countries. In: 6th Nordic-Baltic Conference in Regional Science, Sept 2000. Proceeding Report, Riga, pp 293–296

Robeyns I (2004) The capability approach: a theoretical survey. Journal of Human Development

Room GJ (1995) 'Poverty and Social Exclusion: the New European Agenda for Policy and Research'. In: G. Room (ed) Beyond the Treshold: the Measurement and Analysis of Social Exclusion, pp.1–9, The Policy Press, Bristol

Room GJ (1999) 'Social exclusion, solidarity and the challenge of globalization', International Journal of Social Welfare, 8:166–174

Room G (2000) Globalisation, social policy and international standard-setting: the case of higher education credentials. International journal of social welfare 9:103–119

Rogowski R (2004) The European Social Model and Coordination of Social Policy. An overview of policies, competences and new challenges at the EU level. Paper delivered at the TLM.NET Conference, Quality in Labour Market Transitions: A European Challenge, Amsterdam 25–26 November 2004

Rosenbrock R, Gerlinger T (2004) Gesundheitspolitik. Eine systematische Einführung, Bern, Göttingen, Toronto u.a.

Rostgaard T, Fridberg T (1998) Caring for Children and Older People – A Comparison of European Policies and Practices. The Danish National Institute of Social Research, Copenhagen

Sakellaropoulos T, Berghman J (eds) (2004) Connecting Welfare Diversity within the European Social Model. Intersentia, Antwerpen-Oxford-New York

Salt J et al. (1999) Assessment of Possible Migration Pressure and its Labour Market Impact Following EU Enlargement to Central and Eastern Europe. UCL, London

Saraceno C (1997) The importance of the concept of social exclusion. In: Beck W et al. (ed) The social quality of Europe. Kluwer, The Hague

Sarfati H, Bonoli G (eds) (2002) Labour Market and Social Protection Reforms in International Perspective. Parallel or converging tracks? Aldershot, Ashgate

Sauli H, Bardy M, Salmi M (2004) Families with small children face deteriorating circumstances. In: Heikkilä M, Kautto M (eds) Welfare in Finland. STAKES, Helsinki, pp 20–37

Scharpf FW, Schmidt VA (eds) (2000) Welfare and Work in the Open Economy. Volume 1: From Vulnerability to Competitiveness. Oxford University Press, Oxford

Scharpf FW (2002) The European Social Model: Coping with the Challenges of Diversity, Journal of Common Market Studies 40:645–670

Schaub VE (2001) Grenzüberschreitende Gesundheitsversorgung in der Europäischen Union. Die gesetzlichen Gesundheitssysteme im Wettbewerb In: Henke K (ed) Europäische Schriften zu Staat und Wirtschaft. Baden-Baden

Schmähl W (1981) Soziale Sicherung im Alter, w: Handwörterbuch der Wirtschaftswissenschaft, Band 6, Fischer/Mohr, Stuttgart u.a., pp 645–661

Schmähl W (1986) Alterssicherung als verteilungspolitisches Problem, Versicherungs Rundschau, pp 110–136

Schmähl W (2001a) Umlagefinanzierte Rentenversicherung in Deutschland. Optionen und Konzepte sowie politische Entscheidungen als Einstieg in einen Grundlegenden Transformationsprozess. In: Schmähl W, Ulrich V (eds) Soziale Sicherungssysteme und demographische Herausforderungen, Mohr Siebeck, Tübingen

Schmähl W (2001b) Alte und neue Herausforderungen nach der Rentenreform 2001, Die Angestelltenversicherung, 9/01, pp 313–322

Schmähl W (ed) (2001c) Möglichkeiten und Grenzen einer nationalen Sozialpolitik in der Europäischen Union. Schriften des Vereins für Socialpolitik, Bd. 281

Schmähl W (2002) Die "offene Koordinierung" im Bereich der Alterssicherung aus wirtschaftswissenschaftlicher Sicht. In: Offene Koordinierung der Alterssicherung in der Europäischen Union. Internationale Tagung am 9. und 10. November in Berlin, Verband Deutscher Rentenversicherungsträger, Sonderausgabe der "Deutschen Rentenversicherung", DRV-Schriften, Band 34, pp 108–121

Schmähl W (2003a) Dismantling the Earnings-Related Social Pension Scheme – Germany Beyond a Crossroad, Zentrum für Sozialpolitik, Universität Bremen, ZeS-Arbeitspapier Nr. 9/03

Schmähl W (2003b) Erste Erfahrungen mit der 'Offenen Methode der Koordinierung': Offene Fragen zur 'fiskalischen Nachhaltigkeit' und 'Angemessenheit' von Renten in einer erweiterten Europäischen Union, Zentrum für Sozialpolitik, Universität Bremen, ZeS-Arbeitspapier Nr. 11/03

Schmähl W (2005) Generationengerechtigkeit als Begründung für eine Strategie nachhaltiger Alterssicherung in Deutschland. In: Huber G et al. (ed) Einkommensverteilung, technischer Fortschritt und struktureller Wandel. Marburg, S. 441–459

Schmähl W, Rische H (eds) (1997) Europäische Sozialpolitik

Schmähl W, Horstmann S (eds) (2002) Transformation of Pension Systems in Central and Eastern Europe. Elgar

Schmid G, Schömann K (2004) Managing Social Risks Through Transitional Labour Markets: Towards a European Social Model. Pp. 37. TLM.NET Working Paper No.2004-01. Amsterdam: SISWO/Insititute for the Social Sciences

Schmidt A (2003) Europäische Menschenrechtskonvention und Sozialrecht. Baden-Baden

Schmitter PC, Bauer MW (2001) A (modest) Proposal for Expanding Social Citizenship in the European Union, Journal of European Social Policy 11:55–65

Schneider M (2003) Ökonomische Bewertung des Leistungseinkaufs in Europa. In: Klusen N (ed) Europäischer Binnenmarkt und Wettbewerb – Zukunftsszenarien für die GKV. Baden-Baden, pp 161–179

Schneider M (2004) Gesundheitsversorgung – zu welchem Preis?, Die Krankenversicherung, 56. Jg., Juli, pp 173–176

Schneider M, Biene-Dietrich P, Gabanyi M, Hofmann U, Huber M, Köse A, Sommer JH (1995) Gesundheitssysteme im internationalen Vergleich – Ausgabe 1994, BASYS, 1995

Schreyögg J (2003) Eine internationale Bestandsaufnahme des Konzeptes der Medical Savings Accounts und seine Implikationen für Deutschland, Zeitschrift für die Gesamte Versicherungswissenschaft, Jg. 92, Nr. 3
Schulenburg M Graf von, Greiner W (2000) Gesundheitsökonomik. Tübingen
Schuler R (1988) Das Internationale Sozialrecht der Bundesrepublik Deutschland, Grundlagen und systematische Zusammenschau des für die Bundesrepublik Deutschland geltenden Internationalen Sozialrechts. Nomos
Schulte B (1997) Soziale Sicherheit in der EG, Verordnungen (EWG) Nr. 1408/71 und 574/72 sowie andere Bestimmungen. 3rd ed., Beck
Schulte B (2002) Der Europäische Binnenmarkt und das Gesundheitswesen. Europäische Rechtsprechung und stationärer Sektor, DKG-Forum Europa "Chancen für die Krankenhäuser im Binnenmarkt", Vortragsmanuskript
Schulte B, Barwig K (eds) (1999) Freizügigkeit und Soziale Sicherheit, Die Durchführung der Verordnung (EWG) Nr. 1408/1 über die soziale Sicherheit der Wanderarbeitnehmer in Deutschland. Nomos
Schulte B, Zacher HF (eds) (1991) Wechselwirkungen zwischen dem Europäischen Sozialrecht und dem Sozialrecht der Bundesrepublik Deutschland. Duncker und Humblot
Schulte O (2003) Supranationales Recht. In: v. Maydell B, Ruland F (eds) Sozialrechtshandbuch. 3rd ed, Nomos
Schulz O (2003) Grundlagen und Perspektiven einer Europäischen Sozialpolitik. Heymanns
Sen AK (1985) Commodities and Capabilities. North Holland, Amsterdam
Sen AK (1992) Inequality Reexamined. Harvard UP, Cambridge
Sen AK (1999) Development as Freedom. Oxford University Press, Oxford
Shalala DE, Reinhardt UE (1999) Viewing the US Health Care System from Within: Candid Talk from HHS (Health and Human Science), Health Affairs May/June, Vol.18, no.3:47–55
Siebert H (1998) Pay-as-You-Go versus Capital-Funded Pension Systems: The Issues. In: Siebert H (ed) Redesigning Social Security, Mohr, Tübingen, pp 3–33
Sieveking K (ed) (1998) Soziale Sicherung bei Pflegebedürftigkeit in der Europäischen Union. Nomos
Silver H (1994) Social exclusion and social solidarity: three paradigms, International Labour Review 133 (5-6):531–578
Sinn H-W et al. (2001) EU-Erweiterung und Arbeitskräftemigration. Wege zu einer schrittweisen Annäherung der Arbeitsmärkte, ifo – Institut für Wirtschaftsforschung, Max-Planck-Institut für ausländisches und internationales Sozialrecht, München
Sipilä J (ed) (1997) Social Care Services: A Key to the Scandinavian Welfare Model. Aldershot, Avebury
STAKES (National Research and Development Centre for Welfare and Health) (2004) Statistical Yearbook on Social Welfare and Health Care 2004. Helsinki
Statistical Office of Estonia (2001) Social Trends 2. Tallinn, Ofset Ltd
Statistical Office of Estonia (2002) Estonia and the European Union 2002. Tallinn
Statistical Office of Estonia (2004) Estonia in Figures 2004. Tallinn
Statistisches Bundesamt (2003) Bevölkerung Deutschlands bis 2050, 10. koordinierte Bevölkerungsvorausberechnung. Wiesbaden
Statistisches Bundesamt (2005) www.destatis.de
Stewart K (2003) 'Monitoring social inclusion in Europe's regions', Journal of European Social Policy, 13/4, 335–356
Straubhaar T (2001) East-West Migration: Will it be a Problem?, Intereconomcs, July/August 2001, pp 167–170
Strang A, Schulze S (2004) Integrierte Versorgung: Mit neuen Partnern über alte Grenzen, Gesundheit und Gesellschaft, Ausgabe 10, pp 32–37

Streeck W (1995) From Market Making to State Building? Reflections on the Political Economy of European Social Policy. In: Leibfried S, Pierson P (eds) European Social Policy. Between Fragmentation and Integration, The Brookings Institution, Washington D.C., pp 389–431
Sykes R (1998) The future of social policy in Europe? In: Alcock P (ed) Developments in European Social Policy. Convergence and diversity. The Policy Press, Bristol, pp 251–264
SZW (2002) Verkenning Lecensloop (Monitoring the Life Course). The Hague: Ministry of Social Affairs and Employment, The Hague
Taipale V (2004) Lapsuus tietoyhteiskunnassa. [Childhood in the knowledge society, in Finnish.] Yhteiskuntapolitiikka 69:401–406
Takala P (2000) Lastenhoito ja sen julkinen tuki. [Childcare as a social service] Tutkimuksia 101. STAKES, Helsinki
Taylor-Gooby P (2002) The Silver Age of the Welfare State. Perspectives on Resilience. Journal of Social Policy 31:597–621
Taylor-Gooby P (2003) Introduction: Open Markets versus Welfare Citizenship: Conflicting Approaches to Policy Convergence in Europe. Social Policy & Administration 37:539–554
Taylor-Gooby P (2004) Open markets and welfare values. Welfare values, inequality and social change in the silver age of the welfare state. European Societies 6:29–48
The Government of Estonia (1999) The Concept of the Children and Family Policy, Tallinn (http://www.riik.ee/pere/inglise.htm)
Therborn G (1995) European Modernity and Beyond. The Trajectory of European Societies 1945–2000. SAGE Publications, London
Therborn G (2004) Between Sex and Power: Family in the World 1900–2000. SAGE Publications, London
Torfing J (1999) Workfare with Welfare: Recent Reforms of the Danish Welfare State. Journal of European Social Policy 9:5–28
Townsend P (1979) Poverty in the United Kingdom. Allen Lane, Harmondsworth
Tsakloglou P, Papadopoulos F (2002) Aggregate level and determining factors of social exclusion in twelve European countries, Journal of European Social Policy 12:211–225
Ulrich V, Schmähl W (2001) Demographische Alterung in Deutschland: ein Überblick. In: Schmähl W, Ulrich V (ed) Soziale Sicherungssysteme und demographische Herausforderungen. Tübingen, pp 1–19
UNDP Human Development Report, Estonia (2003) Eesti Inimarengu Aruanne (In Estonian)
UNDP – United Nations Development Program (2004) Human Development Report 2004 (http://hdr.undp.org/reports/global/2004)
UNFPA – United Nations Population Fund (2002) Population Ageing and Development. Social, Health and Gender Issues. Number 3. Population and Development Strategies, Report of an Expert Group Meeting. UNFPA, New York
Välimäki A-L, Rauhala P-L (2000) Lasten päivähoidon taipuminen yhteiskunnallisiin muutoksiin Suomessa. [Children's day care yields to societal changes in Finland, in Finnish, a referee article] Yhteiskuntapolitiikka 65:387–405
Van Creveld M (1999) The Rise and Decline of the State, Cambridge University Press, Cambridge
Van Oorschot W, Abrahamson F (2003) The Dutch and Danish Miracles Revisited: A Critical Discussion of Activation Policies in Two Small Welfare States, Social Policy & Administration, 37/3, pp 288–304
Van Parijs P, Vanderborght Y (2001) From Euro-stipendium to Euro-Dividend, Journal of European Social Policy 11:342–346
Vandenbosch K (2001) Identifying the poor, Using Subjective and Consensual Measures. Aldershot, Ashgate

Vandenbroucke F (1999) 'The Active Welfare State: a European Ambition'. Den Uyl lecture, Amsterdam

Vandenbroucke F (2001) 'Open Co-ordination on Pensions and the Future of Europe's Social Model', Closing speech at the Conference 'Towards a New Architecture for Social Protection?' (Oct.), Leuven

Vandenbroucke F (2004) Foreword: Sustainable Social Justice and 'Open Coordination' in Europe. In: Esping-Andersen G (ed) (2002) Why we need a New Welfare State. Oxford University Press, Oxford

VDR, BMA, Max-Planck-Institut für internationales und ausländisches Sozialrecht (ed) (2002) Offene Koordinierung der Alterssicherung in der Europäischen Union, DKV-Schriften, Bd. 34, pp 21–26

Verband der Angestellten-Krankenkassen e.V. (VdAK) (2004) Ausgewählte Basisdaten des Gesundheitswesens. Siegburg

Verhoeven H (2004) Allochtonen op de Vlaamse Arbeidsmarkt. In: Van Hootegem G, Cambré B (eds) Over Werk(t) in de Actieve Welvaartsstaat, Acco, Leuven/Voorburg, 401–415

Verschraegen G (2002) Human Rights and Modern Society: A sociological analysis from the Perspective of Systems Theory, Journal of Law and Society, 29:58–281

Vihalemm P, Lauristin M, Tallo I (1997) Development of Political Culture in Estonia. In: Lauristin M, Vihalemm P, Rosengren KE, Weibull L (eds) Return to the Western World. Tartu University Press, Tartu, pp 197-210

Vleminckx K (2004) In search of the developmental welfare state. A theoretical exploration. Unpublished paper.

Vleminckx K, Smeeding T (2001) Child Wellbeing, Child Poverty and Child Policy in Modern Nations. The Policy Press, Bristol

Wagner P (1994) A Sociology of Modernity. Liberty and Discipline. Routledge, London

Wagner G, Meinhardt V, Leinert J, Kirner E (1998) Kapitaldeckung: Kein Wundermittel für die Altersvorsorge, DIW Wochenbericht Nr. 46, Berlin

Wennberg L (2004) Social Assistance for Solo Mothers in Sweden, Finland, Norway, and Denmark. A comparative legal study. Umeå Studies in Law No 8/2004. Umeå University, Umeå

Weinbrenner S, Busse R (2004) Europäische Gesundheitssysteme im Vergleich, Die Ersatzkasse, Oktober, pp 384–389

Wille E (2000) GKV Reformbedarf bei der Beitragsgestaltung, Wirtschaftsdienst 5, pp 263–265

Weidenfeld W (ed) (2001) Nizza in der Analyse. Bertelsmann Stiftung

Weihe-Lindeborg L (2000) Europe of the Regions – a Growing Reality. In: Structural Change in Europe – Innovative Cities and Regions. Hagbarth Publications, Bollschweil, pp 6–15

Whelan CT, Layte R, Maitre B (2002) Multiple Deprivation and Persistent Poverty in the European Union, Journal of European Social Policy 12 (2):91–105

Whelan BJ, Whelan CT (1997) In what sense is poverty multidimensional? In: Room G (ed) Beyond the threshold. The Policy Press, Bristol

Wiegand D (1993) Das europäische Gemeinschaftsrecht in der Sozialversicherung. 2nd ed., Asgard

Willaschek M (1998) Was ist 'schlechte Metaphysik'? In: Wenzel UJ (ed) Vom Ersten und Letzten. Fischer, Frankfurt am Main, pp 131–151

Willers D (1994) Sozialklauseln in Internationalen Handelsverträgen. In: Bundesministerium für Arbeit und Sozialordnung u.a. (ed) Weltfriede durch soziale Gerechtigkeit. 75 Jahre Internationale Arbeitsorganisation. Baden-Baden, pp 165–178

Williamson JB, Pampel FC (1993) Old-Age Security in Comparative Perpective, Oxford University Press, New York-Oxford

Wilthagen T (1998) Flexicurity: A new paradigm for labour market policy research. WZB Discussion Paper FS I 98–202, Berlin

Wilthagen T, Tros F (2004) The concept of 'flexicurity': a new approach to regulating employment and labour markets. Transfer 10,2:166–186

Wilthagen T, van Velzen M (2004) the road towards adaptability, flexibility and security. Brussels, European Commission/DG Employment. Thematic Review Seminar on "Increasing adaptability for workers and enterprises"

Wismar M, Busse R (1999) Auswirkungen der europäischen Binnenmarktintegration auf das deutsche Gesundheitswesen. In: Bellach B-M, Stein H (eds) Die neue Gesundheitspolitik der Europäischen Union, München, pp 83–98

Wissenschaftlicher Beirat beim Bundesministerium der Finanzen (2000) Freizügigkeit und soziale Sicherung in Europa, Gutachten, Schriftenreihe des Bundesministeriums der Finanzen, H. 69

World Bank (1994a) Poverty in Poland. Washington D.C.

World Bank (1994b) Averting the Old Age Crisis. Policies to Protect the Old and Promote Growth, Oxford University Press, Washington D.C.

World Bank (1995) Understanding Poverty in Poland. Washington D.C.

World Bank (2001) Social Protection Sector Strategy: From Safety Net to Springboard, OUP, Washington D.C.

Wrobel RM (2000) Estland und Europa: Die Bedeutung des Systemwettbewerbs für die Evolution und Transformation von Wirtschaftssystemen. Tartu Ülikooli Kirjastus, Tartu

Zacher HF (ed) (1978) Sozialrechtsvergleich im Bezugsrahmen internationalen und supranationalen Rechts. Duncker und Humblot

Zerna C (2003) Der Export von Gesundheitsleistungen in der Europäischen Gemeinschaft nach den Entscheidungen des EuGH vom 28. April 1998 in den Rechtssachen "Decker" und "Kohll". Peter Lang

Zijderveld AC (1997) The Democratic Triangle. The Unity of State, Society and Market. In: Zijderveld AC (ed) The gift of Society. Enzo Press, Nijkerk, pp 11–24

Zimmermann KF, Bauer TK, Bonin H, Fahr R, Hinte H (2001) Arbeitskräftebedarf bei hoher Arbeitslosigkeit. Ein ökonomisches Zuwanderungskonzept für Deutschland. Berlin

Zimmermann H, Henke KD (2001) Finanzwissenschaft. Eine Einführung in die Lehre von der öffentlichen Finanzwirtschaft. 8th Edition, München

Żukowski M (1994) Pensions Policy in Poland after 1945: Between 'Bismarck' and 'Beveridge' Traditions. In: Hills J, Ditch J, Glennerster H (eds) Beveridge and Social Security. An International Retrospective, Oxford University Press, Oxford, pp 154–170

Żukowski M (1996) Das Alterssicherungssystem in Polen – Geschichte, gegenwärtige Lage, Umgestaltung. Zeitschrift für ausländisches und internationales Arbeits- und Sozialrecht 10(2), pp 97–141

Żukowski M (1997) Wielostopniowe systemy zabezpieczenia emerytalnego w Unii Europejskiej i w Polsce. Między państwem a rynkiem [Multitier old age security systems in the European Union and Poland. Between state and market], University of Economics in Poznan, zeszyty naukowe seria II, zeszyt 151

Żukowski M (2003) Country Study Poland: Pensions. In: GVG, 2003, pp 39–58

ZUS (1992) Rocznik statystyczny ubezpieczeń społecznych 1985–1990 [Statistical Yearbook of Social Insurance 1985–1990], Zakład Ubezpieczeń Społecznych, Warsaw

Zweifel P (2004) Ökonomische Überlegungen zur Rationierung im Gesundheitswesen. In: Zeitschrift für die gesamte Versicherungswissenschaft, Heft 4, pp 583–601

In der Reihe *Wissenschaftsethik und Technikfolgenbeurteilung* sind bisher erschienen:

Band 1: A. Grunwald (Hrsg.) Rationale Technikfolgenbeurteilung. Konzeption und methodische Grundlagen, 1998

Band 2: A. Grunwald, S. Saupe (Hrsg.) Ethik in der Technikgestaltung. Praktische Relevanz und Legitimation, 1999

Band 3: H. Harig, C. J. Langenbach (Hrsg.) Neue Materialien für innovative Produkte. Entwicklungstrends und gesellschaftliche Relevanz, 1999

Band 4: J. Grin, A. Grunwald (eds) Vision Assessment. Shaping Technology for 21st Century Society, 1999

Band 5: C. Streffer et al., Umweltstandards. Kombinierte Expositionen und ihre Auswirkungen auf den Menschen und seine natürliche Umwelt, 2000

Band 6: K.-M. Nigge, Life Cycle Assessment of Natural Gas Vehicles. Development and Application of Site-Dependent Impact Indicators, 2000

Band 7: C. R. Bartram et al., Humangenetische Diagnostik. Wissenschaftliche Grundlagen und gesellschaftliche Konsequenzen, 2000

Band 8: J. P. Beckmann et al., Xenotransplantation von Zellen, Geweben oder Organen. Wissenschaftliche Grundlagen und ethisch-rechtliche Implikationen, 2000

Band 9: G. Banse et al., Towards the Information Society. The Case of Central and Eastern European Countries, 2000

Band 10: P. Janich, M. Gutmann, K. Prieß (Hrsg.) Biodiversität. Wissenschaftliche Grundlagen und gesellschaftliche Relevanz, 2001

Band 11: M. Decker (ed) Interdisciplinarity in Technology Assessment. Implementation and its Chances and Limits, 2001

Band 12: C. J. Langenbach, O. Ulrich (Hrsg.) Elektronische Signaturen. Kulturelle Rahmenbedingungen einer technischen Entwicklung, 2002

Band 13: F. Breyer, H. Kliemt, F. Thiele (eds) Rationing in Medicine. Ethical, Legal and Practical Aspects, 2002

Band 14: T. Christaller et al. (Hrsg.) Robotik. Perspektiven für menschliches Handeln in der zukünftigen Gesellschaft, 2001

Band 15: A. Grunwald, M. Gutmann, E. Neumann-Held (eds) On Human Nature. Anthropological, Biological, and Philosophical Foundations, 2002

Band 16: M. Schröder et al., Klimavorhersage und Klimavorsorge, 2002

Band 17: C. F. Gethmann, S. Lingner (Hrsg.) Integrative Modellierung zum Globalen Wandel, 2002

Band 18: U. Steger et al., Nachhaltige Entwicklung und Innovation im Energiebereich, 2002

Band 19: E. Ehlers, C. F. Gethmann (eds) Environment Across Cultures, 2003

Band 20: R. Chadwick et al., Functional Foods, 2003

Band 21: D. Solter et al., Embryo Research in Pluralistic Europe, 2003

Band 22: M. Decker, M. Ladikas (eds) Bridges between Science, Society and Policy, 2004

Band 23: C. Streffer et al., Low Dose Exposures in the Environment, 2004

Band 24: F. Thiele, R. E. Ashcroft (eds) Bioethics in a Small World, 2004

Band 25: H.-R. Duncker, K. Prieß (eds) On the Uniqueness of Humankind, 2005

Außerhalb der Reihe sind ebenfalls im Springer Verlag die Übersetzungen der Bände 5 und 18 erschienen: „Environmental Standards. Combined Exposures and Their Effect on Human Beings and Their Environment“ (Streffer et al., 2003) sowie “Sustainable Development and Innovation in the Energy Sector” (Steger et al., 2005).

MIX
Papier aus verantwortungsvollen Quellen
Paper from responsible sources
FSC® C105338

If you have any concerns about our products,
you can contact us on
ProductSafety@springernature.com

In case Publisher is established outside the EU,
the EU authorized representative is:
Springer Nature Customer Service Center GmbH
Europaplatz 3, 69115 Heidelberg, Germany

Printed by Libri Plureos GmbH
in Hamburg, Germany